THE BRITISH COAL TRADE

THE
BRITISH COAL TRADE

by

H. STANLEY JEVONS

A reprint with an introductory note

by

Baron F. Duckham

AUGUSTUS M. KELLEY, PUBLISHERS
NEW YORK 1969

Published in the United States of America by
Augustus M. Kelley, Publishers, New York

SBN 678 05559 9

Library of Congress No 68–58858

This book was first published in 1915 by
Kegan Paul Trench Trubner & Co Ltd

This edition published 1969

PUBLISHER'S NOTE The original edition of 1915 is here
reproduced in full, with the exception of a large fold-out
map of the coalfields of the British Isles, forming the fron-
tispiece, which was based on the Geological Survey Map
of the Royal Commission on Coal Supplies, 1905, with
certain alterations and additions. This has been omitted
for reasons of economy.

Printed in Great Britain by
Redwood Press Limited Trowbridge Wiltshire

INTRODUCTORY NOTE TO 1969 EDITION

Herbert Stanley Jevons was the famous son of a famous father. Born on 8 October 1875 to William Stanley Jevons (1835–82) and Harriet Ann, *née* Taylor, he could claim a lineage of intellectual distinction on both the paternal and maternal sides.* His father was a considerable scholar of international reputation who occupied the chairs of Political Economy and Logic, and Mental and Moral Philosophy at Manchester from 1866 to 1876 and the chair of Political Economy at University College, London, 1876–80. W. S. Jevons is assured of an honourable place in the canon of English economic thought in the nineteenth century and can justly be considered one of the shapers of that discipline. He was a prolific writer, not merely on economics (especially currency and finance), but also in the fields of logic, railways and the state, and labour. His own important work on the coal industry, *The Coal Question* (1865), quickly established itself as a classic and went through two further editions in 1866 and 1906. The elder Jevons had an original and fertile mind. Looking at his voluminous output it is hard for one to believe that he died at the comparatively early age of forty-seven, leaving his son—then aged seven—to the care of his mother. The boy was educated at Giggleswick School in Yorkshire and subsequently went up to University College, London, completing his academic training at Trinity College, Cambridge, and the Geological Institute, University of Heidelberg.

Jevons's first appointment was as a demonstrator in petrology at Cambridge (1900–1), a post he relinquished for a lectureship in

* His mother was the daughter of the lively J. E. Taylor, founder of the *Manchester Guardian*.

mineralogy and geology at the University of Sydney. In 1904–5
he returned to Britain, by way of an extended tour into south-east
Asia, the United States and Canada, to take up a lectureship in
political science at the University College of South Wales and Monmouthshire, Cardiff. His success in this new intellectual field was
considerable. He edited occasional papers for the university's
Department of Economics and Political Science, published his
Essays on Economics in 1905 and was elected Fulton professor. But
he was no detached academic and soon became actively interested
in housing reform throughout the South Wales coalfield. In fact,
so important did this work become to him that in 1911 he resigned
his chair to devote more time to it. As his obiturist in *The Times*
noted, his 'mild and gentle manner, and soft caressing voice, were
deceptive characteristics, hiding the personality and character of a
man who felt deeply on every human question, and could attack
with ruthless denunciation the wrongs he felt should be righted.'

Although he had published in 1909 a short paper on foreign trade
in coal, it was in the coalfields of the Welsh valleys that he became
seriously involved with the industry and its human as well as economic
problems. The fruit of his interest and concern was the present
book, in which he trod some of the steps his father had marked out
two generations earlier. Curiously, his activities after 1915 were
to lead him far from the British coal mining industry. In 1914 he
accepted a chair in economics at the Indian University of Allahabad
and exchanged this for a professorship in the same subject at Ran-
goon in 1923. Many of his later books and articles thus deal with
Indian or Commonwealth matters. After resettling in England in
1930 he took a prominent part in the work of the Royal Statistical
Society and was a council member from 1932 to 1937. His involve-
ment with human rights did not diminish and found particular outlet
in his founding of the Abyssinia Association following Mussolini's
cowardly attack on that country in 1935. During the exile of
Haile Selassie, Jevons was often of personal service to the emperor
and in 1954 was invested with the Order of Commander of the

Star of Ethiopia. He took a keen interest, too, in the wider issues of war and peace and, characteristically, was chairman of the Bombing Restriction Committee from 1943 to 1945. He died in London aged seventy-nine on 27 June 1955.

The British Coal Trade, as the author was at pains to point out in his preface, was virtually completed before the outbreak of the first world war. It was written at a time when the industry was—to quote 1913's figures—producing nearly 290 million tons of coal per annum and exporting over a third of this enormous output. With a labour force over a million strong and a militant federated trade union (from 1908 representative of all districts) claiming the loyalty of rather more than half, coal mining dominated the industrial scene. Production had roughly doubled between 1880 and the last full year of peace. Despite some diminishing returns on new labour added to mining, almost any observer of the period would naturally have expected even higher levels of both output and exports to follow. A massive showdown between employers and employees could not, it is true, be discounted, especially in view of the Triple Alliance and vague rumblings of syndicalism in less responsible union circles. But many, like Jevons, were reasonably optimistic of a settlement to the industry's chronic labour problems. Nationalisation, given sensible leadership on the part of both government and the Miners' Federation, did not seem unattainable to 'advanced' thought by 1915. Indeed, progressive government control of the pits during the war, culminating in the Act of February 1918 which, while it remained in force, brought virtual nationalisation without the name, was seen by some as confirming the need for, and likelihood of, real changes taking place in mine and royalty ownership once peace came.

In any case—and whatever the clash between labour and capital brought—scarcely anyone could foresee the rise of alternative fuels to coal on an appreciable scale. Though it was apparent that some coalfields had cost-advantages over others and that such differentials would tend to increase, no one anticipated that for much of the

1920s and 1930s the whole industry would be in the doldrums. It is, of course, easy for us now to smile at the seeming naivety of some of Jevons's beliefs—it would be unfair to call them predictions—about the future course of coal mining. Output, which he supposed might rise to a peak of 902 million tons by about the year 2101, never again reached the record for 1913. Exports, too, so far from doubling within a generation or so, were in actuality well past their peak* and eventually, after the second world war, dwindled into complete insignificance. Even expectations for individual coalfields were to be spectacularly wrong. Kent, it was assumed by Jevons, would soon require 'some hundreds of miles of railways . . . great docks' and produce ten million tons of coal by 1925. Only about a fifth of this output was ever achieved in any one year. Public ownership itself, recommended by the left wing sections of the Sankey Commission of 1919, had to wait until 1947. Although Jevons was not one of those who expected nationalisation to be necessarily either quick or easy, even he presumably did not anticipate so long a delay.

Part of the explanation of the coal industry's stagnation and decline belongs to the run down of other basic coal-using industries after 1918. Fuel economies, too, in steel furnaces, gas retorts, electricity generating stations and so on also inhibited demand-growth. German production affected some of our export markets, but more ominous in the long run was the expansion in supplies of petroleum fuels. Jevons, like his contemporaries, had no way of foreseeing the impressive finds of oil which speedily made his assessment of the probable impact on demand for coal unrealistic. Nor could he guess the rapidity with which petroleum products were to widen their use and application. Though the turn to electrical power rather than steam, or to petroleum rather than coal must not be exaggerated for the inter-war years, it is yet true that the mono-

---

* Export performance 1922–4 was, of course, exceptional because of factors, particularly the French occupation of the Ruhr, affecting our rivals.

poly of coal was decisively broken in the interval between Versailles and Munich.

We have learnt little from history, however, if we allow ourselves the dubious amusement afforded by hindsight. Predictions of fuel needs even within our own generation have more than once been made to look ridiculous by faulty assessments of the pace of technological progress. Nor does one have to return to the dark austerity and power shortages of the late 1940s to find what have become wholly unrealistic forecasts of the requirements of the present. Readers may not need reminding of the over-optimism assigned to the coming role of nuclear power-stations in the mid and late fifties. Planners and 'experts' have often a fallibility incomparably more pathetic and potentially more dangerous than that of the honest scholar who ventures, unwisely perhaps, into inventing the history of the future. Few men were better qualified to speculate on probable developments in the coal industry in 1915 than H. S. Jevons. We can still derive instruction from his closing chapters if we are prepared to read them with sympathetic insight rather than patronage.

But by far the greater part of the work is descriptive and diagnostic. Jevons may not always fully satisfy us with his analysis of the economic and social problems of the mines and their workmen, but he is stimulating and free from narrow preconceptions. His survey of the industry on the eve of the Great War is absorbingly interesting and still valuable. In particular, one might instance his chapters XIV–XXII, covering the nature of the miner's employment, improvements made towards his greater safety, the complexity of his scales of pay in an industry subject to considerable price fluctuations, his unions, housing, habits and so on. These are all topics which still merit and receive the attention of scholars, but it would be difficult to think of a better introduction to them than that provided in Professor Jevons's book. Certainly it remains essential reading for anyone interested in the British coal industry at its productive zenith and in the contemporary viewpoint of a well-qualified commentator. And for good measure there is a capital account, within the space available, of the methods of extracting coal at a time when 'machine-mining' had just emerged from the experimental stage.

It is, of course, a book for historians (amateur or professional) and not a book by an historian. Jevons was a man of great natural talents and wide training, but he did not pretend to be writing a history of the coal trade. Not surprisingly, then, a few errors inevitably crept into the short sections giving historical background to the subjects being discussed, though they are generally mistakes of a minor nature and away from the main thesis of the work. Iron smelting with coke pre-dates 1735 and, as at least a few of the proverbial schoolboys know, Watt did not invent the steam engine in 1782 (p 4). Goole does not stand on the Humber—though a 'Humber port' (p 67)—nor did the Newcomen engine reach Scotland (or anywhere else) in 1690! (p 153). While it is true that in Yorkshire and the [North] Midlands 'the conditions are such as to make [coal-cutters] specially advantageous' (p 211), it was not quite accurate to state that 'most of the machines' were employed there (p 69 and p 211). In fact it was noticeably the older fields, reduced to working thinner seams, which responded most rapidly to machine-cutting. In 1913 Scotland had 876 cutters and the North-East 644 to Yorkshire's 407 and Nottinghamshire's and Derbyshire's 277. (North Staffordshire had admittedly 452 cutters, but Jevons was thinking predominantly in terms of the newer Midlands fields.) There are one or two other statements with which some trade union historians might take issue, but in general the author is at his best when writing of the miners themselves, of their unique communities, way of life and struggle for a more egalitarian society.

Jevons knew the coalfields of Britain well—none more so than that of South Wales—and recognised the cultural differences which separated them as much as those they shared. His comments on the various districts are full of insight. In the older fields, such as Scotland or Durham and Northumberland, division of labour was usually very advanced. Despite wretched housing (especially in Scotland) and other disruptions to family life (such as double and treble shift working) miners had often more self-respect than those in the newer areas and had learnt to assume a more constructive role in the amelioration of their own physical and spiritual environment. Friendly societies, co-operative societies and medical clubs were well supported in addition to the local trade union lodge. In parts of the Midlands and South Yorkshire Jevons found a somewhat rougher, less literate type of pitman with, on the whole, cruder forms of relaxation. But vague generalisations are resisted and, as the author graphically points out, conditions in the same coalfield could exhibit wide differences. Not all Welsh miners were pillars of the non-conformist religion, though the place of the chapel was central in the shaping of social behaviour and reaction in many mining communities. It was, however, noticeable in all districts that the standard of overt civilisation frequently reflected the local 'price lists' (scales of remuneration).

Collieries offering the best cutting rates naturally attracted and kept the most responsible hewers. Because such scales were in practice relatively inflexible, there was a tendency for remuneration at these normally larger and better managed collieries to be consistently higher than the average. The quality of life in the associated townships was correspondingly superior. High wage communities tended to boast 'a large Workmen's Institute with fine hall, occupying a commanding position', had 'numerous chapels, solidly built', and produced men who were able to save and even—in some cases —buy their own houses. 'Respectability' was valued and 'in the streets one [saw] many serious and intelligent workmen of the best kind'. On the other hand, where earnings were lower over long periods, the mining towns and their inhabitants wore the marks of degradation and demoralisation. Sometimes the very fact of physical isolation was enough to ensure 'no library, no evening classes, and no institute where men can see newspapers and periodical literature, or meet for social and business purposes'. Much current research is being pursued nowadays into the historical sociology of alienation and it would be wrong to suggest that Jevons offered a satisfactory analysis of the miner's condition. None the less he was an intelligent, contemporary observer, and if his notion of respectability unashamedly reflected his solid middle-class background, his sympathy and sense of justice went some way towards seeing society from the pitman's point of view. More than most of his generation he recognised and valued the richness of this traditional working-class culture. It is now so nearly a thing of the past that we too should be thankful for the account of it which is given in this book.

This present edition is a simple reprint of that of 1915. The length and nature of the book largely preclude any attempt at up-dating. Its value, as has been stressed, lies in its picture, on a broad canvas, of a great industry half a century ago. The expert will know where to look for further reading. Others may be referred to Jevons's own bibliography at the back of this work and to the basic list of books and articles (to 1914) in the 1969 reprint of R. L. Galloway's *A History of Coal Mining in Great Britain*. Developments since 1915 have already an extensive literature and the appended items form only a small part of it.

University of Strathclyde   BARON F. DUCKHAM

Suggestions for further reading:
The following works range from academic studies of considerable
objectivity to blueprints for, and tracts about, the industry which,
in some cases, are written from a particular political viewpoint.
They do nevertheless represent a fair cross-section of non-technical
literature on modern coal mining published since 1915. Space has
forbidden the inclusion of references to most articles in periodicals
and to parliamentary papers.

Abercrombie, P. and others, *The Coal Crisis and the Future: a study
of social disorders and their treatment* (1926)

Arnot, R. Page, *The Miners: Years of Struggle* (1953)

Bulman, H. F., *Coal Mining and the Coal Miner* (1920)

Court, W. H. B., 'Problems of the British Coal Industry between
the Wars', *Econ Hist Rev*, XV (1945)

Court, W. H. B., *Coal* (History of the Second World War, 1951)

Dickie, J. P., *The Coal Problem* (1936)

Edwards, N., *History of the South Wales Miners' Federation* (1938)

Foot, R., *A Plan for Coal* (Mining Assoc of Great Britain, 1945)

Heinemann, M., *Britain's Coal* (1944)

International Labour Office, *World Coalmining Industry* (Geneva,
1938)

Jones, J. H., Cartwright, G., and Guérault, P., *The Coal-Mining
Industry: an International Study in Planning* (1939)

Lubin, I., and Everett, H., *The British Coal Dilemma* (New York,
1927)

NCB, *Plan for Coal* (1950)

NCB, *Investing in Coal* (1956)

Neuman, A. M., *Economic Organisation of the British Coal Industry*
(1934)

PEP, *Report on the British Coal Industry* (1936)

Raynes, J. R., *Coal and its Conflicts* (1928)

Redmayne, R. A. S., *The British Coal Industry During the War*
(Oxford, 1923)

Redmaye, R. A. S., *Men, Mines and Memories* (1942)

Rowe, J. W. F., *Wages in the Coal Industry* (1923)

Shurick, A. T., *The Coal Industry* (1924)

Smart, R. C., *The Economics of the Coal Industry* (1930)

Stewart, W. D., *Mines, Machines and Men* (1935)

Thomas, I., 'The Coal Mines Reorganisation Commission', in Robson, W. A., (ed) *Public Enterprise* (1937)

Townshend-Rose, H., *The British Coal Industry* (1951)

Watkins, H. M., *Coal and Men: an Economic and Social Study of the British & American Coalfields* (1934)

THE BRITISH COAL TRADE

THE
BRITISH COAL TRADE

BY

H. STANLEY JEVONS
M.A., B.Sc., F.S.S., F.G.S.

PROFESSOR OF ECONOMICS AT THE UNIVERSITY OF ALLAHABAD,
AND FORMERLY FULTON PROFESSOR OF ECONOMICS AND
POLITICAL SCIENCE AT THE UNIVERSITY COLLEGE
OF SOUTH WALES AND MONMOUTHSHIRE

*With Twenty-three Illustrations in Black and White
and two Maps*

LONDON
KEGAN PAUL, TRENCH, TRÜBNER & Co., Ltd.
BROADWAY HOUSE, 68-74, CARTER LANE, E.C.
NEW YORK: E. P. DUTTON & COMPANY
1915

PREFACE

THIS book is intended to be a popular account of
the coal mining industry and of the coal trade of
the British Isles, in which special attention is paid
to the economic and social aspects. Whilst on the
technique of mining or selling coal, and on the geology
of the Coal Measures I have written for the lay
reader, there is a good deal of information in some of
the chapters which will be new, I believe, to students
of economic and social science—at any rate it
has not previously been collected in one book.
Many of the facts which I have obtained orally
from persons engaged in coal mining, or by my own
observation, are not generally known and have not
been published before.

The growing part which the coal trade and the
miners are destined to play in the economy and the
politics of this country has always been before me
as defining the object and scope of this book.
Stricter and more extensive legislative regulation
of the mining industry will be demanded, and the
nationalisation of mines will be seriously proposed
and discussed. No adequate discussion of such
measures by the public or in Parliament is possible
whilst there is widespread ignorance of the modern
methods of working coal, and of the conditions

of employment of miners. My purpose has been to provide knowledge of facts and a point of view. It is always the lives and personality of the workers in an industry which appeal to me as the most important aspect of it, as indeed they are to the nation ; so I have grudged no space for the chapters dealing with miners' lives and their organisations.

I collected materials dealing with the history of coal mining, the finance of mining enterprises, mineral royalties and coal leases, a comparison of American and continental mining methods with our own, export duties in coal, the retail coal trade, the nationalisation of mines, and other subjects, but want of space forbade my treating them adequately—I also rejected them in some cases because they were too technical, and in others because I had determined to avoid controversy. The book has had to be written in my spare time, so that although I have been engaged on it for some four or five years, there has been no time for me to make minute verification of all my information. I can hardly hope, therefore, that there are not a few errors or misstatements ; but I believe they are not many, as I have read carefully the earlier written portions without finding anything serious amiss. One great difficulty arises from difference of usages and technical names in the principal coalfields, which are so confusing that both the employers and the miners' leaders are constantly troubled by them when they meet together in national conferences. I must also point out that almost the whole of the book was written before the outbreak of the Great European War ; and some of my rather confident predictions may be vitiated, or at least postponed, by this tremendous and unexpected calamity.

My thanks are due to so many persons for information and assistance that it is hard to name them. In particular I must express my indebtedness to Professor Galloway, whose kindness I have much appreciated. Colliery owners and managers have always answered my questions with the greatest readiness and frankness. Amongst the miners' representatives I am indebted to Messrs. Robert Smillie, J.P., President of the Miners' Federation of Great Britain, Thomas Ashton, J.P., William Brace, M.P., Tom Richards, M.P., J. Winstone, J.P., George Barker, Vernon Hartshorn, J.P., Hubert Jenkins, Edward Gill, T. I. Mardy Jones, and several others. To Mr. David Evans, of Cardiff, I am indebted for much of the information connected with the sale of coal and with shipping ; and I have had much extremely valuable assistance in collecting information and statistics and in seeing the book through the press from Mr. Edgar L. Chappell, Secretary of the South Wales Garden Cities Association. Other friends also have given generous help.

To the Business Statistics Publishing Co., of Cardiff and Newcastle-on-Tyne, who issue the South Wales Coal Annual and the North Country Coal and Shipping Annual, I am also indebted, not only, in common with the general public, for these excellent annuals, of which I have made extensive use, but also for kindly allowing me the use of their proof sheets of the issue for 1915, and for permitting me to reproduce several valuable illustrations. I wish also to thank the Powell Duffryn Steam Coal Co. and the Ashington Coal Co. for kindly providing me with several photographs.

CONTENTS

CHAPTER		PAGE
I.	INTRODUCTION	1
II.	COAL AND COAL SEAMS	11
III.	USES OF COAL	31
IV.	THE ENGLISH COALFIELDS	58
V.	THE WELSH, IRISH AND SCOTTISH COALFIELDS	93
VI.	THE KENT COALFIELD	155
VII.	NEW SINKING AND DEVELOPMENT	176
VIII.	METHODS OF MINING COAL	202
IX.	PREPARATION OF COAL FOR MARKET	219
X.	THE BYE-PRODUCTS OF COAL	232
XI.	ECONOMICS OF THE COAL TRADE	257
XII.	THE SALE OF COAL	291
XIII.	AMALGAMATIONS	314
XIV.	METHODS OF PAYING WAGES	331
XV.	SAFETY IN MINES	365
XVI.	MINING LAW AND INSPECTION	404
XVII.	MINERS' TRADE UNIONS	445
XVIII.	SLIDING SCALES AND CONCILIATION BOARDS	489
XIX.	ABNORMAL PLACES AND THE NATIONAL COAL STRIKE	520
XX.	COAL MINES (MINIMUM WAGE) ACT	587
XXI.	MINERS' LIFE AND WORK	608

CHAPTER	PAGE
XXII. The Housing of Miners	637
XXIII. The Anthracite Coal Trade . . .	660
XXIV. Foreign Trade in Coal	675
XXV. Oil Fuel	694
XXVI. The Coal Question	718
XXVII. The Coal Question—Future of British Coal Trade	742
XXVIII. The World's Coal Resources . . .	772

APPENDICES

I. Collieries of the Northern Coalfield . .	799
II. Collieries of the Yorkshire Coalfield. .	801
III. Welsh Colliery Companies	802
IV. Sale Contract Form	805
V. Extracts from Coal Mines Act, 1911 . .	811
VI. Rules of the Miners' Federation of Great Britain	819
VII. Variations in Percentage Additions to Standard Wages in Different Districts . . .	825
VIII. South Wales Miners' Federation, Wages Agreement of 1910	826
IX. Extracts from Coal Mines (Minimum Wage) Act, 1912	842
X. District Rules for the District of South Wales and Monmouthshire	849
Bibliography of the Coal Trade . . .	857

LIST OF ILLUSTRATIONS

PLATES

FACING PAGE

1. PLAN OF MINE—LONGWALL METHOD OF WORKING
   COAL 204

2. ASHINGTON COLLIERY, NORTHUMBERLAND . . 212

3. ELECTRIC COAL CUTTER 213

4. INTERIOR OF THE COAL WASHERY AT THE BARGOED
   PIT OF THE POWELL DUFFRYN STEAM COAL
   CO., LTD. 228

5. COAL WASHERY AT THE BARGOED COLLIERY OF THE
   POWELL DUFFRYN STEAM COAL CO., LTD. . . 229

6. STEMMING THE SHOT 404

7. PONY HAULAGE 405

8. CLEARING OUT THE CUT 420

9. AFTER THE SHOT 421

IN THE TEXT

FIG. PAGE

1. SECTION THROUGH A PART OF THE COAL MEASURES IN
   SOUTH DERBYSHIRE 14

2. SERIES OF HORIZONTAL BEDS 21

3. RESULT OF GREAT PRESSURES SET UP IN THE EARTH'S
   CRUST 22

FIG.   PAGE

4. OUTCROPS OF COAL SEAMS FOR A SYNCLINE, AN ELON-
   GATED BASIN, AND A TRUE BASIN   .   .   .   23

5. SECTION ACROSS SOUTH WALES (N. TO S.)   .   .   25

6, 7. *Faults* CAUSED BY FRACTURES OWING TO EARTH
   MOVEMENTS   27

8, 9. SEPARATIONS OF COAL SEAMS   29

10. SUPPORTING A MASS OF COAL BY SPRAGS AND PROPS   209

11. GRAPH SHOWING CONDITIONS OF DEMAND AND
    SUPPLY   258

12. GRAPH SHOWING MARKED CHANGE OF PRICE IN A
    SHORT PERIOD OF TIME IN RELATION TO CON-
    DITION OF SUPPLY   261

13. SKETCH PLAN OF HOLBROOK COLLIERY EXPLOSION,
    27TH APRIL, 1913   392

14. SKETCH PLAN OF LODGE MILL COLLIERY, WHERE TWO
    MEN WERE SUFFOCATED ON THE 26TH JANUARY,
    1913   393

15. GRAPH SHOWING VARIATIONS IN FREIGHTS   *Facing*   686

THE COAL TRADE

CHAPTER I

INTRODUCTION

A GREAT industry by which in this country five
millions of people are supported, which has a history
extending over centuries, and which is the foundation of nearly all the manufactures and commerce
of this great Empire—how is such a mighty organisation of human energy to be described and explained
in one small book ? The task is indeed great, and
the author must perforce pick and choose, presenting
to the public a general outline of the whole, and
dealing more fully only with such features as will
be of general interest. There are scientific treatises
dealing with the origin of coal, and there are many
technical treatises dealing with all the methods of
mining and the problems and duties of mining
engineers ; there are even works devoted to the
book-keeping and business sides of coal-mining,
others dealing with the finance and economics of the
industry, and others yet with the sale and oversea
carriage of coal.

But all these books deal with stones and machinery
or facts and figures ; and it is not, I think, such
inanimate matter which really interests the great

public, but rather the human picture of the industry. A great army of men is engaged delving in the bowels of the earth for the " black diamond " ; a quest which kills too many men, and makes million- aires of others. What are these men like ? What is their work ? What is their home life ? What are their organisations and their political ideas ?

Owing to the nature of their occupation the miners are very much a class apart from the rest of the community, and their work is less known and understood by the people at large than it should be. I shall try in this book, therefore, to give a sympa- thetic picture of the miner's work and life, besides dealing with the economic and commercial aspects of the coal trade, and very briefly, by way of intro- duction, with scientific and technical matters. This chapter may serve as a brief summary of the principal subjects dealt with in the book.

Coal occurs in layers or seams, usually from one to ten feet thick ; and it originated from accumula- tions of the tree trunks and other vegetable matter becoming buried and mineralised during the course of long ages. It is found in certain " fields " or districts of which the principal fields in the British Isles are (1) the Northumberland and Durham field ; (2) the Yorkshire—Derbyshire—Nottingham —Lincolnshire field ; (3) the South Wales field, and (4) the Scotch field (Ayr—Lanark—Lothian and Fife). The lesser coalfields are (5) the Lancashire field ; (6) the Cumberland field : (7) the North Wales (Flint—Denbigh) field ; (8) Midland field, which includes the following, often spoken of as separate fields, although they are unquestionably connected underground : North Staffordshire, South Staffordshire, Shropshire (Coalbrookdale), Leicestershire, Warwickshire, and Worcestershire (Forest of Wyre) ; (9) Forest of Dean field ; (10) Bristol and Somerset field ; (11) Kent field. In Ireland the principal coalfield is (12) that of Leinster (Queen's and Kilkenny counties), whilst there is also (13) the Tyrone field in the North.

The Irish coalfields are worked on a very small scale, and their total output is scarcely more than that of one large colliery company in England or South Wales. It is probable that several of these English coalfields are in reality connected by the coal seams extending underground from one to the other (see Chapter IV). The Yorkshire—Nottinghamshire field undoubtedly extends eastward under the greater part of Lincolnshire, almost to the sea, as new borings are continually carrying further eastward the limit of coal proved. Further there is good evidence in favour of the theory that a concealed coalfield lies beneath large parts of Wiltshire, Berkshire, Oxfordshire, and Buckinghamshire.

Coal, although quarried here and there from the earliest ages, was first systematically mined in the neighbourhood of Newcastle and on the Firth of Forth in the fifteenth century, and it was carried to London by coasting boats during the sixteenth century, and known as "sea-coal." It was impossible to work far below the surface, as the mines became flooded out with water, until the invention of steam pumping engines, which began to be used towards the end of the eighteenth century in the Durham and Northumberland coalfield.

Staffordshire collieries were developed in the seventeenth century, and working along the out-crops began in South Wales in the middle of the eighteenth century. The great development of deep mining in South Wales, Scotland and Lancashire dates only from the middle of the nineteenth century, and there was not much trade in the export of coal to foreign countries until this time.

Coal was first used for heating houses, and its use for smelting iron dates from about 1735, and for raising steam from Watt's invention of the steam engine in 1782. At the present time, much the larger part of all the coal mined is used for raising steam, but much is used for gas for lighting purposes, some for conversion to coke, and some for burning in producer-gas engines. For the latter, anthracite is chiefly used, which is practically pure carbon and contains no tar or volatile matter. House-coal, on the other hand, is bituminous ; and when heated gives off much gas, oils, tar and other bituminous matter. Steam-coals are those containing less bituminous matter than " house " and " gas " coals, whilst " hard " and " dry " steam coals approach nearly to the anthracite.

Coal used to be got simply by quarrying at the surfaces, and later by cutting " levels " or " slants " into the hill-side, and working galleries from either side. The sinking of a shaft, and the use of a winding engine came with the exhaustion of coal near the surface. Now there are few levels and slants, except in Wales ; and over 90 per cent. of the total output of the country comes from deep pits. From the bottom of the shaft the workings are opened out radially in all directions along " main roads " and *headings*, and each hewer is assigned a working place on the " face " of the coal. The " face " is constantly extending outwards, assuming normally the shape of an irregular circle round the pit bottom, which may, however, be pushed out in one direction more than another if the coal is there easier to work, or of better quality.

A very large amount of capital must be spent in sinking a pair of shafts and erecting elaborate machinery before any return is obtained by working the coal. It is a somewhat risky investment, as difficulties may be encountered or the quality of the coal may be poor. Many economies are being realised in the working of collieries by amalgamation of a number of collieries in one ownership or at least under one control.

The principal class of workman employed underground is the hewer or collier, who actually " gets " the coal. He is mostly paid by piecework on a complicated scale and he generally earns good wages (30s. to 65s. per week), although they may be reduced by many causes, and fluctuate more or less in accordance with the price of coal. In nearly all mines the use of a safety lamp is imperative, so that the miner works under difficulties, almost in the dark. The other considerable class of underground workers is the hauliers, who fetch the coal from the face to the shaft bottom. They are paid a day wage and earn much less than the hewer.

Trade Unionism is strong amongst miners, although of later growth than amongst most of the great industries of this country. The first or general method of regulating wages was by the *sliding scale*, which determined how wages should vary in proportion to the price of coal. The fluctuations of wages between good trade and bad trade proved too great for the miners under this system ; and it has been replaced in each coalfield by a *conciliation board* composed of equal numbers representing masters and men. Each conciliation board determines from time to time the percentage at which wages in their district shall stand above the base which is separately fixed for every grade of labour in each seam of every colliery. In spite of the elaborate machinery of the Miners' Federations and Coalowners' Associations, and Conciliation Boards, the whole system broke down in 1912, mainly owing to the inadequate and uncertain earnings of men working in abnormal places, or who were otherwise prevented from earning a full day's pay. The result was the National Coal Strike of 1912, which has forced Parliament to establish other big machinery for providing a minimum wage for underground workers.

There is an extensive series of legislative enactments providing for the safety of workers in mines, and especially for preventing explosions of gas or coal dust. These Acts culminated in the Coal Mines Regulation Act of 1911, which imposes stringent conditions on the colliery proprietors and is designed to increase the efficiency of inspection. At present it seems to be inadequately enforced in certain collieries and districts.

The miners are to a large extent a class apart from the rest of the community, although their aloofness from the general public and detachment from other organisations, is showing signs of disappearing. In recent years there has been a great spreading of education amongst the miners, and the standard of life has risen considerably both in regard to the general level of intelligence and in regard to the degree of comfort expected in their home life. The miners include all sorts and conditions of men, drawn, especially in the newer coalfields, from the most varied employments ; but a surprisingly large proportion of the hewers, who are the *élite* of the miners, are serious, earnest men, devoting their spare time to reading or study in classes, or to work on local government bodies, or for the Federation, or in connection with socialist organisations, or with their church or chapel. The hewers, repairers, and a few other grades of workers are highly skilled and earn good wages in the majority of pits ; but the average type of hewer differs very much from one coalfield to another, and is evidently dependent upon the rate of wages earned. In other words, where the hewers have the opportunity of earning high wages with comparatively short hours of labour, as in Durham and Northumberland, and in South Wales, there is undoubtedly evolved in the second and third generations an aristocracy of workmen who are a very great asset to the whole country, and would be more so if they could be better housed and could be brought more into touch with the general current of English life. This is suggestive of the progress which might be possible for the working classes of all trades, if higher wages could be paid to them.

The coal trade is of enormous importance to this country and presents several features of peculiar interest, prominent amongst which is the rapid growth in recent years of our export trade in coal. This continues to grow, particularly from the Tyne ports and from South Wales, in spite of the opening up of coalfields in our own colonies and in so many other foreign countries. When compared with other staple commodities of English commerce, coal is seen to be heavy and bulky in relation to its value ; and it is the only single commodity of the nature of raw material which is produced upon a great scale in this country. Coal, therefore, has a peculiar influence upon our foreign trade, and there is a curious paradox that the export of coal, by reducing inward steamship freight rates and raising the home prices of coal, and so of other commodities dependent on it, at the same time both benefits and injures our home manufactures. The importance of coal as a raw material lies, however, in the fact that it is of service not in one industry only but in all ; and at the present time it is the original source of practically all our artificial heat and light and power. The energy obtained from oil, from water, and from wind is for industrial purposes practically negligible in comparison with the like services obtained from coal. A new industry is arising in the preparing of coal in various forms for the market, and in the manufacture of bye-products which are put to the utmost variety of uses, from driving motor-cars to making scents or delicate shades of colours. The establishment of extensive coking ovens with plant for the recovery and separation of bye-products is a new industry which has come over from Germany ; and those who look into the future may anticipate the time when coke, not coal, will be the usual form of fuel, because the bye products obtained from the bituminous part of the coal will be worth far more than the coal in solid form. Coke of the highest quality may indeed itself become a bye-product produced in such quantities that it becomes a cheaper form of fuel for household and general purposes than most qualities of coal now used ; and this is the only bright point in an outlook which is rather dark or the coal consumer.

The price of coal has not only been rising for many years past, but seems likely to do so in the future. This has been caused partly by the increasing stringency of the regulations for the safety of mines, partly by the increase of wages which has been very properly granted to miners, and partly as the result of the increasing cost and difficulty of working coal in thinner seams, or at greater depths, or in old mines at a greater distance from the shaft —all of which tendencies result from the exhaustion of the coal which is most easily and cheaply worked, and which was naturally, therefore, the first to be worked.

The late Professor Jevons wrote in 1863 upon the *Coal Question*, and foretold serious injury to the trade of this country ; not so much from complete exhaustion of the coal, which would take some hundreds of years, as from the rise of prices incidental to the working out of the coal seams which were cheapest to work. The discovery of extensive coalfields in South Yorkshire, Nottingham and Lincolnshire, in Warwickshire and in Kent, and the probability of coal being found in Wiltshire and Berkshire, which could not be foreseen at that date, may tend somewhat to postpone a serious rise of price which would deal a heavy blow to English industry and shipping, but the danger is nevertheless still with us ; and Englishmen must take heed that in the future, in commercial competition with other nations, we rely less upon exploiting our vast store of natural wealth, and more upon the resources which scientific skill and practical education can place at our disposal.

CHAPTER II

Nature of Coal

THE term " coal " includes those mineral substances
which have resulted from the decomposition and con-
solidation of vegetable matters, and which now occur
in the fossil state. Ordinary coals are always black
in colour, but differ very much in other properties,
coal from some seams being bright or glistening,
from other seams dull black. Sometimes the coal
is broken only with great difficulty with a heavy
hammer ; at other times it is quite brittle, almost
friable, and falls to powder if dropped from a few
feet high, or touched with a pick or hammer. Some
coals break into large, irregular, square or rounded
blocks ; others are especially flaky in character and
contain white laminæ, or films of mineral matter,
which add greatly to the amount of ash remaining
when the coal is burnt.

All of the properties just enumerated are of great
economic importance ; and it is essential for a proper
understanding of the coal trade to have familiarity
with the properties of coal. For example, the cutting price paid to the collier or hewer, when he
wins coal from the face, varies considerably with the

11

hardness and other physical properties of the coal. He must be paid at a high rate per ton for the very hard coal, the lowest rate for coal which breaks up easily into lumps of the right size, and a high rate where the coal is very friable and much of it falls away in dust, which it is useless to send up to the surface. This is particularly the case in South Wales, where the steam coal is required in large lumps, and the hewer is paid for the large coal only. Again, brittleness of the coal affects its handling. Where the coal is brittle, the tracks underground should be well laid, the railway wagons should have spring buffers, and the coal must not be tipped from a great height into the wagons, or from wagons into the ship's hold.   It was especially to avoid breakage of the coal through tipping from a great height at the Cardiff Docks, that the Lewis-Hunter Crane was invented, which takes half a truck-load of coal at a time, and lowers it to the bottom of the hold, where the bottom of the box is opened and the coal is distributed without falling any distance to speak of. " Self-trimming " vessels, the decks of which can be opened all over, are built specially for use with such cranes.

Origin of Coal

There is plenty of evidence in every coal mine from the beautiful fossil remains of various kinds of ferns, tree trunks, and reed plants found associated with the coal, that all of it is the result of accumulations of vegetable materials which have subsequently been covered up by sand, clay, and other deposits, the whole having been converted through long ages to a solid mass of rock. There are two opposing views commonly held as to the exact origin of coal. Some geologists regard many of the seams in this country as having been formed of drifted material which collected in large shallow lagoons or estuaries. Probably the land must have been sinking very slowly into the sea, so that it was possible for a layer of vegetable matter, 20 or 30 feet thick, to accumulate, and this, owing to the very great weight of rocks subsequently formed above it, has been compressed into a coal seam three or four feet thick. On the other hand, there are geologists who lay stress upon the theory of growth *in situ*, supposing that vast forests grew upon low-lying land with trees falling and rotting where they lay, so that layer upon layer of decaying vegetable matter underlay the still growing forest, as we now find in many of the tropical forests of Africa or the Amazon.

There is no need for us to weigh up the relative merits of the two theories—indeed the truth probably lies with both, as many seams provide so much evidence in favour of the one theory, and many seams good evidence in favour of the other theory. A seam of coal behaves like a layer of rock, varying in thickness from one foot up to 30 or even 40 feet ; and from the point of view of coal mining the principal points are the nature of the coal itself, and the thickness and angle of dip of the seam.

The Coal Measures

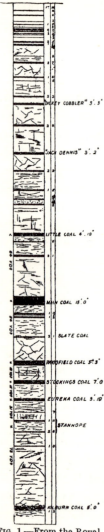

FIG. 1.—From the Royal Commission on Coal Supplies. 1st Report, 1903.

One aspect of the origin of coal is, however, of great practical, and therefore, commercial importance. The strata or layers of sandstone and shale in which coal occurs are called the coal *measures*. They are usually some thousands of feet thick, containing at irregular intervals a number of seams of coal varying from 1 foot to 10, or even 20 feet in thickness, but mostly from 2 to 6 feet thick. According to Mr. Aubrey Strahan the coal measures " are built up of repetitions of a certain definite sequence of deposits, sandstone or conglomerate (sandstone full of pebbles) being succeeded by shale, and shale by coal. By the repetition of this sequence each coal would be overlaid by a sandstone, and as a matter of fact this relation of coal to rock is found to hold in the majority of cases." A typical section through a part of the coal measures in South Derbyshire is shown in Fig. 1. It shows 27 seams of one-foot thickness and

over, and several thinner ones. The total thickness
of all the seams added together usually varies from
60 to 120 feet. In one part of Glamorganshire there
are 48 seams yielding 125 feet of coal.

Economic Products other than Coal

The sandstones are generally hard and bedded,
and can sometimes be used as building stones, if it
is found necessary to bring them out of the pit.
The shales are a soft flaky stone, easily cut with the
penknife, and rather like soft slate. They are only
hardened clay ; and by some years' exposure to
weather on the colliery rubbish tips, they disinte-
grate to a good soil. With the improved brick-
making machinery now on the market it is becoming
a practice to grind up the shale dumped out of the
colliery wagons and convert it into pressed bricks,
the quality of which is often excellent. Both sandstones and shales are dark grey, bluish, or almost
black, due to their containing a small percentage of
carbonaceous matter. There are bands of the sandstone, and more often of the shale, in which the
percentage of carbonaceous matter increases so
much that it is possible to burn the rock or shale in
a furnace with a good draught. Such carbonaceous
stone leaves when burnt such masses of ash that it is
commercially worthless. It is possible to find, however, every stage connecting more or less carbonaceous rocks through a series of very ashy
coals to pure coal. Some of the shales contain a
fair percentage of mineral oil ; and in Scotland a

considerable industry has arisen in the distillation of oil from shale and its subsequent purification.

Fire-clay, because it is almost pure clay (silicate of alumina), will stand a great heat when made into bricks. It is found in the coal measures inter-stratified with sandstones and ironstones, frequently underlying a coal seam. ' Ganister ' is a fine-grained, highly siliceous clay or shale used for lining furnaces, found especially in the lower coal measures of the North of England.

Very important in the past have been the beds of clay-ironstone, which occur in the coal measures of many districts in bands, seams, or nodules. It is hard and bluish when freshly uncovered, but weathers yellow and brown with rust if left exposed. Sometimes the ironstone is black, and is called " black-band," containing in itself sufficient carbonaceous matter to allow it to be calcined in furnaces without admixture of coal or coke. The clay ironstone of the coal measures, especially of Staffordshire and South Wales, used to be the chief source of iron smelted in Britain. Owing to the discovery of the richer or hæmatite ores in Cumberland and elsewhere, which are comparatively free from phosphorus and other impurities detrimental to steel-making, and to the low cost at which high grade ores can now be imported from Spain and other foreign countries, the iron-ores of our coal fields are almost neglected, except for the rich " black-bands " of Scotland and North Staffordshire.

It is curious to note that the industrial development of South Wales began, not so much with the mining of coal as with the working of beds of clay-ironstone occurring in the coal measures there, which was smelted with charcoal in various places. At Dowlais the smelting of iron with charcoal was firmly established by the energy of John Guest, soon after his arrival in 1760, and a few years later he began smelting with coal. The greatness of Merthyr and Dowlais in the iron industry may be attributed to the fact that here were rich beds of ironstone, associated with easily worked coal seams, whilst the two other requisites of iron-making, namely, limestone and furnace sandstone, were also to be had on the spot in any quantity. Now Dowlais ironstone is worked no more, the Welsh steelworks, amongst which Guest, Keen and Nettlefold's is foremost, deriving their ores almost wholly from Spain.

Geological Position of the Coal Measures

In this country the coal measures overlie the mountain limestone, which is so widely spread in the North of England, and again in the South-West. This limestone and the coal measures together form what is called the Carboniferous Series. Beneath it lies the Devonian Series, below that the Silurian Series ; above the Carboniferous lies the Permian Series, then the Triassic Series, next the Jurassic and then the Cretaceous Series. Each " series " is a distinct kind of rock easily recognised as a rule by its appearance, but known with cer-

B

tainty by the distinct kind of fossils it contains ;
and as each series has been laid down upon the one
beneath, the lower must be older than the upper.
The Cretaceous rocks are possibly a million years
old ; but it is not the actual age of the coal measures
which is of practical importance, but the relative
age of the different series of rocks, as evidenced by
the order of succession in which they are found.
For instance, since the mountain limestone always
underlies the coal measures, it is pretty obvious that
when the mountain limestone appears now on the
surface of the ground, the coal measures must have
been washed away and removed altogether by
ages of denudation by rain and rivers or the sea.

In the nineteenth century persons prospecting
for coal by boring sometimes were absurd enough
to start operations by boring through a rock which
any geologist could have told them was older than
the coal measures, so that there was no possible
hope of finding coal beneath it. Now-a-days
mining engineers are consulted who have a good
knowledge of geology, and expert geologists are
often called in when there is a difficulty in locating
the coal.

Excellent examples of the geologist's assistance
are afforded by both the new Kentish coalfield and
by the South Yorkshire and Lincolnshire coalfield.
In each case the coal measures lie under a deep
covering of newer rocks, some thousands of feet
thick. It was entirely by geological reasoning that
the presence of coal measures under the chalk and

other cretaceous rocks of Kent was thought to be probable. Several borings under London and its neighbourhood had revealed rocks older than the coal measures underlying the Jurassic and Cretaceous series ; and it might be that the basin of the Belgian and French coalfields extending from Mons to Calais would extend under the Straits of Dover and under most of Kent, the basin having retained the coal measures between the older and newer rocks. Several boreholes near Dover have proved this surmise to be correct, and coal is now being extensively worked from collieries in Kent, the first cargo of Kentish coal having been exported from Dover in 1913. Again, geologists told us that there must be workable coal seams in South Yorkshire around Doncaster ; and now a number of new and large collieries are working them. We are told also that there is almost certainly good coal under the greater part of Lincolnshire and Nottinghamshire and Warwickshire ; whilst it is probable that coal may be found at some depth under Wiltshire, Oxfordshire, and Buckinghamshire. It is most desirable in the national interests that the Geological Survey should be provided with sufficient funds to conduct a detailed search for the concealed coalfields.

Not only can the geologists locate the coal measures, they can also distinguish different strata or " horizons " in the coal measures themselves, partly by their physical characters and partly by the fossils they contain. The " millstone grit."

for example, which is a hard coarse sandstone, so called from its extensive use in Yorkshire in former times to make grinding stones for wind and water mills, can be easily distinguished from the sandstones occurring higher up in the coal measures. The different bands of shale can be distinguished by their characteristic fossils, which differ somewhat. An expert can thus considerably assist a mining engineer who is trying to locate a particular seam, which is often a difficult matter in a place where the coal measures are 15,000 feet or more thick and it is uncertain what part appears at the surface. It is a case where a few days of an expert's time are far cheaper than cutting out tons of rock in trial levels or shafts, whilst the geologist's assistance is necessary to interpret the results of boring as it proceeds.

Formation of Coal

The strata of the coal measures, like those of the other series, were all laid down horizontally on the floors of great estuarine lakes. Around the British coast at the present day some of the rivers, where they widen out to estuaries, have mud bottoms and mud banks ; others have sandy bottoms and sandy banks. It was so in the coal period. We can find now at many points along our coasts (Norfolk, Devonshire, North Wales, etc.) many remains of sunken forests, the trees being of the same kinds as now grow in this country, proving clearly that the coast has sunk many feet relatively to the sea level during geologically recent times. It was probably after such a sinking of the land surface during the Carboniferous period that the tides washed in quantities of sand, until the estuary gradually silted up and a bar was formed separating it from the sea. Then the only deposits would be the muds brought down by the rivers ; and as the estuary became shallower and shallower, vegetation would take root and would grow thicker and thicker. The tropical forests stretching from the low banks and islands right into the water, would be a confused mass of fallen tree trunks, ferns and creepers, the shores being blocked up by masses of floating deadwood carried down from

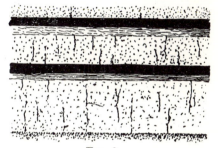

FIG. 2.

the upper reaches of the rivers. These conditions can be seen at the mouths of the great tropical rivers of to-day—the Mississippi, the Amazon, and the Congo.

A further sinking of the land leads to the cycle : sands, muds, vegetation occurring over again, so that the net result is a series of horizontal beds lying as shown in Fig. 2. In the course of ages these beds have become hardened into rock ; but in many districts they still lie nearly horizontal, just as they were laid down, except that they have been raised above the sea level. When the coal

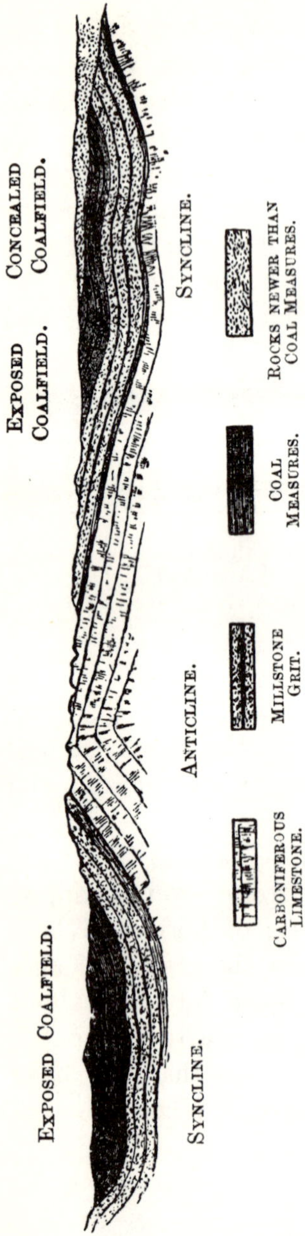

EXPOSED COALFIELD.

CONCEALED COALFIELD.

EXPOSED COALFIELD.

SYNCLINE.

ANTICLINE.

SYNCLINE.

ROCKS NEWER THAN COAL MEASURES.

COAL MEASURES.

MILLSTONE GRIT.

CARBONIFEROUS LIMESTONE.

FIG. 3.

thus occurs in unbroken horizontal seams it is a point of great commercial importance as it can be so much more cheaply worked. As a general rule, however, the coal measures, like other series of rocks, have suffered from what are called "earth-movements." The slow cooling of the earth as a whole means that it contracts in size; and the "crust," or outer layers, become wrinkled, or folded on a huge scale, just like the peel of an orange or skin of a plum, which has shrivelled up through being dried. Great pressures are set up in the earth's crust (and sometimes tension), and the result is that the rocks get bent and folded as shown in Fig. 3.

Where the fold is concave upwards it is called a *syncline*, where convex it is named an *anticline*.

The beds are sometimes forced into the shape of a basin, or raised like a dome. A syncline which is more pronounced in the centre than at the ends, really forms an elongated basin. If the strata shown in Fig. 3 are intersected by a horizontal plane to represent the surface of the earth, the line of intersection of any particular bed with the plane is called its *outcrop*. The outcrops of coal seams for a syncline, an elongated basin, and a true basin, are clearly shown in Fig. 4.

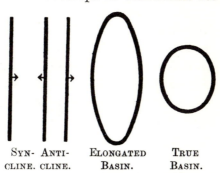

SYN-   ANTI-   ELONGATED   TRUE
CLINE.   CLINE.   BASIN.   BASIN.

Fig. 4.

The Coalfields

We are now in a position to understand the disposition of the coalfields—indeed why coal occurs in fields at all. There were probably coal seams originally deposited over practically the whole of England and South Wales, over much of the South of Scotland and over most of Ireland, excepting only Devon and Cornwall, Central and North Wales, and the mountainous parts of Scotland, all of which districts, with a few smaller islands, were probably elevated land whilst the coal measures were being deposited. At the end of the Carboniferous period, there must have been a thick continuous bed of coal measures lying flat and undisturbed over much the greater part of these isles. Individual coal seams stretch over an area of two or three counties, and perhaps more ; but where one thins out and gradually disappears another takes its place at a higher or lower level.

During the subsequent ages, beginning probably soon after they were deposited, the coal measures have been disturbed by earth movements from their original horizontal disposition. In common with the series of strata above and below them they have been bent and folded into anticlines and synclines, or rather mainly into domes and basins, more or less elongated. It is chiefly the basins and synclines which form the coalfields of to-day.

To understand why the coal measures are now found principally where they have been bent into basins it is necessary to understand what is called " denudation," which is the process by which rocks are " weathered " and the material carried away by the rivers to the sea. If you watch the surface of the roadway during heavy rain you will see innumerable little streams running off towards the gutter, each carrying along sand and mud. Here and there they form beautiful little deltas—almost exact models of the great deltas of big rivers. On steep roads or paths the streamlets cut out little valleys and gorges which cost the roadman some trouble to fill up. Wherever there is open sandy country with not much grass, the same action can be seen going on, and it goes on to some extent, even in turf, although the blades of grass are of course continually arresting the motion of the grains of sand and earth.

Since water always runs downhill, this denuding action can only result in the solid particles being carried downhill, so that earth is always being washed down from a higher level to a lower. This gradually lowers the whole surface level of the hills, and in the course of hundreds of thousands of years

SECTION ACROSS SOUTH WALES (N to S).

FIG. 5.—1 = Old Red Sandstone   2 = Carboniferous Limestone;
3 = Millstone Grit ; 4 = Lower Coal Measures; 5 = Pennant
Sandstone ; 6 = Upper Coal Measures.

a great mountain range is gradually worn down and denuded until it becomes almost a flat plain. Frost plays a prominent part in breaking up rocks and earth, and so does a hot sun followed by cold nights. Yet, of course, sands and clays yield much more easily to denudation than hard rocks of any kinds, as rain plays the greatest part, so that it is not uncommon to find the valleys lying along the out-crop of a soft bed of clay or sand, whilst the hard strata of sandstone or limestone form ridges or ranges of hills along their outcrop. Fig. 5 is a section from north to south across the southern edge of South Wales coalfield, where there are two well

marked ranges of hills due to the mountain limestone and the Pennant sandstone respectively.

Strange as it may seem it is not the arch or
anticline which now forms the highest hills, but
rather the synclines. This is because the beds
slope outwards in an anticline, and because hard
layers of rock are always traversed by innumerable
cracks or " joints " which divide up each layer into
angular blocks. Quarrymen know how to make
use of these joints, always working the beds uphill.
For the same reason the denuding forces of nature
can work best uphill, for the blocks loosened by
water and frost soon slide out of position and the
face of a cliff disintegrates rapidly. On the other
hand, where the beds slope in towards the hill,
every block has a tendency to remain where it is.
The effect is the same in the case of another very
powerful denuding agency known as " soil creep."
On every hill slope the whole of the upper two or
three feet of soil moves down the hill side, by sometimes as much as one or two inches a year, being
removed at the bottom by the river or stream eating
away the banks. Where the beds slope with the
hill surface, blocks are easily wrenched off and
carried away by the soil ; but beds sloping into the
hill tend to stop the soil creep. When the beds are
bent in the shape of an arch or anticline, the blocks
slide away on each side, whilst both sides of a
syncline retain them.

This explains why our coalfields are mostly
basins, where denudation has proceeded far, and

why at the same time they are often hilly or mountainous regions.

Geological Faults

The strata are not only bent by earth movements, they also break. Such fractures are often accompanied by some vertical and lateral movement of one side relatively to the other, and this produces a *fault* (see Figs. 6 and 7). When the broken ends of a coal seam or other stratum tend to separate horizontally from each other as in Fig. 6, the fault is the result of tension in the strata; but when the broken ends come one over the other, it is called a

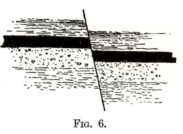

FIG. 6.

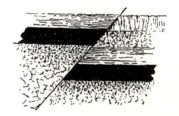

FIG. 7.

reversed or *thrust* fault (see Fig. 7), and is due to lateral compression of the strata. Cases of *thrust* faults are known where a stratum appears to be duplicated, because it has been pushed many hundred feet over itself along a nearly horizontal fault-plane.

Faults are of very great importance in mining, and are the bane of the mining engineer and the colliery proprietor. You may work a seam back

further and further from the bottom of the shaft
and everything goes smoothly and pays well, till,
in a certain direction you find there is no coal whatever—nothing but stone. The engineer is then
confronted with a tremendous problem : where has
the coal gone to, up or down, and how far ?   Often
the first question—up or down—can be decided
from the inclination (or " *hade* ") of the fault and
a knowledge of whether ordinary or reversed faults
are the rule in the district ; also sometimes by a
leader of shattered coal being found along the fault
plane. How far the coal has gone can only be
decided very approximately from examination of
the strata and general geological considerations ;
and in such a difficulty the combined efforts of a
mining engineer, an expert geologist, and an official
who has worked for years in the pit will give the
best results. When the probable location of the
coal is decided, it is necessary to decide how to
reach it : whether to continue the existing heading,
with a different slope, or to drive a new one altogether at great expense. Many a colliery Board
Meeting has begun in perplexity and ended in perplexity, the question being : how many more
hundreds or thousands of pounds can we spend
driving headings through stone to find the coal ?
It is often a huge, but necessary, speculation ; and
many previously successful colliery owners have
come to grief in this manner.

   It is evident that the average cost of mining in
any district must increase considerably if faults are

common. Seams absolutely or nearly flat, and
unbroken by faults, such as are found in the Midlands and South Yorkshire, and are so extensive in
the Middle States of America and in Canada, can
be mined very cheaply, particularly with coal
cutting machines. Where the strata are disturbed
by faults or *rolls* (*bends* or *waves*), there is too much

FIG. 8.— C **=** Coal Seam.   Strata A and B are below it, and stratum
D above it.

FIG. 9.

dead-work in removing stone without coal, and
increased cost in haulage through the roads having
to turn up or down to follow the coal or be specially
cut.

The Belgian coalfield is very much disturbed, and
coal-seams are sometimes found standing vertically
or even overturned. In this country the South
Wales coalfield is the most disturbed. Besides
faults and *rolls*, there are what are termed *wash-outs*,
where the coal either becomes very thin or disappears altogether for a few yards, without there
being any fault. It was thought formerly that this

phenomenon was due to the vegetable deposit having been washed away soon after it was formed by a river branch becoming diverted over it as often happens in great deltas ; and no doubt this is a satisfactory explanation of some cases. In others, however, the disappearance of the coal appears to have resulted from the plane of a thrust fault passing actually along the coal seam or at a very acute angle with it, so that the broken edges of the seam become separated by a few yards along the direction of the coal seam itself, as in Fig. 8. The formation seen in Fig. 9 is often produced where the middle section has been thrust forward between the lower and upper sections.

Speaking generally, in the South Wales coalfield the disturbance of the strata increases in a westerly and northerly direction, until in the anthracite region it becomes intense, a fact which largely accounts for the high price of anthracite as compared with other coals. In many of the anthracite mines there is a constant struggle with faults, and the output of coal may at times become seriously reduced though the same number of men is employed —naturally with disastrous results financially.

CHAPTER III

USES OF COAL

" Coal " to the uninitiated means something pretty definite, but to the dealer in coal it is only the name of a whole class of substances, and, without further qualification means about as much as the words " cloth " or " paper." When you order a ream of paper you specify the size, the thickness, the texture, the colour, whether ruled or unruled ; and it is really, for most purposes, just as necessary to specify carefully the kind of coal required. Coal varies as much as paper, but, unfortunately, it is not easy to recognise its qualities except by actually burning it for trial.

There are two principal ways in which coal is used : by burning it, and by distillation. With the commoner modes of burning coal we are all more or less familiar ; but a few words as to the process of distillation are necessary. All that this means is the heating of coal in a closed steel vessel, usually called a retort, to such a temperature that the various constituents of the coal are decomposed. A great variety of gases and liquors are thus produced, which pass out of the retort by its only outlet into a series of cooling pipes where they are condensed.

There remains behind some kind of coke, which is only reckoned of commercial value when it forms a bright, hard, spongy mass and burns so as to leave little ash. Coke which is soft and powdery, or full of ash, is of little use. The distillation of coal is carried on principally for the manufacture of illuminating gas, when the liquors and coke are bye-products ; but also to a growing extent, apart from gas works, for the sake of getting good quality coke, when the liquors and the gas are bye-products. Thirdly, the chief object of the distillation is occasionally the liquors, the gas and coke being bye-products. The production of oil fuel from coal for the use of motor-cars falls under the last heading, and is of growing importance.

We may, perhaps, regard as a third distinct use of coal its application to smelting iron and other metals, where it is mixed with the molten metal in the furnace.

The chemical analysis of coals shows that they all consist principally of carbon—the purest form of which is lamp-black—and of hydrogen and oxygen, the two gaseous elements of which water is composed, and which enter into almost every organic compound. Coal also contains small quantities of nitrogen and sulphur which disappear in the burning, and also more or less earthy matter which remains behind as ash. The sulphur varies much in quantity, but its presence in many coals is easily detected by the sulphurous character of the smoke ; and, now that the law no longer requires gas companies to extract sulphur compounds from gas supplied to houses, the sulphurous smell of burnt gas is good evidence of the sulphur present in the coal used by the gas companies.

Composition of Coal

The following analyses of the coals taken from the various localities indicated will sufficiently illustrate the differences in ultimate composition of the different classes of coal in use.

	1 Anthra- cite. Emlyn, S. Wales.	2 Smoke- less Steam. Great Western Clly. Rhondda S. Wales.	3 Steam Coal. N. Wal- bottle, North- umber- land.	4 Gas Coal (bitu- minous). Wigan, Lancs.	5 Best House (Bitu- minous). Nailstone, Leicester.
Carbon	91·20	87·56	82·14	80·07	74·91
Hydrogen	2·95	4·09	5·05	5·53	4·98
Oxygen	2·46	3·67	6·80	8·08	14·99
Nitrogen	1·13	0·89	1·18	2·12	1·22
Sulphur	0·80	0·71	1·17	1·50	0·86
Ash	1·46	3·08	3·66	2·70	3·04
	100·00	100·00	100·00	100·00	100 00

The analyses are all made on dried coal, or calculated free of water. Coal, as hewn, usually contains from 0.05 to 12.00 per cent. of water. It will be noticed that the proportion of carbon is greatest in the anthracite and least in the bituminous coal,

C

whilst the volatile constituents, principally oxygen and hydrogen, increase as the carbon decreases. The percentage of sulphur and of ash is very variable, and has no relation to the percentage of carbon.

Whilst the ultimate analysis of coal, which states its composition in the above form, is of the greatest importance to the scientist, it is too tedious and costly a process for general commercial use. An adequate method of analysis for business purposes is that which divides the constituents of coal into four parts : (1) moisture, (2) volatile matter, (3) fixed carbon, (4) ash. The process is quite simple, though considerable care is needed to carry it out accurately. A weighed sample of powdered coal, usually taken from different parts of the mass, is gently heated to slightly over the temperature of boiling water, and cooled in a dry atmosphere, and then weighed again. The difference is the amount of moisture. The sample is next placed in a closed crucible pierced only by a small hole for the escape of gases, and it is heated in a gas furnace to a white heat. The whole of the gas and liquors of a bituminous character are driven off ; and there remains only the fixed carbon, which in some cases would be a good coke. When cold the crucible is weighed, and the difference gives the volatile matter. The crucible is again heated white hot, but this time open, and in a current of oxygen, whereby all the carbon is burnt away, and there remains solely the ash, the amount of which is accurately found by weighing the crucible with it

and without it. The amount of sulphur present
may, if desired, be separately determined by another
process.

The following are a few typical analyses obtained
by this method ; but it should be pointed out that
the fixed carbon thus found is not the same figure
as the content of carbon obtained by ultimate
analysis, because some of the carbon is carried off
in the gases and liquors as volatile matter.

	1	2	3	4	5
	Anthracite. Cawdor, S. Wales.	Smokeless Steam. Great Western Colly. Rhondda, S. Wales.	Steam and Gas. Bearpark, Brancepeth, Durham.	Manufacturing and House (bituminous).	Steam (second quality, bituminous). Arley, near Coventry.
Fixed Carbon .	89·65	77·39	68·44	55·80	47·78
Volatile Matter .	6·30	18·33	27·62	31·15	38·24
Moisture	1·29	1·24	0·84	8·26	8·52
Ash .	2·76	3·04	3·10	4·79	5·46
	100·00	100·00	100·00	100·00	100.00

Another very important test of the quality of coal
is by measuring what is called *calorific value*. This
means determining the actual amount of heat which
it produces whilst being completely burnt. The
test is made by burning a weighed quantity of coal

in a small furnace under a copper bell immersed in a weighed quantity of water. The rise of temperature of the water is measured by a thermometer; and when account is taken of the weights of water and coal used, the increase of temperature indicates the amount of heat developed. As coal is usually wanted for producing heat, it would seem that all buyers ought to determine the calorific value of coals they intend to purchase; because it is really a certain amount of heat, rather than a certain weight of coal, which the manufacturer wants to purchase with every sovereign expended on buying coal. It is, indeed, the custom of most Continental buyers to specify that coals delivered shall be up to a certain standard of calorific value. There are, however, a number of other considerations which affect the price of coal, such as the amount of labour required in stoking, the amount of waste in small coal, dirt, or ash, and the degree of cleanness or smokiness with which it burns.

There is also a peculiar point connected with the utilisation of heat developed in burning coal. Many bituminous coals will show a higher calorific value than Cardiff best steam coals; yet, in practice, the latter has a greater evaporative effect in a boiler furnace than the bituminous coals. Possibly the explanation is that the volatile matter comes off so rapidly from the bituminous coals in the furnace that much of it, including the smoke, is incompletely burned; whereas, the South Wales steam coals contain just the right amount of volatile constituents to allow of their being fully burnt in the furnace chamber itself.

Classes of Coals

The various kinds of coals and their properties in burning are so important that it is desirable to give some detailed particulars. We shall take them in order, according to the proportion of volatile matter, beginning with the most bituminous coals.

(1) *Highly Bituminous Coals.* These contain over 40 per cent. of volatile matter, and burn with a characteristic long and brightly luminous flame. Coals of this class are used to some extent in gas manufacture and for household purposes. They are not good for steam raising in boiler furnaces.

(2) *Bituminous Coals.* This is a very large class, which may be taken to include all those with from 20 per cent. to 40 per cent. of volatile matter. They are used for innumerable purposes, according to their qualities in burning and distillation. Nearly all can be used for house coals, but they differ very much as to brightness of burning, amount of ash, sulphur, and smokiness. The bituminous coals contain much gas, and are largely used by gas companies. Although production of coke is not the gas companies' principal object, yet coke is so important a bye-product that they choose, by preference, the coal which forms a fairly hard coke ; or at least, a mixture of coals is used containing a

large percentage of coal which cakes well on burning. On the other hand, for raising steam, coking-coals are avoided ; for it is obviously difficult to stoke and maintain a uniform bright fire if the coal, as it burns, keeps setting into hard masses. For boiler purposes, the coal wanted is one which burns brightly to detached cinders, which themselves burn almost completely away in the furnace. The coking property is best developed in coals of this class, low in oxygen, and having from 25 to 30 per cent. of volatile matter.

(3) *Semi-bituminous Steam Coals.* The volatile matter is from about 14 to 20 per cent., and this class includes the finest steam coals of this country. As a rule they burn with little smoke, and give a bright hot fire with little caking. The best Welsh steam coals, the so-called " Admiralty " steam coals which are used for the navies of many countries, and for fast mercantile ships, are included in this class. They give an exceedingly hot and quite smokeless fire, and burn freely and completely without raking, so that there is a great saving of labour in stoking. If the usual turning and raking of the fire is done, it is pretty certain the fire-bars will be burned through.

This class also includes excellent coke-making coals, and a kind of gas-coal which is rather poor in illuminating properties. There are household coals also in this class which, however, are not much used outside South Wales ; they are rather difficult to light, and burn with little flame or smoke ; but in a large grate, or closed kitchen range, give a very bright hot fire.

(4) *Dry or Hard Steam Coals.* These contain a comparatively small proportion of volatile matter ; namely, from 8 to 14 per cent. They are rare in Great Britain, except in South Wales. They are free burning, and are very satisfactory for steam raising where there is a good draught. There are but few coking coals in this class, though they may be used in a mixture with class 2 or 3. They are of little use as a household coal, except in closed stoves.

(5) *Anthracite.* This name is strictly applied only to coals containing less than 8 per cent. of volatile matter, though for market purposes many coals, both in South Wales and Pennsylvania, are denoted anthracite which really belong to class (4). True anthracite is used for a great variety of purposes, where the desideratum is a small or moderate-sized supply of heat, with freedom from smoke and dirt. It is of no use for steam raising purposes ; and only burns in an open grate as a dull, flameless fire, which has to be started with bituminous coal. In specially constructed closed stoves it is widely used for heating houses and offices, especially on the Continent. It is also used for central heating furnaces of houses, hotels, etc., and for greenhouses. For commercial purposes it is used for drying malt, for curing rubber in the plantations of Malay, Ceylon, etc. ; and very widely as a fuel in producer-gas plants for internal combustion engines of moderate size.

It should be stated that the above figures giving the percentage of volatile matter of the different classes of coals, assume that the coal is freed from ash, the percentage of ash being deducted before the amount of volatile matter is calculated. If this were not done a coal might be shifted from one class to another, merely through containing a greater or less proportion of ash. It is generally supposed that the amount of ash present in a coal is quite independent of the content of volatile matter, but a suggestion of a connection in the coals of South Wales has been made by Mr. A. Strahan.[1]

Coals are very variable in the quantity of the ash they contain, and the coal lessens in value more than in proportion to the increased quantity of ash ; there is, of course, actual loss of heating power in proportion to the increase of ash, but it is mainly the trouble in keeping fires and furnaces clean which leads to the general dislike of ashy coals.

Consumption of Coal

There are no actual statistics, and very few reliable estimates, obtainable as to the consumption of coal for domestic purposes, and in the various industries of this country. The most reliable estimate is that made for the year 1903 by the Royal Commission on Coal Supplies from information collected from many different sources. The following are the figures :

[1] See Geol. Surv. Memoirs : *The Coals of South Wales.*

COAL CONSUMPTION IN THE UNITED KINGDOM.

		Tons.
Railways (all purposes) . . .		13,000,000
Coasting Steamers (bunkers) . .		2,000,000
Factories		53,000,000
Mines		18,000,000
Iron and Steel Industries . . .		28,000,000
Other Metals and Minerals . .		1,000,000
Brick Works, Potteries ⎫		
Glass Works . ⎬ . . .		5,000,000
Chemical Works ⎭		
Gas Works		15,000,000
Domestic		32,000,000
		————————
Coal consumed in 1903—Grand total .		167,000,000

This estimate, although ten years old, probably still represents accurately the relative proportions in which coal is used for different purposes, as it is only very gradually that an increasing or decreasing proportion is used for any one purpose. It is probable that the consumption has increased in all uses, as the total British consumption in 1913 was 189,693,000 tons. It is worth noting that the principal class of consumers is the factory owners, whilst domestic consumption ranks next, and the iron and steel makers rank third; mines, gasworks, and railways, are the next in order of importance. The following few statistics of consumption are collected by the Board of Trade annually :—

CONSUMPTION OF COAL IN THE UNITED KINGDOM, 1911.[1]

Purpose.	Quantity in Tons.
Production of Pig Iron . . .	19,218,491
Gas Undertakings	15,635,012
Coke Ovens	19,123,625[2]
Locomotives	12,821,641

There are some estimates available of the consumption of coal in other countries, and it is worth while to quote one made for France. The quantities used for different purposes are expressed as percentages of the total production.

COAL CONSUMPTION IN FRANCE.

	per cent.
Metallurgy	17·5
Railroads	13·6
Mines	7·5
Gasworks	7·2
Marine Purposes	2·3
Various Industries	31·4
Domestic Use	20·5

Another more recent estimate is that contained in *Business Prospects*, 1914.[3] This statement analyses the uses to which the 550 million tons of

---

[1] Coal Tables, 1912 (published 1914). The figures for 1911 are given in preference to those for 1912, as in the latter year the consumption was below the average on account of the National Coal Strike.

[2] Amount of coke produced was 10,720,352 tons. To get the above consumption of coal it has been assumed that 100 tons of coal are necessary to produce 60 tons of coke.

[3] Published by the Business Statistics Publishing Co.

coal assumed to be produced in the British and
Continental Coalfields were put.

ANALYSIS OF USES OF EUROPEAN COAL, 1913.

	Percentage.	Equal in tons per annum.
Consumed at Collieries . .	7	38,500,000
Used in Iron, Steel and other Metalliferous Works . .	35	192,500,000
Railways and Tramways .	10	55,000,000
Steamships' Bunks . .	10	55,000,000
Gas and Electric Lighting .	8	44,000,000
General Manufactures . .	15	82,500,000
Domestic Consumption . .	15	82,500,000
Total . . .	100	550,000,000

In the 1915 issue of the same publication is given
an analysis of the uses to which the 190 million tons
of coal retained for home consumption in 1914 were
put.

CONSUMPTION OF COAL IN UNITED KINGDOM, 1914.

	Percentage.	Quantities. Million Tons.
Metallurgical Industries . .	29	55
General Manufactures . .	29	55
Railways	7	13
Gas Making	9	16
Domestic Consumption . .	15	30
Bunkers shipped at ports in United Kingdom . . .	11	21
	100	190

Development of Power by Steam

The remainder of this chapter will be occupied with some description of the great variety of ways in which coal is used to produce power, heat and light ; whilst I shall endeavour to be of some practical service to users of coal, whose name is legion, by indicating the methods and appliances by which economies in consumption can be realised. Some of the facts and suggestions in what follows are based upon the very valuable report of the Royal Commission on the Coal Supplies.

By far the greater part of the power used by factories, mines and works of all kinds—by steamships, electric generating stations, etc.—is obtained by steam pressure raised in a boiler, or series of boilers, heated by a coal fire, the steam as it leaves the engines being condensed rapidly by passing into cold water. Adequate condensation is very important, as the work done by the steam in the engines depends upon the difference of the pressure of the steam in entering and leaving the engine, and if the steam is rapidly condensed its pressure in the exhaust is much reduced. This explains the great, unsightly, black, wooden structures which in recent years have been making their appearance at power stations, and wherever large stationary steam engines are at work. They provide a constant rain of cold water falling about the exhaust outlets from the engines. In railway locomotive engines the steam is exhausted directly into the air, no condenser and low pressure cylinder being possible ; consequently every ton of coal does much less work than in a well equipped stationary engine.

There are two distinct kinds of steam engines. The older is the *reciprocating engine*, in which the steam expands in one or more cylinders, a triple expansion engine being so arranged that the steam can expand in three successive cylinders, leaving finally at quite a low pressure. The other class consists of the *rotary engines*, of which the *steam turbine* is the chief member. This is simply a modification of the principle of the windmill. A continuous blast of steam passes through a tube in which is an axle set with almost innumerable steel blades, upon which the steam impinges, and so turns the axle.

The steam turbine is growing greatly in favour because it works at high speed, and so can be coupled directly to a dynamo for providing electric current, and partly because it can now be made more economical of steam than a reciprocating engine. The great practical difficulty has been that it cannot be made to reverse, so that in the great steamships *Lusitania* and *Mauretania*, as well as the numerous smaller ships fitted with turbines, it is necessary to have a separate engine and screw for reversing the propellers. Other rotary steam engines are not yet of practical importance.

Boilers for raising steam are of three classes. The first includes the old Cornish and Lancashire boilers, in which the boiler consists of a long cylinder of sheet iron or steel, with one, or in the Lancashire, two large tubes passing through it rather below the centre. The fire is built below and in front of it so that the flames pass not only under the boiler itself, but also through the tubes, which are covered by the water. The second class is the tubular boiler, in which the flames and the hot gases pass through a large number of tubes in the boiler, thus giving a very great heating surface. It is necessary, however, to use soft water only in such boilers as any deposit of "fur" on the tubes can only be removed by pulling the boiler to pieces. The third class is called *water-tube* boilers. The water circulates in a mass of parallel tubes, which look almost like organ pipes, and the flames of the furnace have free play all round them. These boilers allow of a high pressure of steam being raised very quickly, and are much used in the navy. For most industrial purposes they have been regarded as too expensive ; but Messrs. Babcock and Wilcox, amongst other firms, are bringing them into prominence for large steam plants.

The stoking of boiler furnaces has usually, until recent years, been carried out by hand, the stoker deftly throwing in the coal and occasionally raking the fire to distribute the coal evenly, get rid of the ash or clinker, and prevent caking. Now, however, various forms of *mechanical stokers* are in use at large stationary steam boiler plants. They are made and used for two purposes : to economise in fuel and save wages in stoking. They also produce a nearly smokeless fire, which is another great advantage where local authorities are active in suppressing the smoke nuisance. There are three distinct principles used in making different patent mechanical stokers ; one contrives by means of a spring box to shoot into the furnace every few seconds a measured quantity of small coal, scattered well over the fire-bars. Another involves turning the fire-bars into an endless chain constantly revolving so that the floor of the furnace is constantly moving inwards. At · the mouth of the furnace, small coal is fed evenly on to the moving bars and is gradually carried forward into it through the fire. A third method pushes the coal up from underneath into the bed of the furnace, so that the coal becomes gradually coked before it reaches the burning surface, which is bright red hot and smokeless. The second method also gives a nearly smokeless fire. There seems little doubt that mechanical stokers will gradually be adopted for all stationary steam boilers, owing to the great economy of fuel and labour, and the absence of smoke. They have not yet been applied to steamships or railway locomotives.

An enormous saving in the consumption of coal might be effected by adopting improved appliances, and the findings of the Royal Commission on the Coal Supplies (1905) on this point are worth quoting :—

" Mr. Beilby has estimated that about 52 million tons of coal are annually converted into steam power at mines and factories in the United Kingdom.

" It is generally agreed that the consumption of coal per indicated horse-power per hour is on an average about 5 lbs. When it should not exceed 2 lbs., and might even be less, the waste and extravagance of our methods of raising steam will be realised. It is true that improvements and economies have been going on, especially during the last twenty-five years, and according to the evidence there is now very little hope of improving upon the best type of modern steam engines ; but it is said that if all steam engines were as efficient as the best, 50 per cent. of the coal now used for steam raising might be saved.

" There seems to be no doubt that a large percentage of waste must be attributed to small consumers of power, and to engines scattered over factories and workshops with long ranges of pipes and small, ineffective boilers. The use of oil and gas engines is increasing, but for greater economy we must look to the general installation of central power stations."

There is no doubt that many manufacturers are not alive to the very considerable savings which might be effected in their coal bills by adopting the latest machinery. On the other hand it must be pointed out that investigation will frequently show that the capital cost of making the change in an existing factory is very heavy ; and that, even if the necessary capital for making such a change is available, the annual saving anticipated may not amount to more than 6 or 7 per cent. upon the necessary capital expenditure. If money for effecting the change could be borrowed at 6 per cent. per annum it would obviously be worth while to take the trouble of making the change if the annual saving were more than 10 per cent. per annum on the capital outlay. It is not merely ignorance, but often various considerations of this kind, which prevents the installation of the latest and most economical steam engines.

Prepared Coals as Fuel

In mining parlance, coal in the same condition as it is worked by the hewer is known as " through and through." It is a mixture of large, small and slack just as it comes from the coal face ; and every householder is familiar with it, as many house coals are still sold unscreened. " Nuts " and " cobbles " are, however, sorts of coals for domestic purposes which have been broken and screened so that they contain no large coal or slack.

In boiler firing the use of carefully sized or prepared coal has largely extended in recent years. For hand stoking, large coal is preferred from which all the small has been removed by screening ; and for the bunkers of steamships such coal is also washed, so as to lessen risks of spontaneous combustion, and because it is cleaner in handling. Mechanical stokers all work with small coal or slack, and the extension of their use is bringing up the price of small coal relatively to large. Many collieries are equipped with elaborate screening and washing

plants, which are described in the later chapter on
" The Preparation of Coal for the Market."

" Patent fuel " is the name commonly used for
large blocks made in hydraulic presses out of small
coal and pitch. They are very popular for use
abroad both for steamship and railway purposes,
but are not much used in this country. Their
advantages and manufacture are fully dealt with in
Chapter X. Smaller blocks, called " briquettes,"
are made for household purposes. They are more
extensively used on the Continent than here, and the
quality is better, as they are made there harder and
cleaner.

There is another interesting method of burning
coal, namely, as a powder. Small coal is dried and
ground to a very fine powder and is then fed through
a special form of burner by a blast of hot air, or, in
another patent, by steam. The result is a hot,
brilliantly luminous flame beneath the boiler. In
Germany this process has been extensively used ;
but in this country it is confined to the calcining
operations in the manufacture of Portland cement,
and is not used for steam raising. Speaking of the
plants in use in Germany and of the experimental
installations in this country, the Royal Commission
on the Coal Supplies reported (1905) :—

" It is claimed that by such appliances the
maximum heat can be got out of the coal and that
small dusty coal as well as peat, lignite, etc., can be
utilised to advantage. It is further claimed that
dirty coal can be used, even if it contains a large

percentage of ash, and, according to one witness, the only limitations are that the coal used must contain not more than 14 per cent. of moisture, and not less than about 21 per cent. of volatile matter."

It is possible that the poorer qualities of coal will be largely burnt by this method in the future. There is, however, another way of burning inferior coals ; by means of powerful forced or induced draught produced by a ventilating fan. Long experience can alone decide the cheapest method.

Internal Combustion Engines

A class of engine which is growing in popularity works upon a totally different principle from the steam engine of either class. Instead of the energy being obtained by the expansion of steam produced in a boiler separated from the engine, the combustion of the fuel, usually gas or oil, takes place actually in the cylinder of the engine. The gas engine was the first of this class. Coal gas taken from the ordinary town supply is passed through an apparatus which admits a certain proportion of air and then passes into the cylinder of the engine, where it is exploded at the right moment to give the piston a kick, and so drive the fly wheel. By an ingenious mechanism just the right amount of gas is used to keep the engine up to the work which it has to do. Such engines are now constructed for use with " *producer-gas*," which is made by drawing air over hot coke or coal so as to partially burn it, forming *carbon monoxide*, which is then admitted, with the necessary air, to the cylinder of the engine and exploded. Usually coke or anthracite is used, but in larger engines bituminous coals give satisfactory results. Such *producer-gas engines* are now made of very considerable size, up to about 5,000 horse-power, and they appear to work very satisfactorily, provided the gas-making plant is carefully attended to and kept in proper working order. Some users have complained of high cost of repair in these engines, but this would seem to be due mainly to careless management or to imperfections in the earlier patterns of engines manufactured. Apart from this, the producer-gas engine shows considerable saving over the steam engine, and seems destined largely to replace it when the technical perfection of construction avoids the expense and delay of repairs, and excepting for purposes where extremely smooth running is a desideratum. The internal combustion engine naturally produces a heavy throbbing, as there is only one explosion to each revolution of the fly-wheel, and for the same reason it is difficult to start. These difficulties can be overcome by using three or four different cylinders ; but of course this adds greatly to the cost of the engine.

The Royal Commission of 1905 came to the following interesting conclusions as regards the possible future of producer-gas engines :—

" According to the witnesses much economy of fuel results from the use of producer-gas plants, but this depends on several conditions, especially their size and their load factor. The fullest economy is obtained in large plants of 4,000 horse-power and upwards, with recovery of bye-products, in which case the cost of coal is balanced by the value of the bye-products : without recovery of the bye-products it does not pay to put down plant for bituminous coal of less than, say, 100 horse-power, but in this country some are in use of 20 horse-power. Up to at least 100 horse-power anthracite or coke plants are the most economical, but as to plants beyond 100 horse-power the opinion of witnesses differ, some preferring anthracite plants up to 250 horse-power.

" As for the quality of the coal which can be used in producers of the modern type, any coal which does not cake excessively will suit if it contains sufficient nitrogen recoverable as ammonia. Coal with a large percentage of ash can be used, and according to Mr. Crossley, who has already got good results from coal with 30 per cent. of ash, there is no reason to think that 50 per cent. of ash would be prohibitive. It would be difficult to exaggerate the importance of the development of these gas-producing processes, which are said to have rendered practicable the utilisation of inferior coal and have thus enormously increased the available resources of the country."

So far we have noticed two forms of the internal combustion engine, both of which depend upon coal as the ultimate source of energy. It is this type of engine, however, which has made the extensive use of oil possible as a motive power. The air is passed through the oil in a *carburettor*, thus producing an explosive mixture of oil vapour and air, which is admitted to the cylinder and ignited by an electric spark. The stationary oil engine was extensively used where coal gas was not available. Now, however, the producer gas engine is largely taking its place as being less costly in use.

The possibilities of oil fuel in permitting the construction of small portable engines has opened up several entirely new spheres for the application of mechanical power to locomotion, and has thus created entirely new industries. The more important types are the motor-car and other vehicles for road traction, the motor-boat, and oil-driven agricultural machinery ; and finally the aeroplane. It is interesting to note that in all these developments oil is to a very small extent supplanting coal. In road traction and for agricultural uses the motor is displacing horses, and the aeroplane is an entirely new use of mechanical power. It is only in the case of railway locomotives and ocean-going ships that oil is generally used as a fuel to generate steam, and not by way of internal combustion. Cases where Diesel engines are employed are an exception to this statement, as they provide examples of very large internal combustion oil engines used in ships. In the navy the superior convenience of oil—as detailed in a later chapter— gives preference to it, even if it should cost more than coal for the same work. For commercial purposes, however, oil must compete with coal on

equal terms as to cost ; and there seems much
reason to doubt whether the use of oil can be much
extended without seriously raising its price, which
would in itself at once check any further extension
of use. There has already been a conversion of
many oil-driven locomotives on the Russian railways
back to coal-heated engines, owing to the rise of
price of oil from the Baku oil-fields.

Coal Gas

The use for purposes of illumination of gas
obtained by heating coal in retorts is now over a
century old. It is, however, within the last forty
years that gas has come to be used extensively for
domestic heating and for power. Prior to the
invention of the incandescent mantle, about 25
years ago, the illuminating power of gas depended
entirely upon the constituents which burned with
a luminous flame ; and it was often necessary to
enrich the gas by passing it through heavy oils,
which would give the gas a bright yellow flame.

The advent of incandescent mantles has meant a
great economy in the consumption of gas ; and it
is possible to obtain ten times the amount of light
from the same consumption of gas, as when it is
burned in the ordinary fish-tail burner. The
tendency in recent years has been for gas companies to lower the price of gas to the consumers,
owing to the competition of electric light ; and this
end has in many places been attained without any
sacrifice of profits by one or more of three modes of

economy : (1) by installing new and largely automatic plant ; (2) by abandoning the elimination of
sulphur from gas, which was always an expensive
process and is now no longer required by the Board
of Trade ; (3) by a more sparing use of valuable
oils, etc., for enriching the gas for illuminating
power, this being now of less importance to most
consumers.

The invention of the *high-pressure gas* burner
which gives, with a suitable mantle, an intensely
brilliant light, suitable for outdoor lighting and
shops at very low cost, yields the palm to gas again
as against electricity—for a time at least. The
compressing plant is too expensive for any but large
consumers, so it is only where the gas is supplied
at high pressure in special mains that it can be used.

Gas companies or local authorities have laid
ordinary gas mains throughout every town and
suburb, large or small ; so that, if the price is low
enough, gas is most extensively used for a great
variety of purposes. It is doubtful whether gas
may not yet prove a cheaper means of transmitting
power over long distances than electricity ; and it
would doubtless be used for very large power plants
if it could be had at 8*d*. to 1*s*. per 1000 cubic feet.
With the development of coke-manufacture in bye-
product recovery plants, as described in chapter X,
a great quantity of good gas will become available
at a comparatively small cost ; and companies may
be formed for collecting this from the collieries and
distributing it through pipelines to various manu-

facturing districts. In the future, perhaps, coal will be rarely seen on railway trucks except for export ; the whole of it leaving the inland collieries in the form of briquettes or of coke, or oils and gas, or of electric current generated from the gas of the coking furnaces.

CHAPTER IV

General Location

THE term *Coal Field* means any area of the country in which productive coal measures appear at the surface so that coal may be worked where it crops out on the surface, or by sinking pits to no great depth. Where the coal measures occur overlain by newer strata, and their presence can only be inferred by geological reasoning and ascertained by boring, the area is termed a *concealed coal field*. Naturally mining began in coal fields of the first class, which, for clearness we may sometimes call *exposed* or *visible coal fields*, so we shall consider them first and then refer to the concealed coal fields.

It is to be observed, in the first instance, that all the exposed coal fields lie to the west of a line running north and south through York to Nottingham, thence to Warwick, and from Warwick to Bath and on to Portland. None of them lie within 90 miles of London, which means that the great centres of industry in this country are divorced from London, the great centre of commerce and the greatest aggregate of population. Many broad acres of purely agricultural country lie between them.

58

Whether this condition will continue is doubtful, however ; for the concealed coalfields lie mainly to the east of the line named and seem to extend much nearer to London. Coal may before long be found within 30 miles of London, either in Kent or under the Chiltern Hills ; and the neighbourhood of such new collieries, which would provide a supply of cheap fuel, whilst being also in close proximity to the great markets of London, would be a most favourable location for factories and works of many kinds. London would then begin a new and startling growth.

If anyone will examine a railway map of Great Britain, like that supplied with Bradshaw's Railway Guide, he can pick out with ease the principal coal-fields of the country by the plexus of railways built over each coalfield. The valuable but bulky mineral cannot be economically carried except by rail or water, so that every coalfield must be traversed by numerous railway lines with branches to tap every colliery. I believe that the eastern half of the coalfield of South Wales and Monmouthshire contains a greater total length of railways than any other area of the same size chosen in any part of the world, excepting possibly London, with its tubes.

Durham and Northumberland Coalfields

Geologically speaking, this is one coalfield, of which the larger part lies in the county of Durham. On the western margin the coal measures rise steeply against the Pennine Chain.   To the east in Durham they dip under newer strata, but in Northumberland they extend to the sea coast and several miles beyond.   From pits sunk near the coast the coal is worked for a distance of more than two miles out to sea.   The coalfield is nearly 60 miles in length from north to south, and its width varies from about 5 to 30 miles.   It is, roughly speaking, of triangular form, the apex being on the sea coast near the mouth of the river Coquet.   The area of the visible coalfield is about 590 square miles, that of the part overspread by Permian rocks 125 square miles, while the area beneath the sea is roughly estimated at 136 square miles.   The under-sea coal is worked by old collieries at Seaham, Ryhope, Monk Wearmouth, and by newer mines at Horden and Easington.   The Durham and Northumberland coalfield, which in structure is a trough, is not much troubled by faults and other irregularities, while the seams vary little in thickness, and are only slightly inclined.   About 25 out of 60 known seams are capable of being worked.

As has already been pointed out this coalfield is the oldest in the country, having been worked for centuries ; and for this reason a greater proportion of its resources have already been exploited than is the case in other coalfields ; moreover, the production is greater than that of any other field except the Yorkshire coalfield, which covers a much larger area.

The whole coalfield is well and almost exclusively

served by the North-Eastern Railway, which in many ways is more enterprising than any other line in handling coal for export. The company have introduced trucks of 30 and 40 tons capacity to replace the old 10 and 12 ton wagons, and they use very powerful locomotives. They are now experimenting with electric haulage for mineral traffic as well as passenger traffic. Several of the larger colliery companies [1] have their own private lines to the docks, and the North British Railway serves some of the Northumberland collieries. The principal ports of shipment taken in order from the north southwards are as follows, with the total coal (only) shipped at each port in 1912 :—

	Tons.
Blyth	3,399,450
Newcastle-on-Tyne . .	6,418,067
South Shields . . .	5,292,077
Sunderland	2,786,178
West Hartlepool . . .	1,305,229

The docks at Newcastle and South Shields are fitted with the most elaborate and efficient loading apparatus in the form of gigantic tips and cranes.

There being a central electricity generating station, a large number of collieries have adopted electrical power both for surface and underground machinery, there being 111,000 horse-power of electrical apparatus in operation in 1912, which is a figure only surpassed in British coalfields by South

[1] For a list of the principal companies of the Northern Coalfield see Appendix I.

Wales.   Coal cutting machines are not extensively used.

The following table shows the growth of output, number of persons employed, and total foreign shipments of coal and coke as cargoes, and of coal in ships' bunkers, from the ports of the coalfield :—

| | Output. | | Number of Persons |
Year.	Durham.	Northumberland.	Employed.
1870	21,773,275	5,840,264	
1880	28,063,346	6,850,162	
1890	30,265,241	9,446,032	
1895	31,133,253	8,694,651	135,627
1900	34,800,719	11,514,521	152,553
1905	37,397,176	12,693,885	159,942
1910	39,431,598	13,121,691	212,350
1911	41,718,916	14,682,427	216,733
1912	37,890,404	13,381,641	218,937
1913	41,532,980[1]	14,819,284[1]	226,817[1]

The reserves of coal as estimated by the Royal Commission on Coal Supplies in 1904 are :—

	Durham. Tons.	Northumberland. Tons.
Under land . .	4,401,087,700	4,253,401,300
Under sea . .	870,028,600	1,256,224,200
	5,271,116,300	5,509,625,500
Quantity raised to end of 1912 . .	353,049,400	120,954,400
	4,918,066,900	5,388,671,100

[1] Provisional figures.

The total production of coal in the years 1904 to 1912 was 474 million tons ; and if it is assumed that the future output remains stationary at this rate, more than 250 years must elapse before the point of exhaustion is reached. It is probable that in many of the oldest districts no very considerable further supplies are available, and this fact explains why it is that the housing conditions are so much worse in these areas than in those portions of the coal-field where the exploitation of coal began at a later date. The dwellings of the people are old and insanitary, but owing to the approaching exhaustion of the coal supplies in such districts and the total absence of alternative industries, little or no action is being taken to provide the people with dwelling accommodation of an improved character.

As might be expected in so old and highly developed a coalfield, wages are at a high average level, and there is an exceptional degree of division of labour in the collieries. Double shifts working at the coal face have been the custom for generations, and three shifts have been largely introduced since the Eight Hours Act. The hewers work only seven hours in some of the collieries and $7\frac{1}{2}$ hours in others ; but all other grades work 8 hours. The hewers customarily work only five days a week, and some only four days, a custom which has prevailed for a long time. The high rate of cutting prices possible with a normal double shift system makes the earnings good even with short hours.

The Durham and Northumberland miners are

some of the finest manual workers in the country. Many are well educated men with a keen interest in local affairs or national politics. The housing conditions are not so good as might be expected. A large proportion of the cottages are owned by colliery companies, and are old and inconvenient, if not actually insanitary. The flat system prevails to a large extent. Social life is much interfered with, and home life is made a burden to the women folk by the double and treble shift system, men going and returning from the house at all times of the day and night.

The Yorkshire and North Midland Coalfield

The Yorkshire and North Midland coalfield is the largest, and in many ways the most important coalfield of the United Kingdom. It extends from Leeds to Nottingham, a distance of about 60 miles, and includes considerable portions of the counties of Nottingham, Derby, and York. The area of the proved coalfields is about 1,376 square miles, while a further area of 760 square miles in the valley of the Trent contains reserves which can probably be worked but which have been only partly proved. The number of known seams in the northern end of the coalfield is 16 and the aggregate thickness of coal is 52 feet ; in the central portion there are 18 seams containing 40 feet of coal, while in the southern part there are 21 seams with 50 feet of coal. The strike of the beds is about north and south, and the measures dip gently towards the east, where they are overlain by Permian or other formations.

The future extension of the coalfield must be towards the east to the unproved coalfields to which reference has already been made. Great developments are already taking place to the east and south of Doncaster, and it is anticipated that this portion will shortly become the wealthiest of this coal-field. There is some uncertainty as to the limits of extension towards the east, and the geology of the district has not yet been certainly ascertained.

The principal and best known seam in Yorkshire is the Barnsley Bed, which is found at its best in the vicinity of the town from which it takes its name. In some parts this seam is from 7 to 10 feet in thickness, and it has been worked continuously throughout its whole length from Nottingham to Barnsley. Other well-known seams are the Parkgate or Deep Hard, the Flockton or Deep Soft, the Silkstone or Black Shale, and the Warren House. The coalfield as a whole is comparatively free from faults and disturbances of the strata, and intrusive igneous rocks are unknown. The coal can be cheaply worked as the strata are only moderately inclined.

Most of the coal of this area is used for manufacturing and household purposes, but some seams provide good gas, steam, and coking coals. Nearly 250,000 persons are employed in the 630 mines in this important coalfield, and during 1912 an output of nearly 70,000,000 tons was recorded.

Coal has been worked in the country around Leeds, Halifax, Huddersfield, and Sheffield for at least 200 years, but only in a small degree. It is only in comparatively recent times that the coalfield has become one of great importance. Its rise is synchronous with the Industrial Revolution, and the development of the steam engine, and the iron and steel manufactures. The woollen and worsted industry, which is the staple trade of the West Riding of Yorkshire, began to adopt improved machinery about 150 years ago ; but this at first was driven by hand or horse power—and later very extensively by water power—so that the mills are scattered up and down the valleys beside the mill-dams. It was only in the first half of the nineteenth century that steam power began to be generally adopted by the textile mills as an auxiliary to water power ; and the gradual replacement has been going on ever since, many of the mills still using water power, though generally for less than five per cent. of their total requirements. In the district around Leeds, Wakefield, Sheffield, and Rotherham, the great industry of iron and steel manufacture arose during the eighteenth and nineteenth centuries, and the machine-making industry has taken firm root in all the large West Yorkshire towns. Largely, of course, it is textile machinery that is manufactured, spinning frames and looms being supplied to all parts of the world. Iron smelting has been carried on chiefly in the westerly half of the exposed coal-field, where the beds of ironstone,

occur, and the bars are sent westward and northward to the textile towns for manufacture into
machinery.

The coalfield is served by a large number of
railways, including five of the principal English
railway systems. These are : The London & North-
Western, the Great Northern, the Midland, the
Great Central, and the Lancashire & Yorkshire.
The North-Eastern Railway also taps the north-
eastern part of the coalfield, whilst the Hull and
Barnsley Railway is also of importance. The Great
Central Railway is paying especial attention to
developing the eastern extension of the coalfield,
both near Doncaster and to the south of that town.

As already indicated, the greater part of the production has always been for consumption in manufacturing and for household purposes in the immediate
neighbourhood, or the central part of England ;
but there has for a long while existed a small export
trade from Goole on the Humber, and from Hull.
In recent years strenuous efforts have been made
to increase the export trade ; and with great success,
as the Yorkshire coals are now strong competitors of
the North-Eastern and South Wales steam coals in
most of the European and distant markets. Special
depots for coaling ships with Yorkshire coals have
been established at Brixham on the south coast of
Devon, and at the Suez Canal and elsewhere. The
result is that the total exports from the Humber
ports have doubled in the last fifteen years, whilst
those in the United Kingdom as a whole have only

increased by about 50 per cent. At the same time the total exported is less than half that of Durham and Northumberland, and under one-third of the exports of South Wales.

The principal ports are all situated on the Humber or at its mouth on the Lincolnshire coast. Boston is the only port tapping the extreme south of the coalfield around Nottingham ; but it is not yet of much importance. The exports in 1912 were as follows :—

Ports.	Coals (only) shipped in 1912. Tons.
Goole	1,106,121
Hull	3,646,872
Grimsby	2,016,363
Immingham (newly opened)	
Boston	182,917

The following table shows the growth of output, number of persons employed, and total foreign shipments of coal, including coke and coal for ships' bunkers, from the Humber ports :—

COAL PRODUCTION IN YORKSHIRE COALFIELD.

Year.			Output.	Number of persons employed.
1870	.	.	18,665,039	—
1880	.	.	29,810,033	—
1890	.	.	39,656,836	—
1895	.	.	40,671,683	154,582
1900	.	.	52,118,521	174,542
1905	.	.	54,913,651	195,504
1910	.	.	66,757,894	241,767

Year.			Output.	Number of persons employed.
1911	.	.	. 67,922,358	245,111
1912	.	.	. 65,980,984	250,983
1913[1]	.	.	. 74,195,168	271,219

It will be observed how very rapidly the output of this great coalfield has increased in recent years, the output of 1910 being more than $3\frac{1}{2}$ times as great as that of 1870. This is, of course, mainly due to the opening of the coalfield eastwards under the newer strata in the district around Doncaster. The collieries which are being sunk here are planned on a gigantic scale, the takings of individual companies[2] running to from 12 to 20 square miles, two or three pits being thus worked in conjunction with one another. The total thickness of workable coal in the eastern extension is not great, making large areas and large scale machinery desirable. Electrical machinery is being adopted extensively in the newer mines, and many of the undertakings are on a very large scale. In 1913 there were 23 companies with a yearly output ranging from 1,000,000 to 2,679,000 tons.

Coal-cutting machines are extensively used in this coalfield, 601 being reported as in use in 1912, which is nearly one-fourth of the total for the United Kingdom. The undisturbed nature of the coal measures is very favourable to the use of machines.

[1] 1913 figures provisional only.
[2] A list of the principal companies is given in Appendix 2.

The reserves of coal in seams of 12 inches and upwards, as estimated by the Royal Commission on Coal Supplies in 1904, are :—

West Yorkshire	8,367,385,600
South Yorkshire . . .	10,770,620,700
Derbyshire and Nottinghamshire .	7,360,725,100
	26,498,731,400
Less Coal raised to end of 1912 .	559,918,300
	25,938,813,100

The wages paid in Yorkshire coal mines are distinctly lower than in the North-Eastern, Scottish, or South Wales coalfields. Whilst this may be partly due to the coal being more easily worked, it is undoubtedly mainly due to the fact of there being available such extensive and well-paid employment for women in the textile industries of the district ; because it is a well-known fact that where the men's earnings are supplemented by the women also obtaining regular employment, there is a tendency for men's wages to be reduced. In determining the supply of labour, the wages of the family are really the effective factor ; and where the necessary family maintenance is found partly by women, the men's labour comes to be paid at a lower rate.

As regards the regulation of the percentage on wages by the Conciliation Board, the whole of the coalfield belongs to the Federated Districts. There are, however, three principal miners' trade unions :

the Yorkshire Miners' Association with about 90,000 members, the Derbyshire Miners' Association (40,000 members), and the Nottinghamshire Miners' Association (35,000). The Yorkshire miners first began to combine effectively as a result of the great lock-out of 1858 ; and they were in the forefront of the fight for the right of appointing checkweighers in 1859–60.

The social conditions cannot be said to be quite so satisfactory as in Durham and South Wales. Although the housing accommodation is not, in most parts, actually deficient, the cottages are usually only small brick boxes put up in monotonous rows by jerry-builders. The miners live in dirty, ill-made, noisy streets under a murky atmosphere ; and many of the cottages being built back to back there is no through ventilation and no privacy. The Yorkshire miner is on the whole rather a rougher type than his confrères of Durham, Scotland, or South Wales ; and, speaking broadly, his recreations are less intellectual, and his interest in local affairs and politics are not so marked. Yorkshire miners have been great devotees of cock-fighting and rabbit coursing ; whilst the pit-boys are rather notorious as a wild and rough lot. The same characteristic is true to a less extent of the Lancashire and Cheshire and Midland coalfields ; and a possible explanation is that where coal mining is carried on alongside other flourishing industries, the men of superior intelligence and morale can have their choice of employment, and prefer the skilled mechanical, manufacturing and textile trades to going underground. In Durham, South Wales, and Scotland there are no other industries to speak of offering skilled employment and high wages for men, so there is no selection ; and practically the whole male population goes underground. The men of intelligence and strong character then give a tone to the whole. This is true to a large extent of Derbyshire miners also.

The housing conditions in the neighbourhood of the large new pits which are being sunk in South Yorkshire are being very greatly improved by the movement for building model colliery villages on garden city lines, referred to in Chapter XXII.

Lancashire and Cheshire Coalfield

The Lancashire and Cheshire coalfield is separated from the coalfield of Yorkshire by the Pennine Chain anticlinal. The coalfield has an area of 217 square miles of exposed coal measures, the greater part of which lies to the north of Manchester ; and the field is a trough or basin, well defined in most directions by outcrops of older rocks. The southern edge is concealed beneath Permian formations, and the extension of the field in this direction is now being proved, and is likely to bring the total area of the field up to nearly 600 square miles. There are in the northern portion 15 seams with over 40 feet of coal, in the south-eastern parts 19 seams with 70 feet of coal, and in the south-western part 21 seams with 75 feet of coal. The portion of the coalfield lying in Cheshire is small, and the output in this county is less than 2 per cent. of the total production of the coalfield.

Some of the deepest workings in the British Isles are found in this region at the Pendleton Colliery near Manchester which has a total depth from the surface of 3,483 feet. The number of seams varies in different parts from 15 to 31, and a considerable number of these are workable. In some collieries, indeed, as many as 10 seams are being worked at the same time. On account of the numerous faults, and the steep inclination of the measures, however, mining operations are rather difficult and costly. The best known seams are the Arley Mine, the Wigan Cannel Coal, Ravenhead, Rusheypark, and St. Helen's Main.

The coalfield is traversed by the London & North-Western Railway, Lancashire and Yorkshire, and other lines, which convey the coal to the manufacturing towns where it is consumed. The immense growth of the cotton spinning and weaving industry, and of the machinery manufacture of Lancashire, has been greatly facilitated by the abundant supply of cheap coal on the spot. It is, perhaps, partly a case of action and reaction. The price of coal is low partly because wages are low, and this from the same cause as already noted for Yorkshire, viz. : that there is so much well paid employment for women. The amount of coal exported is comparatively small, the total exports during 1912 for the four north-western ports of Liverpool, Manchester, Preston, and Runcorn amounting only to 674,234 tons.

The following figures give the output of coal and numbers employed, and show that there has been no growth of production in recent years, and even a slight tendency to decrease of output :—

COAL PRODUCTION IN LANCASHIRE AND CHESHIRE.

	Output. Tons.	Numbers employed.
1895 . . .	22,764,171	87,227
1900 . . .	25,542,317	87,944
1905 . . .	24,247,514	93,896
1910 . . .	23,766,377	104,659
1911 . . .	23,980,814	103,959
1912 . . .	23,063,198	105,748
1913[1] . . .	24,627,515	109,021

The stationary production is probably due to the fact that only a few new mines are being sunk to the deeper measures, and that any new output from this source is counterbalanced by the closing of small collieries which have worked out their takings of the upper seams. Account must also be taken of the fact that the Lancashire miners, like those of South Wales, used to work 10 hours per day, and the coming into force of the Eight Hours Act in 1909 undoubtedly cut down production. This fact, the requirements of legislation, and the working of deeper mines, account for the great increase in the number employed in spite of a stationary or decreasing production.

[1] Provisional figures.

Cumberland Coalfield

On the opposite side of the Pennine Chain from the Durham and Northumberland field lies the small coalfield of Cumberland. It is about 25 miles long from north-east to south-west, and about 6 miles wide at its widest point. On the east it is bounded by outcrops of the Millstone Grit and Carboniferous Limestone series, which here, as in Northumberland, contain a few thin seams of coal; on the west it is bounded by the Irish Sea, under which the workable coal measures probably extend for several miles; and to the north-west by Triassic strata, under which the coal-measures pass. The Solway Firth occupies a syncline, the mountain limestone coming up again on its northern shore, and it is quite probable coal measures extend under most of its floor, and will be worked from the coast between Silloth, Maryport and Workington. The coal-measures also probably extend eastward under the Trias as far as Carlisle, and they have been proved to the north of that town at Canonbie. The total area of the exposed and concealed coalfield cannot be less than 250 square miles, little of which is yet worked.

The middle series of the coal measures contains seven workable seams, of which the most important is the " Main Band," 9 feet thick. There are no coals in the upper series, and those of the lower series and of the mountain limestone, are thin and of poor quality. The coals are all bituminous, suited for steam, gas and household purposes; whilst the small coal after washing makes good blast-furnace coke.

A feature of this coalfield is the extent of the workings under the sea, which occur all along the coast. At Whitehaven the coal is being worked under the sea at a distance of four miles from the coast. The output of this coalfield is an increasing one; and it is likely to grow in importance. At present there is practically no export of coal to foreign countries; but there is some coast-wise shipment to Barrow, Liverpool, Manchester, the Lancashire coast towns, and the Isle of Man. The level of wages in this coalfield is a medium one, somewhat higher than Lancashire and the Midland fields, but below Durham, South Wales, and most of Scotland.

OUTPUT OF CUMBERLAND COALFIELD.

	Tons.
1870	1,408,235
1880	1,680,841
1890	1,740,413
1895	1,883,592
1900	2,022,327
1905	2,151,060
1910	2,174,855
1911	2,296,254
1912	2,133,563

The number of persons employed in the coalfield in 1912 was 10,742.

The reserve of coal is estimated at about

1,507,000,000 tons, allowing for the amount raised since the date of the Royal Commission on the Coal Supplies. In the estimate made for the Commission the coal was assumed to extend and to be workable five miles under sea from the coast, but only along twelve miles of coast line. The undersea coal will probably eventually be workable under a much larger area when the price of coal justifies the cost.

Bristol and Somerset Coalfield

This coalfield occupies a basin elongated in a north and south direction, extending from Wickwar in Gloucestershire at the north end, for twenty-five miles southwards to near Frome. The principal parts of the field lie between Bristol and Bath, and in the neighbourhood of Radstock to the south. The succession of the strata corresponds roughly with that of South Wales, but all parts of the coal measures tend to be thicker in the Bristol district. Looking at the geology of the field broadly we may presume that the Somerset basin is the result of two synclinal folds crossing one another nearly at right angles. There are, in fact, two series of folds in this region, one running east and west, and another north-north-east to south-south-west. The South Wales coalfield is a great syncline of the east and west series, its termination eastward being due to an anticline of the other series. On the east of that anticline comes the syncline of the north-north-east and south-south-west series, and this brings in the Forest of Dean coalfield as a continuation of the

South Wales syncline. The next north-north-east and south-south-west syncline to the east produces the Bristol and Somerset coalfield where it crosses the next east and west syncline immediately south of the South Wales coalfield, a syncline partly covered by Triassic rocks in South Glamorgan but largely covered by the Bristol Channel. This Bristol Channel east and west syncline is one possibly of the greatest importance, as it may extend right across England in the Carboniferous rocks under the covering of newer strata, and be responsible for the Kent coalfield. If this be so it may, indeed, have produced coal basins under Wiltshire, Hampshire, Berkshire and Surrey (south of Croydon), but this remains to be proved by numerous deep borings.

Whilst the Bristol and Somerset field as a whole may, as above stated, be regarded as one basin, the colliery operations and geological surveying have shown that it is subdivided into a number of smaller basins by earth-movements later in age than the coal-measures, but completed before the deposition of the Triassic strata which overlie them. At the close of the long period intervening between the Carboniferous and Triassic areas, the coal-measure basins stood out as hills or elevated plateaux,[1] and the Triassic marls and pebble-beds were deposited around them and eventually covered them. There are probably concealed extensions of the coalfield in several directions—to the west under the Severn, to the south beyond the Mendip Hills, though

[1] See Chapter II.

probably at great depth, and eastwards in the neighbourhood of Bath.

The coals of the Somerset and Bristol field are of very varying quality. The best steam coals have given satisfactory results in ocean-going steamers, and there are found excellent house-coals, good gas coal, and ordinary bituminous coking and manu- facturing coals, for all of which there is a very fair demand in the district. The railways serving this coalfield are the Great Western and Midland.

The output of coal in Bristol and Somersetshire has grown somewhat, though not rapidly :—

	Output.
1870	525,000
1880	757,800
1890	921,870
1895	841,490
1900	1,046,792
1905	957,442
1910	1,208,046
1911	1,161,023
1912	1,072,356

The output of one large colliery company in South Wales or Durham often surpasses the whole output of this county.

The number of persons employed in 1912 was 6,448.

The reserve of coal was estimated by the Royal Commission on Coal Supplies, 1904, and allowing for coal subsequently worked may be put at about 4188,0 00,000 tons.

The level of wages is low in this coalfield, and it was one of the stumbling-blocks of the Miners' Federation of Great Britain when, in the early stages of the minimum wage agitation, they sought to establish a national minimum for coal hewers at the same figure for all coalfields. There were loud protests from the coalfields of Forest of Dean and Somerset, the miners fearing that their employers were right in saying that many of the mines would have to close down if wages were raised to such an extent. The cause of the low wages of the district must be a matter of surmise, but amongst other possible reasons some weight must be given to the facts that this comparatively small coalfield is surrounded on all sides by agricultural districts in which there is usually a surplus of youthful labourers, that Bristol is not growing fast as an industrial centre, and thus is not creating a great demand for coal or for men's labour in other employments, and that the mining has so far been mainly at shallow depths, the hewers requiring less skill and intelligence than in most deep mines.

Devonshire Culm and Lignite

It will be news to many that Devonshire produces coal—of whatever kind. In fact it produces two kinds—*culm* from the culm-measures of mid-Devon and the west coast, and *lignite* from Bovey Tracey, near Chudleigh and Newton Abbot.

The culm-measures are contemporaneous with the middle and upper coal-measures, but are very different in appearance. They are slates and shales with but few sandy beds. They contain seams of anthracite which is in Devonshire called *culm*, and this has for centuries been worked in a small way near Bideford. In 1856 the amount produced was 5,036 tons, but it appears to have been even then on the decline, and there is no record of production since 1879, when the figure given is 600 tons.

The lignite beds of Bovey Tracey are interesting as the only example in this country of coal not of carboniferous age. It occurs in Tertiary beds, which lie above the Triassic, Jurassic and Cretaceous series, and are probably not one-tenth the age of the coal measures. This is evident from its different appearance, as it is only partially fossilized. It is dark brown in colour and friable, and soft enough to take an impression from the finger nail. It feels lighter than coal, and absorbs moisture. The total output at Bovey Tracey in 1853 was 18,633 tons, but the workings, which were established at the beginning of the eighteenth century or earlier, gradually declined, the output being only 1,383 tons in 1869. The decay of the industry was due, as in the case of culm, to the cheapening of coal in the locality by improved means of transit.

In other countries the mining of lignite, or *brown coal* as it is widely called, is a most important industry. There are large deposits in North Germany, and other European countries, and in North America, India and Japan, Tertiary coals are extensively mined, they being in some cases more mineralised than typical brown coal. The output of lignite in European countries is as follows :—

TABLE OF LIGNITE PRODUCTION ABROAD.

	1911. Tons.	1912. Tons.
Germany . .	70,469,000	79,634,000
Austria . .	24,859,000	25,861,000
Hungary . .	8,024,000	—
France . .	697,000	739,000[1]
Bosnia . .	757,000	—
Italy . . .	548,000	653,000[1]
Spain . . .	248,000	—
Servia . .	263,000	—
Bulgaria . .	241,000	—

In Germany the price of lignite is only about 2s. 6d. per ton at the pit's mouth.

Forest of Dean Coalfield

As explained above, the Forest of Dean occupies a small, irregular oval basin formed by the intersection of the South Wales east and west syncline with a syncline running north-north-east and south-south-west. It is a perfect specimen of a basin, the margin being completely defined by the outcrop of the massive millstone grit. The area covered by coal measures is 34 square miles, and there are 16 workable seams over one foot in thickness, the best known seam being the Coleford Highdelf coal in the Lower Series.

[1] Provisional figures.

The seams at present worked yield a rather soft bituminous coal suitable for household, manufacturing and gas-making purposes. The lower seams, which are being developed by deep mining, afford second-class steam coals.

Coal has been worked in the Forest of Dean from very ancient times ; but always, until the past few years, by working from the outcrop down the dip of the beds towards the centre of the basin. The mines have, therefore, generally become waterlogged at no great depth, their output not justifying the very powerful steam pumps which would be necessary to keep down the water.

There is a very interesting system of mineral rights in force in this coalfield which has come down from feudal times, modified only by recent Acts of Parliament. The whole of the Forest is Crown property both as to surface and mineral rights, and it is administered in London by the Office of Woods and Forests, and locally by two Deputy Gavellers, one of whom has the duty of issuing mineral leases. These, however, are of a peculiar form and are called *gales*. Just as owners of surrounding lands have certain defined rights of use of neighbouring common lands, so the freeminers of the Forest of Dean are entitled to a gale or lease of a defined area, from which they may work mineral on payment of a small royalty to the Crown.

A freeminer is a man who has been born within the Hundred of St. Briavels, is over twenty-one years of age, and who has worked a year and a day in a coal or iron mine in the Forest of Dean. Having fulfilled these conditions, he is entitled to be registered in the gaveller's books, kept at the Crown Office at Coleford, and it is part of the deputy gaveller's duty to ascertain that these conditions have been fulfilled before registering a freeminer.

A freeminer is entitled to apply for and have granted to him an area of coal or iron ore, locally termed a gale, upon such terms as the gaveller considers fair and reasonable, subject to the payment of the rents and royalties and the carrying out of the working of the gale in accordance with the rules and regulations appended to the award of coal and iron mines of 1841.

A freeminer cannot be the registered owner of more than three gales. The grant is in fee simple, subject to a rent or royalty payable to the Crown, which they can revise every twenty-one years. The gales vary in size from a few acres up to 700 or 800 acres. The largest gale granted to the freeminers in one was the United Deep, which approached 2,000 acres.

The method of granting the gales is as follows :— When a gale is surrendered or becomes forfeited for non-compliance with the conditions of the grant, fourteen days' notice is given in the local newspapers of such surrender or forfeiture, and the advertisement states that application for the re-grant of the gale will be received at the Crown Office upon a certain day between the hours of ten and five. All applications received upon this day are treated as being simultaneous.

The application having been made, it is open to any of the applicants to request that the gale be granted, and if such a request is made the applicants either agree together amongst themselves to nominate one of their number to receive the grant on behalf of the whole number of applicants, or a lottery is held to decide to whom the grant is to be made. A copy of the grant is kept in a book at the Crown Offices at Coleford.[1]

The amount of actual mining done by the freeminers themselves is small, as they have not usually possessed the necessary capital, or cared to risk it, in workings on the scale which is necessary now that most of the gales on the outcrop are exhausted to waterlevel. Hence the freeminer has become a concessionaire who sells or leases his rights to a capitalist who opens a colliery in the usual way. Difficulties arose when galees attempted to dispose of their interests as regards the deeper measures, as single gales were rarely large enough for working economically by a deep pit, whilst the Forest rules prescribed barriers of coal to be left unworked between adjoining gales. In occasional cases the gaveller could grant the right of removing a barrier ; but the fact of the mine-owner having to deal through different galees made the negotiations

---

[1] This account of the rights of the freeminers is based upon a statement by Mr. W. Forster Brown to the Royal Commission on Coal Supplies, 1905 ; Report, Part X., p. 362.

complex and often fruitless, with the result that the deep measures remained unworked. By a special Act of Parliament passed in 1904, however, provision was made for leasing large areas of the principal seams in the lower series ; and these facilities have led to a number of areas being taken up in which deep pits have been sunk. These developments are creating quite a revolution in the Forest of Dean, greatly increasing the employment, raising wages and making the provision of further housing accommodation an urgent necessity. The number of men employed in the coalfield in 1912 was 8,524.

The output of coal has been increasing for more than half a century, the figures for the coalfield being as follows :—

	Tons.			Tons.
1851	335,000	1895	.	872,000
1860	587,000	1900	.	1,050,000
1870	835,000	1905	.	1,388,476
1880	754,000	1910	.	1,477,150
1890	853,000	1911	.	1,436,200
		1912	.	1,567,701

The available reserve of coal remaining unworked at the present date (end of 1913) according to the estimate of the Royal Commission on Coal Supplies, is about 248,000,000 tons.

The coalfield is served by two railways: the Great Western and the Midland, and a little coal is shipped in small coasting steamers at Newnham and Lydney.

The wages earned by the miners are comparatively low, and in striking contrast to those of South Wales, which is no doubt due to the slow growth of demand for labour, together with a falling demand in the agricultural districts surrounding. The deep mines are now paying rather better wages. The trade union is the Forest of Dean Miners' Association; and it is a member of the Miners' Federation of Great Britain. The Board of Conciliation for the Forest of Dean, which has existed since 1895, consists of eleven representatives of the owners and an equal number appointed by the Miners' Association. Its object is to settle disputes at individual collieries which affect the whole of the workmen of any grade in that colliery or one seam of that colliery. It does not regulate the percentage addition to the standard wages, that duty being performed by a standing Joint Committee, there being a sliding scale still in force.[1] The designations Conciliation Board and Joint Committee have unfortunately been interchanged in their meaning in the Forest of Dean as compared with the rest of Great Britain, where the body which regulates the percentage addition to wages is always called the Conciliation Board, and that which settles disputes at individual collieries the Joint Committee. The Joint District Board under the Minimum Wage Act (1912) decided that the minimum wage should be the standard day-wage at each colliery for each class of workmen plus 30 per cent., giving an actual minimum for

[1] See Chapter XVIII.

most workmen of from 3s. 11d. to 5s. per day.   This
compares with 4s. 9d. to 6s. 10½d. in South Wales.

The Midland Coalfield

In the Midland counties of Staffordshire, Warwickshire, Worcestershire, Leicestershire and Shropshire,
occur a number of coalfields more or less detached
from one another and of varying sizes, the largest
being the North Staffordshire and the South
Staffordshire coalfields. The North Staffordshire
or Potteries coalfield is roughly triangular in shape,
and covers an area of about 110 square miles.
If the Cheadle district, which forms a detached
basin on the east, be included, the area is 128 square
miles. The coalfield is traversed by numerous
faults ; and on account of the irregularity of the
strata, the seams of coal are frequently inclined
at very steep angles, and in some instances are nearly
vertical. There are 36 seams of 1 foot 6 inches and
over in thickness in North Staffordshire with an
aggregate of 144 feet of coal. Most of the seams
now worked average about 6 feet in thickness, the
coal produced being suitable for house, steam and
manufacturing purposes, and some also for coking
and gas manufacture. The Cheadle portion of the
North Staffordshire coalfield contains 17 seams with
an aggregate thickness of 65 feet.

The South Staffordshire coalfield, between Cannock
Chase and Clent Hills, has a total area of about 150
square miles, including the extensions under newer
strata yet unproved. The number of seams varies

from 6 to 11, and in thickness from 1 to 30 feet ; the aggregate thickness is about 65 feet. The celebrated 30-feet seam, which occurs in the neighbourhood of Dudley, and is known as the " Tenyard " or " thick coal," is really a combination of several thinner seams, which a few miles away occur separated by intervening sands and shales. This thick seam has been a source of great wealth to the district, but it is now almost entirely worked out except in a few old and flooded mines which may some day be drained. The bulk of the South Staffordshire coals are suitable for household use and for smelting and manufacturing, but there is a total absence of true steam coal. The coals on the southern margin are of poor quality, and for this reason have been considered unprofitable to work.

The Warwickshire coalfield lies between the towns of Coventry, Nuneaton and Tamworth, and is about 60 square miles in area, including the concealed portion so far proved. There are 10 seams varying in thickness from 1 to 16 feet with an aggregate thickness of 40 feet. The coal measures lie in a broad basin almost flat in the centre and rising rapidly on the eastern and north-western edges. This coalfield is fortunate in being comparatively free from serious faults and other irregularities. It is probable that the coalfield extends further to the south and west under the Permian and Triassic strata than has yet been proved. Pits are now being sunk around Coventry, and it is possible the coal may be proved to extend southwards to Kenilworth and even to Warwick.

The Leicestershire coalfield is only about 30 square miles in area in its visible portions, but in addition there is a further area of about 55 square miles which has been proved, but has not been developed. The centre of the coalfield is at the town of Ashby-de-la-Zouche. The number of seams varies from 22 to 33, of a thickness ranging from 1 to 16 feet, and the aggregate thickness is about 94 feet. The principal seams are the Slate, Ell, Rider, and Twoyard coal. In some collieries the seams are very near to one another and can consequently be worked together and with comparative cheapness.

The Shropshire and Worcestershire coalfields are many in number and include Coalbrookdale, Shrewsbury, Le Botwood, the Forest of Wyre, Dryton and Clee Coalfields. These coalfields extend irregularly from Shrewsbury to the River Teme, and they are mainly irregular basins, with a tendency to a predominantly eastward dip of the strata. The total area is about 98 square miles. There are generally seven seams with an aggregate of 15 feet. In the Shrewsbury coalfield, however, only three seams are known and the total thickness is only 6 feet. The Le Botwood and Dryton coalfields have been abandoned for some years on account of the fewness and thinness of their seams.

Midland coals are for the most part dry, and of a bituminous character, suitable mainly for household purposes and for smelting and manufacturing ;

there is also a considerable quantity of good gas and coking coal in North Staffordshire. The Midland coalfields contain numerous seams of ironstone, and it was the existence of this mineral in the district that mainly led to the development of Birmingham, Wolverhampton, and other big towns in the " Black Country " region. In North Staffordshire, the coal seams are interspersed with layers of excellent clay, and this association of coal and clay was responsible principally for the development of the pottery industry in this locality.

The amount of unworked coal in the Midlands, including seams of 1 foot or over, to a depth of 4,000 feet, has recently been estimated as follows[1] :—

	ESTIMATE OF COAL RESERVES.			
	No. of Seams.	Thickness.	Area Sq. Miles.	1000 tons.
North Stafford- shire . .	30 to 36	1 to 10 Agg. 148	110	7,372,401
Cheadle . .	17	1 to 6 Agg. 65½	18	
South Stafford- shire . .	6 to 11	1 to 30 Agg. 65	149	
Warwickshire .	10	1 to 16 Agg. 40	56	1,468,425
Leicestershire .	22 to 33	1 to 16 Agg. 94	84	2,534,285
Shropshire & Worcestershire	7	Agg. 15	96	370,767
			513	

[1] Aubrey Strahan in *The Coal Resources of the World.*

The extent of coal production in each of the counties in this coalfield and the total production of the coalfield as a whole are given in the following tables :—

COAL PRODUCTION IN MIDLAND COUNTIES, 1912.

County.				Output in Tons.
Staffordshire	.	.	.	13,696,886
Warwickshire	.	.	.	4,577,758
Worcestershire	.	.	.	516,090
Leicestershire	.	.	.	2,765,103
Shropshire	.	.	.	772,205
				22,328,042

GROWTH OF COAL PRODUCTION IN THE MIDLAND COALFIELD.

Year.				Output. Tons.	No. of Persons employed.
1895	.	.	.	17,593,163	62,178
1900	.	.	.	20,878,348	70,299
1905	.	.	.	20,323,761	74,590
1910	.	.	.	22,953,165	87,539
1911	.	.	.	23,232,265	89,160
1912	.	.	.	22,328,042	91,551
1913	.	.	.	24,075,950	96,423

CHAPTER V

THE WELSH, IRISH AND SCOTCH COALFIELDS

THE SOUTH WALES COALFIELD

Extent and Structure of the Coalfield

THE South Wales coalfield is in many respects the principal coalfield of the United Kingdom, for besides the exceptional quality of the mineral it is the largest continuous coalfield, covering nearly 1000 square miles. It extends from Pontypool in Monmouthshire westward as far as St. Bride's Bay on the west coast of Pembrokeshire, a distance of nearly seventy miles, whilst its width at its widest point is about eighteen miles, and it occupies the greater part of the counties of Glamorgan and Monmouth, and portions of Brecknock, Carmarthen and Pembroke. The coalfield forms an elongated oval trough, or *syncline*, lying in a depression of the Old Red Sandstone rocks, the dip of the measures being from the outside towards the centre. Along most of the southern boundary of the coalfield the strata dip much more steeply northward than the same strata at the northern edge of the coalfield dip to the southward. This is shown clearly in the cross section in Fig. 5, chapter II, p. 25, and it is a factor which has had a considerable influence on the development of the coalfield. Owing to the difficulty of working steep measures, and the expense of mining at the depth to which the measures quickly descend on the southern boundary, deep mining has not been undertaken in the southern part of the coalfield until quite recent years. Coal has for long been worked on the northern boundary of the coalfield, both by levels and slants on the northern outcrop and by pits sunk two or three miles to the south of the outcrop.

A kind of arch, or anticlinal ridge, in the strata runs east and west through the coalfield, bringing the important seams of steam coal comparatively near the surface in a central east and west zone. When this was realised about the middle of the nineteenth century the minerals throughout this area were quickly leased for the sinking of large collieries. It is in the area of this anticlinal ridge, and between the ridge and the northern boundary, that all the important and well-known steam coal collieries of South Wales are situated. The measures also curve slightly upwards in an easterly and westerly direction along the length of the field, so that the lower coal seams, which are in parts of the central portion too deep to be worked, are rendered available over most of the coalfield.

The surface of the coalfield is carved into a series of deep and narrow valleys by the forces of denudation. The pedestrian climbing to the top of almost any of the hills dividing them sees at once that all the hills are of about the same level, forming a great moorland plateau, the area of the valleys being less than that of the flat-topped hills, which are composed mainly of the hard Pennant sandstone. A map showing the rivers and railways resembles closely a gridiron, or a series of gridirons, the valleys being generally nearly parallel.

The coal seams of the upper series crop out on the mountain sides, and these are largely worked by levels and slants, forms of mines which are quite unusual in other coalfields. The general direction of the valleys is from north to south, but some of them, especially in the western half of the coalfield, run more from north-east to south-west. Some of the valleys are quite short, their entrances lying close to the sea ; but others are from 20 to 40 miles long. The western portion of the coalfield is deeply penetrated by the Carmarthen and Swansea Bays, under which exists a large quantity of workable coal, a small part of which is already being worked.

Coal Measures

The coal measures vary from about 4,000 to 12,000 feet thick, and consist of three series :—

(1) *Upper Series*, consisting of shales, fire clay, sandstone, and several seams of more or less bituminous coal.

(2) *Pennant Grit*. This is a thick band of massive sandstones, with some shales and conglomerates at the base. It contains a few seams of good coal in the western portion of the field only.

(3) *Lower Shale Series.* This series contains the largest number of seams, and the best steam coals. Near the northern outcrop it also contains bands of ironstone, the existence of which accounts for the many ironworks in this region. At the present time, however, these ores are not mined to the same extent as formerly, as the iron manufacturers are able to obtain supplies cheaper from abroad. The Millstone Grit, with beds of shale and impersistent coal seams, forms the base of this series.

The field is bounded on all sides, except in Swansea and Carmarthen Bays, by outcrops of mountain limestone, which form striking and beautiful ranges of hills, running mainly east and west.

Character of Coal

The coal varies much in character, passing from anthracite through the well-known steam coals to bituminous house coal. Welsh steam coal is famous the whole world over ; there are numerous varieties, but the kinds in demand for Admiralty purposes are regarded as superior to any other coal now being produced in considerable quantities in any part of the world. The Monmouthshire bituminous coals are also of very good quality, being hard, large, clean, and well suited for use in all kinds of climates, on account of their resistance to atmospheric influences. The semi-bituminous coals are much used for ships' bunkers.

Anthracite, or " stone coal," is mined in the north-western and western districts of the field. The exact limits of the anthracite field cannot be clearly defined, for the seams of anthracite are continuous with those of the dry steam coal and bituminous steam coal found in the southern and central areas of the coalfield. There is a gradation in two ways ; for in the anthracite coalfield, as the lower coal measures are penetrated, each seam is found to be more anthracitic than the one above, whilst each particular seam varies in the horizontal direction, becoming more anthracitic as the northern and north-western border of the coalfield is approached. The purest anthracite is found in a narrow strip along the Gwendraeth valley towards Kidwelly. The coal in the narrow strip of coal measures in Pembroke-shire is also almost entirely anthracite.

A big fault running north-east and south-west along the Vale of Neath divides the South Wales coalfield into an eastern and western portion, the downthrow side being on the north and west. The anthracite seams, which become less pure, that is to say, more bituminous, as they approach this fault, can hardly be said to exist as anthracite to the east of it.

From the foregoing it will easily be understood that the above classes of coal can be worked from the same mine ; for example, the upper seams may yield house-coal, and the lower seams a bituminous steam coal ; or sometimes the upper seams are hard steam coal and the lower seams anthracite.

The best house coal occurs in seams like the No. 2 Rhondda, which are of small extent, as they are outliers or detached areas of coal, lying often high up in the mountains where they crop out on the hillside, and have in past centuries been worked by quarrying. They are now worked by levels and slants.

The relative proportions of the three principal classes of coal were estimated by Sir W. T. Lewis for the Royal Commission on Coal Supplies (1904) to be as follows :—Bituminous, 30·42 per cent. ; Steam, 47·31 per cent. ; Anthracite, 22·27 per cent. The number of workable seams varies over the coalfield. In Monmouthshire twelve are known with an aggregate thickness of about 42 feet ; in some parts of Glamorgan, however, there are more than 40 seams with a total thickness exceeding 120 feet, while in Pembrokeshire again the number of seams diminishes to about 18 with 30 feet of workable coal at the utmost. The strata in the latter region where the coal is all anthracite are very much disturbed.

South Wales has practically a monopoly of anthracite in the United Kingdom, and exports it largely to the Continent.[1] As the demand for this class of fuel grows, as it is certain to do, mining development in the anthracite area must increase very considerably. There are at present a number of collieries working anthracite, but most of them are comparatively small. There are only seven pits employing more than 500 men each, and only three

[1] See Chapter XXIII. : *The Anthracite Coal Trade.*

colliery companies employing more than 1000 men at all their pits. This compares with figures of from 1,500 to 3,000 men in many of the single pits of the steam coal area, and of 5,000 to 12,000 men each employed by the largest steam coal colliery companies.[1] It is doubtful whether the anthracite field is not too small, and the strata too disturbed for collieries to be worked upon a very large scale.

History

Coal was won in quarries and small slants along the outcrops at least four hundred years ago; and numerous mineral leases for 21 years, three lives and other periods, granted by the Earl of Pembroke from 1611 to 1677 on the estate which has now descended to the Marquis of Bute are extant.[2]

It was not until the eighteenth century, however, that mining can be said to have become an industry in the district. A great impetus to coal-mining was given by the use of coal for smelting iron which was commenced about 1770 by the first John Guest, at Dowlais. At this time coal was exported from the Merthyr and Dowlais district, but only in small quantities on the backs of ponies and donkeys, which travelled in trains over the mountain paths, each carrying 130 pounds, both into Herefordshire and down to Cardiff, where a little was shipped in

---

[1] A list of the principal South Wales colliery companies and the number of men employed by each is given in Appendix 3.

[2] Charles Wilkins: *History of Iron, Steel and other Trades in South Wales*, 1903, p. 19.

coasting boat.   Iron was taken down to the port
in wagons, two tons at a time, drawn by four horses
over frightfully rough roads.   The making of the
road down the Taff valley in 1767, coupled with
the use of coal for smelting, was a great stimulus
to mining.

The greatest stimulus, however, was provided
by the opening of the Glamorgan Canal from Merthyr
to Cardiff in 1798 ; and its extension a little later
to Aberdare opened up that valley, which had already
been the seat of iron works and coal mining, though
subsidiary to Merthyr and Dowlais.   The canal was
built of very narrow gauge, and although for many
years it easily paid its maximum dividend of 8 per
cent., it could not meet the needs of the district ;
and further developments only became possible
with the opening of the Taff Vale Railway in 1841.
There has been a very striking and steady growth of
the coal traffic on this railway, as shown by the
following table :—

TAFF VALE RAILWAY MINERAL TRAFFIC.

	Tons.			Tons.
1841 . .	41,669	1883 . .		8,614,715
1842 . .	114,516	1890 . .		10,812,942
1843 . .	152,100	1893 . .		11,342,905
1853 . .	874,362	1903 . .		16,168,838
1863 . .	2,772,011	1912 . .		14,475,068
1873 . .	4,527,641	1913 . .		19,392,267

From 1841 onwards begins the great era of
expansion in the Taff Basin, which includes most of

the middle part of the South Wales coalfield—the
Rhondda, Aberdare, and Merthyr valleys. The
coal mining industry in South Wales has been constantly hampered in its growth by the want of
adequate transport and shipping facilities, and is so
at the present day.

Whilst the development of the Merthyr (Upper
Taff) and Aberdare valleys was proceeding in the
first half of the nineteenth century, interesting
developments took place in the Monmouthshire
valleys. As in Glamorgan, it was on the northern
outcrop that coal was exploited first on a large scale ;
but in Monmouthshire also, principally for the iron
manufacture. There are four parallel valleys in
North Monmouthshire, named Rhymney, Sirhowy,
Ebbwy, and Blaina. Each is a deeply cut, narrow
valley, several miles long, the sides rising up steeply
to a peaty moorland plateau which rises northwards
from 1000 feet to about 1,300 feet above sea level.
At the very top of each valley were found the rich
crops of coal with seams of ironstone and limestone
close at hand. Works were established in each
valley, which, after several changes of ownership,
came in the 30's, 40's and 50's of last century, to
be vested in large limited liability companies, one
for each valley. Thus there are respectively in the
four valleys : the Rhymney Iron Company, the
Tredegar Iron & Coal Company, the Ebbw Vale
Steel & Iron Company, and the Nantyglo & Blaina
Ironworks Company. Working many miles away
from existing towns, these companies formed com-

plete industrial colonies. They owned or leased the whole of the land and minerals in their areas for miles ; they had to build many hundreds of cottages for their workpeople, and to provide them with all the necessaries of life in the companies' stores. The truck system flourished in these valleys till fifty years ago, and survived long after in a modified form, the company owning one or more stores, though not excluding free shops. The Rhymney Iron Company still owns and works a brewery which dates from early in its history.

The vicissitudes of the iron and steel trade have led to great changes in these companies. Improvements of methods involving enormously expensive plant and the importation of higher grade foreign ore, together with American competition, have rendered the iron and steel business unprofitable to all of these South Wales companies except the Dowlais, and perhaps the Ebbw Vale company ; and all of them (except the Nantyglo & Blaina) are now largely or exclusively colliery proprietors working and selling steam, gas and house coals on a large scale. Each company has in miniature gone through the transformation which has affected the coalfield as a whole.

In the middle of the nineteenth century it was the great ironmasters of Dowlais, Merthyr and Aberdare who controlled the coal trade of South Wales, although their coal mines were to them merely subsidiary undertakings. In the sixties, however, the serious fall in the price of iron led the ironmasters to turn their attention more and more to the sale of coal as a source of profit, a policy which they have ever since continued. They at once found it profitable, because their mines were more fully developed than those which had been recently started in the Aberdare, Mid-Taff and Rhondda Valleys solely for the sale of coal, and because they paid lower wages. The economic situation, which repeated itself again and again, as each new section of the coalfield was opened up, and is paralleled in other coalfields, is well described by Alexander Dalziel.[1] Public recognition of the excellence of Welsh smokeless steam coal having begun, as he points out, about 1835 to 1840, the sinking of pits to produce coal for sale in the open market was greatly stimulated, and the Aberdare Valley was the first to be thus developed. Dalziel gives the following figures of the production of coal in the valley :—

							Tons.
1844 .	.	.	.	.	.	.	71,031
1856 .	.	.	.	.	.	.	1,173,459
1870 .	.	.	.	.	.	.	2,342,792

and proceeds—

" It is evident that the normal population was utterly inadequate, and, therefore, capitalists, in order to develop their properties, found it necessary to attract workmen from other districts ; but, for the accomplishment of this design, they were

---

[1] *The Colliers' Strike in South Wales* (1871). *Western Mail* Offices, 1872, pp. 14–18.

obliged to offer inducements in the shape of higher wages than were then current in the neighbouring districts of Merthyr, although the description of work and the seams in which that work was to be done were identical in every respect.

" One reason which materially affected the higher rate of wages paid was the inadequate number of dwelling-places, causing the workmen to walk, morning and night, to and from the pits, several miles. This, with the distance of meat and vegetable markets from the houses of others who resided near mines in out-of-the-way places, was alleged by the workmen as a cause of extra expense and therefore a justification for the higher rates of wages. The owners of every new colliery acting upon this principle of offering inducements to obtain labour, several anomalies in the rates of wages were created, and they have not been since reconciled.

" It is the general belief, that the difference between the scales of wages current at the pits belonging to the ironmasters in the Merthyr district and at the pits belonging to the colliery owners of the Aberdare valley, amounts to about 15 per cent. At a later stage this will be more particularly dwelt upon. It is sufficient for the purpose at this point to have stated in a cursory manner the primary cause for the differences in the Aberdare valley.

" Next we speak of the Rhondda valley, where a process of the like nature took place. The American war gave a great stimulus to mining enterprise, and capitalists directed their attention to the minerals underlying that beautiful valley.

" On reference to the returns for the assessment of the poor's rates in the parishes of Llantwit Vardre, Llantrissant, and Ystradyfodwg, it will be found that the entire production of coal was in the years—

1856	.	.	.	205,200 tons.
1870	.	.	.	1,858,826 ,,

" The difficulty of obtaining labour in the Rhondda was fully as great as that experienced in the development of the Aberdare valley, and it became necessary, on the part of the employers, to establish even higher rates of wages than were current at Aberdare. Ensuing from this a distinction as between the payments in the Aberdare and the Rhondda valley pits was caused, to the extent of about 10 per cent. and as between the Rhondda and the ironmasters' pits at Merthyr, of about 25 per cent.

" As it was in the Aberdare, so in the Rhondda— it came about that in later times the employers erected ample dwelling accommodation—villages sprang up, markets were established, and all the advantages, including regularity of employment, which workmen were supposed to possess at the pits belonging to the ironmasters, were really obtained by the workmen in the Aberdare and Rhondda valleys. Therefore, according to the laws of political economy, it might fairly be supposed that a general equalisation of wages would gradually have taken place ; but this did not occur. The race for labour in the Aberdare and Rhondda valleys continued ; emigration removed rapidly the surplus population ; new pits continued to be opened until the production of coal not only overtook but gradually exceeded the demand. The call for labour had been so regular that no one had the inclination to attempt an adjustment or equalisation of rates of wages. The general overtrading of the country resulted in a monetary crash (1866), and a season of depression set in.

" We now arrive at a period when a serious complication arose in the mining commercial interests. With the iron, as with the coal trade, the time of depression meant loss—continuance meant disaster.

" The ironmasters having the command of an enormous supply of labour at the lower scale of wages and possessing large mineral properties fully developed, found that the production and sale of steam coal would considerably ameliorate their monetary position. They cast into the markets increased quantities of coal. For the first time the proprietors of the Aberdare and Rhondda pits found real and formidable competitors in the ironmasters, who, by reason of lower royalties and the lower rates of wages, were able to produce coal at a cheaper rate and thereby undersell, which they did to a serious extent."

The result was a split between the ironmasters and the colliery proprietors pure and simple, the latter reducing wages 10 per cent. whilst the former would not reduce more than 5 per cent. The result was rapid organisation of the workmen, and the great strike of 1871 in the Aberdare and Rhondda valleys. This was the beginning of extensive trade union organisation amongst the miners of South Wales, and the changes of wages made in 1871 were the first in which the system of alteration by adding to or deducting from the total of wages at the old rate on the pay-ticket was adopted, by which the old rate became a standard rate. This system has prevailed ever since, a new standard having been adopted in some parts (e.g. Rhondda) in 1877, and in other parts in 1879, which was a year when wages fell to a very low level.

The gradual transference of the control of the destinies of South Wales from the iron trade to the coal trade was completed in the 70's with the development of the Rhondda and the opening up of the Ely and Ogmore valleys. Henceforth the export of coal was to be the mainstay of the greater part of the coalfield. Only in the neighbourhood of Swansea, where the metallurgical industries (copper, tin, spelter, and lead) were firmly established and the rolling of tin plates was becoming a great industry, was coal export a secondary trade.

The new trade in the export of coal could not grow, however, without facilities of cheap transport which were always and still remain, vital to it. The first Marquis of Bute, with great foresight and at considerable financial risk, constructed the first dock at Cardiff, now called the Bute West Dock, which was opened in 1839. Previously coal was shipped from staithes on the banks of the Taff and in a small basin, but only small ships could be accommodated. The new dock was a necessity for the foreign export trade.

About the same time John Nixon, to whose energy and enterprise the South Wales coal trade owes much, was active in persuading French coal users of the merits of South Wales coal, and with such success that a big trade was soon built up. The first Bute Dock soon proved insufficient to handle the growing trade, and a larger dock called the Bute East Dock was constructed and opened in 1859. Even whilst this dock was in progress of construction it became evident that further accommodation was required, and a group of coal owners commenced a new one at Penarth which was taken over before completion by the Taff Vale Railway Company. Although the Marquis of Bute built another and much larger dock, he could not keep pace with the requirements of the rapidly growing trade. The story goes that an influential deputation of coal-owners, merchants and shipowners waited upon the Marquis and his agent, pointing out the urgent need of greatly extended dock accommodation. The Marquis and his advisers were by no means anxious to take the responsibility of finding a further vast sum of money, and a none too cordial interview ended with the taunt :—" Gentlemen, if you must have more accommodation, build your own dock." And they took him at his word, but it was not at

Cardiff but nine miles away at Barry that a great dock was built and an extensive railway constructed to feed it by tapping the principal valleys. Barry now ships as much coal as the whole of the Cardiff and Penarth docks, including the great Alexandra Dock opened by the Bute company in 1909.

A new and urgent demand for more dock accommodation at Cardiff is now being pressed, as there are frequently serious delays in getting coal shipped, which involves wagons being held up full of coal. When unable to get empty wagons a colliery is obliged to stop for a day or two with disastrous results to both owners and workmen.

Although the building of great docks began earlier at Cardiff than at the other South Wales ports, the latter have not lagged far behind. As more coal is exported from South Wales and Monmouthshire than from any other coalfield, and the proportion of the total output exported is as high as 63 per cent., it is not surprising that the history of the development of the coalfield since 1850 is largely a history of the building of its docks and railways. The Monmouthshire coalfield has developed, and can now develop, only as facilities are provided at Newport, and the middle of the coalfield is dependent on the development of Neath and Port Talbot, whilst the west depends wholly on the Swansea docks.

Railways, Ports and Collieries

The South Wales coalfield is served by quite a large number of railways, mostly local. Of the great railways, by far the most important is the Great Western. The main line from London to Fishguard, after passing through the Severn tunnel, runs through Newport (Mon.), Cardiff, Port Talbot, Neath and Llanelly, skirting the north of Swansea, that is to say, it runs right along the coast throughout the length of the coalfield and through every coal-shipping port. It has numerous branches running up various valleys in Monmouthshire, Mid and West Glamorgan and Carmarthenshire, and the main line taps not only these but also all the local railways, as it cuts across the mouths of all the valleys. The local railways carry coal direct to the docks, but practically all the coal leaving South Wales by land must pass over the Great Western Railway. The other great railway systems which tap the coalfield are the London and North-Western, which runs from Abergavenny to Merthyr along the north outcrop and on to Swansea and also up the Monmouthshire valleys; and the Midland, which comes down from the north to Swansea.

Of the local railways the principal is the Taff Vale, whilst the Rhymney and Barry Railways are next in importance and also serve the valleys above Cardiff. There are also in Monmouthshire the Brecon and Merthyr Railway; in the Taff valley the Cardiff Railway; in Mid and West Glamorgan the Port Talbot Railway, the Rhondda and Swansea Bay Railway, the Neath and Brecon Railway, the South Wales Mineral Railway; and in Carmarthenshire the Burry Port and Gwendraeth Valley

Railway and others. There are also several joint lines. The multiplicity of railways has not tended to economy in working, and the bad passenger services with numerous stations and want of connections are a great hindrance to business. It takes longer, for example, to go from Cardiff to some of the anthracite mines in Carmarthenshire than from Cardiff to Liverpool or through London to Margate. On the other hand, colliery proprietors welcome the numerous railways, there being often two or more in each valley, which gives a colliery a choice of routes and shipping ports, and more important still, competitive rates.

South Wales coal is shipped from six different ports, all on the north shore of the Bristol Channel. From east to west they are: Newport (Mon.), Cardiff, Barry, Port Talbot, Swansea, and Llanelly. For customs and official purposes (including statistics) Barry is included in the Port of Cardiff, but it now ships more coal than Cardiff and Penarth docks together.

The following were the exports from the principal ports during 1912 and 1913 :—

	1912. Tons.	1913. Tons.
Cardiff (including Barry)	17,243,494	16,054,752
Newport (Mon.) . .	3,976,142	3,840,132
Swansea . . .	3,001,541	3,022,900
Port Talbot . .	1,736,639	1,652,511
Llanelly . . .	145,483	221,652

Of the coal shipped from Cardiff and Barry in 1912 the average price of 7,000,000 tons was more than 16s. per ton, whilst a further 4,300,000 tons was priced between 15s. and 16s. per ton. No other port in the world can show such a large export of high-priced coal.

The above figures include only coal shipped as cargoes. They take no account of the large amount of coal put in ships' bunkers and of the important export trade in coke and patent fuel. The following figures for the South Wales ports as a whole illustrate the character of the trade, its growth in recent years, and its relative importance to the trade from all British ports :—[1]

---

[1] The figures are taken from the *South Wales Coal Annuals* (Business Statistics Publishing Co.).

	Coal		Coke		Patent Fuel		Bunker Coal		Total Shipments	
	Tons	% of U.K.	Tons	% of U.K.	Tons	% of U.K.	Tons	% of U.K.	Tons	% of U.K.
1891	12,175,219	35	74,801	10	693,572	94	2,126,088	19	15,069,680	32
1900	18,457,238	42	112,918	11	998,290	97	2,646,364	22	22,214,810	38
1902	19,446,643	45	89,220	13	1,047,153	99	3,430,413	23	24,013,429	40
1904	20,798,807	45	94,236	12	1,228,926	99	3,982,853	23	26,104,822	40
1906	23,400,518	42	128,476	16	1,367,558	99	4,210,293	23	29,106,845	38
1908	25,166,334	40	156,870	13	1,432,611	99	4,514,061	23	31,269,876	37
1910	25,215,203	40	116,112	12	1,466,737	99	4,224,105	21	31,022,257	37
1912	26,098,584	40	93,451	9	1,558,072	99	4,026,549	22	31,776,656	37
1913	29,785,530	41	135,475	11	2,031,148	99	4,830,124	23	36,782,277	38

It may be of interest to examine the destinations of South Wales coal exports, for which purpose I quote the official figures [1] for Cardiff and Swansea :—

DESTINATIONS OF EXPORTS IN 1912.

	Cardiff. Tons.	Swansea. Tons.
Russia	466,339	12,546
Norway, Sweden, Denmark	175,028	201,108
Germany	26,416	191,758
Belgium, Holland . .	206,815	921,463
France	2,470,433	1,584,146
Channel Islands . .	9,388	28,265
Portugal	531,454	19,328
Spain	638,222	113,193
Canary Isles . . .	706,271	817
Italy	3,589,835	515,966
Egypt	1,393,931	42,418
Tripoli, Tunis, Algeria .	410,936	57,619
Malta, Gibraltar . .	385,523	—
Upper Mediterranean . .	674,562	46,778
West Africa . . .	368,281	277
Rest of Africa . . .	104,241	5,810
Chile	216,280	6,661
Brazil	1,302,658	24,914
Uruguay	663,399	3,739
Argentine . . .	1,979,442	6,709
Ceylon	236,270	—
Straits Settlements . .	7,891	7,337
Dutch East Indies . .	38,141	—
Far East	73,805	—

[1] Parliamentary White Paper : *Coal Shipments*, 1912. *Cd.* 6845, Price 6d.

It will be seen that France, Italy, Egypt, the Argentine and Brazil are the largest buyers of Cardiff coal. These are all countries with growing industries and busy railways, but without any adequate local supplies of coal. Being obliged to import, the high qualities of Cardiff coal lead to its being largely used, though the same countries also import largely from Yorkshire and Durham. Cardiff coal goes to every country in the world which possesses a seaport ; and many of the smaller figures are either omitted or grouped in the above table. Compared with Cardiff the Swansea figures bear a very different proportion to one another, which is to be accounted for by the fact that more than half the Swansea exports are anthracite, going chiefly to the North European countries.

Increasing quantities of coal are being sold for export and further patent fuel works and coking ovens erected, so that the export figures are certain to go on increasing. Whether they increase rapidly or not depends very much upon the measures taken to provide further dock accommodation, for without adequate loading facilities the trade cannot increase much, and capital invested in mines and works will not be fully utilised.

Output of Coal and numbers employed

During recent years the mining industry in South Wales has developed very considerably ; and it is now growing faster than ever before. The following table gives the output for South Wales and for the

United Kingdom at various periods during the last half century :—

COMPARATIVE OUTPUT OF COAL IN SOUTH WALES
     AND THE UNITED KINGDOM.[1]

Year.	Output of South Wales Coalfield.	Output of the United Kingdom.	Proportion of S. Wales output to the output of U. Kingdom. Per cent.
1855	8,552,270	64,453,070	13·3
1860	10,255,563	80,042,698	12·8
1865	12,656,336	98,150,587	12·9
1870	13,594,064	110,431,192	12·3
1875	14,173,143	133,306,485	10·6
1880	21,165,580	146,969,409	14·4
1885	24,347,856	159,351,418	15·3
1890	29,415,035	181,614,288	16·2
1895	33,040,114	189,661,362	17·4
1900	39,328,209	225,181,300	17·4
1905	43,203,071	236,111,150	18·3
1910	48,699,982	264,433,028	18·4
1911	50,200,727	271,891,899	18·5
1912	50,116,264	260,398,578	19·2
1913	56,830,072	287,411,869	19·7

From this table it will be seen that not only has there been a steady and enormous increase in the production, but this increase has been greater in proportion than the increase for the United Kingdom as a whole. In 1855 South Wales produced 13·3 per cent. of the total output of the country; in 1913 this proportion had increased to 19 per cent. This great activity has, of course, resulted in an immense increase in mining employment, and in

[1] Figures from *South Wales Coal Annual*, 1915.

the population of the coalfield. From 1905 to 1910 the increase of output was only 12 per cent. as against an increase of 26 per cent. in the number employed. This is accounted for largely by the operation of the Eight Hours Act ; and to some extent, also, by the increase of surface plants, such as coal-washeries, coking ovens and bye-product plants.

The following table, extracted from the Annual Report for the year 1912 of Dr. Atkinson, His Majesty's Inspector of Mines for the South Wales division, gives particulars of the numbers of persons employed above and below ground in each of the South Wales counties :—

County.	No. of Mines at work.	Below ground.	Above ground.		Total No. employed Below and Above Ground.
		Males Total.	Males Total.	Females Total.	
Brecon .	25	2,942	600	7	3,549
Carmarthen	71	9,539	2,501	—	12,040
Glamorgan .	398	129,311	22,375	90	151,776
Monmouth .	122	49,450	8,208	60	57,718
Pembroke .	6	339	108	5	452
Total in 1912	622	191,581	33,792	162	225,535[1]
Total in preceding year .	649	188,349	32,383	155	220,887

The figures show a total increase over the preceding year of 4,648 persons employed, which is equivalent to 2·1 per cent.

[1] This represents 20·7 per cent. of the total number employed in the United Kingdom.

Reserves of Coal

A very careful investigation was made by the Royal Commission on Coal Supplies, which reported in 1904 as to the quantities of coal remaining unworked in South Wales and Monmouthshire. The amount unworked was estimated separately in seams of different thicknesses in each county as follows :—

Division of the Coalfield.	In seams 12 inches to 24 inches in thickness.	In seams 24 inches and upwards in thickness.	Total in seams 12 inches and upwards in thickness.
	Tons.	Tons.	Tons.
Monmouthshire .	559,215,203	2,184,292,871	2,743,508,074
East Glamorgan .	1,892,457,896	9,177,902,572	11,070,360,468
West Glamorgan, Carmarthen and Brecon. . .	2,880,933,337	11,468,402,458	14,349,335,795
Pembrokeshire .	63,486,075	109,097,738	172,583,813
Total of Coalfield	5,396,092,511	22,939,695,639	28,335,788,150

Of the above total of 28,336 million tons, much the greater part is less than 4,000 feet deep from the surface, only 1,864 million tons lying at more than 4,000 feet depth, which is generally assumed to be the limit of depth in working, at any rate so far as reasonable cost is considered to impose a limit. Seams over 2 feet in thickness are easily worked ; and seams from 12 inches to 24 inches thick will be increasingly worked in the future with machine coal-cutters wherever the quality of coal is good.

The reserves of coal were also classified according to kinds or qualities and geographical location with the following results :—

	Monmouth-shire and East Glamorgan.	West Glamorgan, Carmarthen, Brecon and Pembroke.	Total of the Coalfield.	Per cent.
	Tons.	Tons.	Tons.	
Bituminous	4,483,486,542	4,135,202,423	8,618,688,965	30·42
Semi-bituminous	5,393,724,590	—	5,393,724,590	
First-class Steam	3,936,657,410	—	3,936,657,410	47·31
Steam (Western Division)	—	4,076,424,971	4,076,424,971	
Anthracite	—	6,310,292,214	6,310,292,214	22·27
Total of Coalfield	13,813,868,542	14,521,919,608	28,335,788,150	100

Ten years have elapsed since the above estimates were made,[1] and during that period about 492 million tons have been worked in South Wales and Monmouthshire. This reduces the present figure for the total reserve to 27,843 million tons. The South Wales coalfield is now producing at the rate of nearly 60 million tons per annum, and the rate of production is pretty sure to increase within 30 or 40 years to about 100 million tons per annum. The annual production will probably still go on increasing gradually for many decades, and then remain stationary, and ultimately fall off slowly.

[1] The estimates were made in 1904 : this chapter was written in 1914.

Some few of the oldest takings may be worked out completely in 30 or 40 years ; and the older districts of Merthyr, Aberdare, and the Rhondda may be practically entirely worked out 100 years hence. Most of the coalfield, east, central, and west, will last at least 200 years, however, and perhaps 300 years. The following figures show that so far, in spite of our great mines, and an army now reaching nearly 100,000 hewers, we have done little more than scratch the surface of this huge coalfield :—

	Million Tons.
Probable original amount of coal, say, 300 years ago (on basis of estimate of Royal Commission of 1904)	30,400
Less worked from beginning of mining operations, about	2,550
Leaving unworked	27,850

The amount so far worked is only 8·3 per cent. of the amount vouchsafed us by Nature.

Wages

The level of wages in South Wales is higher on the whole than that of any other British coalfield. There appear to be at least three principal reasons to account for this fact. (1) In the first place there always has been, and still is, a continually growing demand for labour through the opening of new collieries, a demand which slackens only during brief periods of severe trade depression. New pits are always being sunk in sparsely populated valleys, and it is necessary to offer a good price list in order to get the necessary labour. In face of the steadily growing demand of the new collieries the general body of employers cannot successfully oppose the general percentage increase of wages in accordance with the rise of the price of coal. The geographical position of South Wales makes it more isolated from large centres of population than any of the English coalfields, so that the supply of labour which flows from the agricultural districts of South and Mid Wales, Gloucester, Hereford, Devon and Somerset, and from towns like Bristol, and to some extent Birmingham and Plymouth, is quite insufficient from the colliery proprietor's point of view. (2) The workmen are highly organised in the South Wales Miners' Federation, and are served by skilful leaders. Thus, wherever new price lists are being negotiated, even in the well populated part of the coalfield, rates are now obtained which are almost as high as in the outlying districts. Through the Conciliation Board the utmost advantage has been taken of the increasing prices of large and small coal. (3) A further reason for the high wages in South Wales is unquestionably the relatively high price of the coal which its superior qualities command in the world's markets. These high prices have made the trade remunerative and have enabled colliery proprietors to open new mines, even though this could be done only by paying high wages. The point is that the high value of the product creates such a demand for it as enables the producers to

pay high wages. If the coal had not been so valuable, the development of the coalfield would have
proceeded more slowly, limited by the supply of
labour which would have been available at a lower
wage rate.   There is another cause which is probably
contributory to the high wages in the fact that
colliery companies in South Wales rarely make
provision for the housing of their workers.   In the
North of England the colliery companies have built
houses out of their own funds, and let them at
nominal rentals to the miners.   In South Wales
the men have had to provide their own houses
by forming building clubs, through which they buy
their houses by instalments.   A fair proportion of
the colliers are, therefore, paying about £1 per
month in capital instalments in addition to rent of
22s. to 28s. per month.   The cost of living in
regard to groceries, vegetables, etc., is also very
high in most of the mining valleys.

Mining Accidents

The same causes which have given Welsh coals
their superiority are also responsible for having
made mining in this coalfield more costly and more
dangerous than in the other coalfields of the United
Kingdom.   The Welsh coal is dry and fiery, and
owing to the dryness of most of the mines the fine
coal dust is a constant source of danger.   Loose
jointed coal, and loose or rotten roof, are also more
frequent in South Wales than elsewhere, so that
there are numerous accidents, frequently fatal, from

falls of the face of coal, as well as from falls of the roof. Most of the collieries, too, are worked upon a large scale, about 80 mines employing more than 1000 persons, and the largest over 3,000 ; consequently, when disasters, such as explosions or floodings do occur, a larger number of lives is lost than would be the case with the older and smaller mines.

The following table gives the average death-rates in the different coalfields of the United Kingdom for the years 1901–1910, the death-rate being expressed as the number of deaths per 1000 of those employed. It cannot be said that a death-rate 70 per cent. higher than that of Yorkshire or the North of England is satisfactory ; and it is hard to believe that the Home Office is doing its duty by the workmen in South Wales in not increasing the rigour of the inspection to correspond with the more dangerous nature of the coal measures. It may be added that the death-rates in quite recent years, culminating in the terrible disaster at Senghenydd in 1913, are not at all reassuring :—

Death Rates from Accidents in Collieries in
the Various Coalfields.

Name of Coalfield.	Average (1901–1910).
South Wales	1·78
Midland	1·29
Lancashire	1·57
Yorkshire	1·03
Northumberland and Durham .	1·08
North Wales	1·42
Scotland	1·49
Other Coalfields in Great Britain	1·72
Ireland	1·66
Total for United Kingdom . .	1·35

General Characteristics of the Welsh Miners

Although there has been a large immigration of
Englishmen and Irishmen, and to some extent of
Italians and Spaniards, to the Welsh mining valleys,
the population is still predominantly Welsh. The
short wiry build and the dark hair of the Welsh
miners proclaim them a race distinct from the
English. Welsh is the every-day language in the
majority of their homes ; and a stranger may constantly hear it in the streets, and in trains or trams.
In temperament also the Welshman is distinct
from the Englishman. There has been throughout
the nineteenth century, and still is amongst the
older miners, an extraordinary religious fervour.
They are practically all Nonconformists, chiefly
Methodists or Baptists ; and the numerous chapels

to be seen in the valleys have been built chiefly out of the savings of the miners, sometimes actually with their own hands in their spare time. During the past fifteen years, however, there has been a distinct falling off in attendances at the chapels ; and the younger generation is growing up imbued mainly with socialistic and political aspirations. To many of them the " war against capital " has become almost a religion ; and in the Rhondda and adjoining districts—particularly in the better paid, fully developed parts of the coalfield—have grown up several schools of thought in socialism and syndicalism which have already to some extent made themselves felt in English life in the National Coal Strike of 1912, and are likely to have an increasing influence in the British Labour Movement, and eventually in Parliament. The Welsh miner of the older valleys is temperate, industrious, thrifty and intelligent. Since the Eight Hours Act came into force there has been considerably increased activity in evening continuation classes, the Workers Educational Association, and in local government election campaigns, the miners now controlling many of the local District Councils.

The peculiar Welsh temperament, essentially imaginative, intellectual, and impatient, manifests itself continually in industrial disputes. Local disputes at the collieries with strikes lasting a few days are of constant occurrence, and the causes often appear comparatively trivial, at least, for such a drastic remedy. The men often strike, too, in de-

fiance of the order of their Miners' Agent and of
their district committee. They are just as impatient
of irksome restraint by their own leaders as
they are of any arbitrary action on the part of
the management. A very good illustration of the
intense feeling of community or class interest
which has developed amongst the Welsh miners is
afforded by a recent incident during the unofficial
strike of railwaymen on the Great Western to the
west of Cardiff. A train was due to take some 800
miners from one coast town up to their work
at the pits ; but the engine driver had not come
out on strike as was anticipated. As the persuasion
of two or three of his comrades failed to induce him
to strike the whole 800 miners resolved not to go to
work and promptly went home. I have heard it
said by colliery proprietors in other parts of the
country that they will not employ the most skilled
miners from South Wales, useful as they might be,
because they are so imbued with socialistic principles,
and try to persuade the other workmen to think
they ought to have the management of the mine in
their own hands.

Housing in the Coalfield

Housing conditions in the coalfield are not at all
satisfactory, as building enterprise has been quite
inadequate to provide for the great inrush of population which has taken place during recent years.
In the older districts of Merthyr, Aberdare, and the
lower Rhondda, many of the cottages are small, old,

and insanitary. The extensive building which has taken place in these and the newer districts during the past forty or fifty years is of comparatively good quality. The cottages are almost exclusively of the usual five-roomed type—parlour, living-room, and scullery on the ground floor, and three bedrooms upstairs. In many of the cottages built in recent years there is a large back kitchen in addition, which is usually in a large back extension. If this is built over it either allows for larger bedrooms, or more usually for four bedrooms. Practically all the building of recent years has been carried out under bylaws adopted by the local authorities, and fairly well enforced. Hence, the sanitary condition is pretty good. On the other hand, the bylaws in operation have produced a deadly monotony of design in the cottages and streets ; and have led to a great deal of unnecessary expense in both the buildings and streets, a fact which is partly responsible for the present shortage of houses. We shall see, however, that very great progress has been made in the rational and scientific ordering of the conditions of life of the industrial workers, if we read the following account of the sanitary conditions of Merthyr as it was in 1848 :—[1]

" The interior of the houses is, on the whole, clean . . . It is those comforts which only a governing body can bestow that are here totally absent.

[1] See article in the *Westminster Review*, about 1848, by G. T. Clark, quoted by Wilkins in his *History of the Iron, Steel and other Trades*, p. 305.

The footways are seldom flagged ; the streets are ill-paved, and with bad materials, and are not lighted. The drainage is very imperfect ; there are few underground sewers, no house drains, and the open gutters are not regularly cleaned out. Dust bins and similar receptacles for filth are unknown ; the refuse is thrown into the streets. Bombay itself, reputed to be the filthiest town under British sway, is scarcely worse ! The houses are badly built, and planned without any regard to the comfort of the tenants, whole families being frequently lodged—sometimes sixteen in number— in one chamber, sleeping there indiscriminately. The sill of the door is often laid level with the road, subjecting the floor to the incursions of the mountain streams that scour the streets. The supply of water is deficient, and the evils of drought are occasionally felt. The colliers are much disposed to be clean, and are careful to wash themselves in the river, but there are no baths, or wash-houses, or even water pipes. In some of the suburbs the people draw all their supply from the waste water of the works, and in Merthyr the water is brought by hand from springs on the hillsides, or lifted from the river, sometimes nearly dry, sometimes a raging torrent, and always charged with the filth of the upper houses and works. It is fortunate that fires are rare, for it seems to be the custom among the miners to keep a certain quantity of gunpowder under their beds in a dry and secure place ! "

The quality of the houses in South Wales, taken

as a whole, is probably better now than in any of the other principal coalfields, except perhaps the new South Yorkshire field. South Wales suffers chiefly from the shortage of houses, leading to considerable overcrowding, owing to a large percentage of cottages intended for one family, having to serve for two families, and to too many lodgers being taken in for the accommodation available. In 1911 the number of families occupying apartments in the county of Glamorgan alone was 21,322, an increase of 9,740 over the corresponding number in 1901. During the intercensal period of 1901–1911 the increase of population in Glamorgan was 30·34 per cent., and the increase of dwellings only 24·57 per cent. It is probable that at the present time a shortage of at least 20,000 houses exists in the mining counties of South Wales. The colliery companies have done, and are doing, little in the way of building, which has been left mainly to the workmen, who form building clubs, as described in the chapter on Housing.[1]

Mining operations are at present seriously hindered by the difficulty which developing collieries, situated in remote valleys, experience in obtaining sufficient labour. The houses in the locality being entirely occupied, the men have to travel often 15 or 20 miles by rail; or to walk three or four miles over the hills, which is a great hardship in winter, as they have to be at work at 6 a.m. The building clubs are falling into desuetude; and there appears to be

[1] See Chapter XXI.

no intention on the part of colliery companies to undertake extensive building operations at their own expense. There are, therefore, only two possible solutions of the difficulty, namely, extensive building by local authorities, and more liberal grants to public utility societies, in the formation of which the colliery companies and the miners could be usefully associated.[1]

Trade Unionism

Trade Unionism has grown to great strength amongst the Welsh miners during the past fifty years, thanks mainly to the untiring efforts of two or three pioneers, prominent amongst whom is the Right Honourable William Abraham, M.P., familiarly known as " Mabon." The South Wales coalfield is naturally divided into districts by the great ranges of hills. A principal valley and its tributary valleys constitute a natural district in which the men have facilities of intercourse during leisure hours. To reach the neighbouring district may involve a walk four or five miles over the mountains, or a journey of about two hours by train, perhaps down to Cardiff, or some other coast town, and up another valley. From the commencement, therefore, Trade Unionism has been organised by districts, and the miners still cling tenaciously to the district organisation, a recent proposal to abolish the districts having been thrown out by a large majority on a ballot vote. Each district has its own committee and

[1] See Chapter XXI.

officers, its own funds and its own rights and methods of dealing with disputes within its area. Only when it has tried its best and failed to negotiate a settlement in an important dispute, or wants external aid for a big strike, does it appeal to the Executive Council of the South Wales Miners' Federation, in which all the districts are united. Each district appoints and maintains its own miners' agent, a paid full-time officer whose duties are manifold, the principal being to organise the district and to represent the men in negotiations with the colliery managers in regard to all manner of disputes. Under present conditions very much depends upon the personality of the agent. If he is a good organiser, energetic and firm in negotiations, he secures good terms for his men in all disputes, and keeps them loyal to the Union, whilst strikes are rare and never unofficial, because he has the men well in hand. On the contrary, an agent who is a poor organiser, and slack or weak in negotiations, does not keep the confidence of his men, and a state of industrial unrest ensues which is bad for all concerned. Unofficial strikes of a few days' duration become frequent, whilst the membership of the Union dwindles away. It would seem that the system of electing agents is partly at fault, for a fluent speaker may capture his audiences with oratory, but have few of the business qualities required for his work.

The South Wales Miners' Federation is a federation within a federation, as it is a member of the Miners'

Federation of Great Britain, which is dealt with fully in Chapter XVII. The essential parts of the South Wales Miners' Federation are the Executive Council and the Conference. The Council meets frequently in Cardiff, and appoints the workmen's members of the joint Conciliation Board, and nominates members for the joint District Board under the Minimum Wage Act. Each of the twenty-one districts elects a member of the Executive Council, and it is the custom to elect the miners' agent of the district, as he can attend meetings without losing time at the colliery, and he is naturally in touch with all the business of the district. The Federation employs a whole-time secretary, assistant secretary, and some clerks, and the business is now all conducted from Cardiff. This, however, is a recent innovation, the secretary and treasurer having previously lived in different remote valleys from which the business was conducted, there being merely a small office in Cardiff for the Executive Council to meet in. Great efforts are being made by the officers and Council to carry out a policy of centralisation, but they receive little encouragement from the workmen, who manifest some degree of suspicion of paid officials.

The Conference consists of two delegates from each of the lodges throughout the coalfield. It meets annually in the Cory Hall, Cardiff ; and from time to time special meetings are held when its sanction is required to some important proposal, such as the use of the Federation funds to support a sectional strike in a particular district, or any alteration in the constitution or powers of districts, or in the status of the officers of the Federation, or in the relations with the Miners' Federation of Great Britain, or its connection with the Labour Party and political questions generally. As there are 444 lodges in the coalfield, the Conference is a very numerous body, and there is naturally always a sharp division of opinion on every question.

The South Wales Miners' Federation has always been weakened by want of funds, the accumulated funds of the Federation never having exceeded £1 per member, and having been entirely depleted by the National Strike of 1912, from which it is now slowly recovering. The funds of the districts vary very much, some having considerable amounts accumulated, others none at all. A proposal of the Executive Council to increase the contribution to the Federation was recently negatived in a general ballot vote, and there seems little prospect of a strong centralised union for the whole coalfield coming into being at an early date. Such a strong central body, supervising and controlling the whole business and negotiations from Cardiff, could not only be a great benefit to the workmen but in many ways also to the Trade as a whole. Inconsistent action in neighbouring districts, vacillating policy, and divided counsels on the Executive of the Federation have been features of the miners' Trade Unionism for many years, and the important strikes

of 1893, 1898, and 1910 were largely brought about
or intensified by differences of opinion between the
miners and their leaders and among the leaders
themselves. The miners have undoubtedly in
many places suffered from real grievances, against
which they have repeatedly struck work, but many
of these grievances would never have become so
serious if they had been handled by an energetic
centralised union. Furthermore, it is unsatisfactory
that the employers themselves get very unequal
terms in different districts. In well organised districts price lists tend to be higher and the men's
privileges are more completely enforced than in the
badly organised districts, where many of the
employers have at times been able to do much as
they liked and thus have an unfair advantage in
being able to undersell their competitors whose
mines are in well organised districts.

Whilst not understanding organisation, the Welsh
miner is yet very enthusiastic for his Union, and
the workmen will rally to the call of their agent in
seeing that every workman is a member of the Union
and up to date in his contributions. It is customary
to have " show card " days, when every workman is
obliged to produce his Union card and show that he
is up to date in his contributions. If he is not in
compliance he is not allowed to descend the pit.
If any number of workmen persistently refuse to
join the Union, the whole body of miners, many
hundreds in number, will go on strike until the
whole of the men have joined. There have been

cases where seven or eight hundred men have remained on strike a week or longer in order to compel
three men to join the Union. In some of the
valleys the old-fashioned "white shirt parade" is
still in vogue. Bands of women carry white shirts
in which they garb the unhappy non-unionists
whom they fetch from their homes. They are
forced through the streets in a crowd of hostile
miners, amidst ignominious epithets, and sometimes
pelted with mud and eggs, etc. These practices are
now deprecated, and strikes on the non-unionist
question which have replaced intimidation have
lately become so frequent that many of the employers are not anxious to employ non-unionists if
they can avoid it.

Politics of the Welsh Miners

The miners of South Wales are keenly interested
in religious and economic questions, and the political
aspect of these subjects claims much of their attention. The older type of miner is rather conservative
in temperament, both in industrial and political
matters ; but being a strong Nonconformist he
usually supports the official Liberal party. The
younger miners are keen advocates of socialism, and
an extreme group which has a large number of
adherents in the Rhondda are syndicalists. The
socialists are organised as branches of the Independent Labour Party, and a very active propaganda
has been successfully carried on in the mining
valleys during the past ten years. Whilst the

oldest miners' leaders are of the Liberal complexion, the younger members of the Executive Council are socialists, and they include many of the ablest miners' leaders. It is probable that in the future the labour candidates selected to fight the parliamentary constituencies in the coalfield will be increasingly of the moderate and responsible socialist type. Perhaps owing to the political cleavage amongst the miners themselves, their contribution to the Federation's political fund is poor, and the Federation is, therefore, somewhat handicapped in providing sufficient registration agents, and is not likely in the near future to have more than four or five representatives in Parliament, although there are fifteen constituencies in the coalfield.

The syndicalist section of the miners looks with disfavour upon political action, seeking to gain its ends by perfecting industrial organisation and by transferring the control of industrial undertakings from the employers to the workmen, rather than by parliamentary representation and industrial legislation. The syndicalists wish to be able to elect the manager, overmen and firemen of every colliery, and to divide the whole profits of the colliery amongst the workmen after paying four or five per cent interest to the capitalists. At the present time, however, the supporters of this movement in South Wales seem to be more keenly interested in destroying the Labour Party than in obtaining their own special object. The syndicalist movement is in many ways an extremely interesting one, and it is further referred to in a later chapter.[1]

The North Wales Coalfield

A small coalfield, but one of growing importance, exists in the North of Wales in the counties of Denbigh and Flint. The total area of this coalfield in its visible portion is estimated at 103 square miles, of which 56 square miles are in Denbighshire. The fields are bounded on the west by the Millstone Grit and Mountain Limestone ; in the east, however, they are overlain by Permian rocks. Much of the coal in this eastern district towards and over the English border has already been proved, and it seems probable that considerable extensions of deep mining will take place in this easterly direction. The coalfield is crossed by numerous faults, one of which, named the " Wrexham fault " was formerly regarded as the boundary of the coalfield on the east side ; during recent years, however, the seams on the eastern or downthrow side of the fault have been proved to be workable, and Wrexham itself is likely to become a large centre of the mining industry.

In Flintshire there are from 12 to 14 seams of workable coal with an aggregate thickness of 58 feet; in Denbighshire the number of seams averages about 16, but the aggregate thickness varies from 36 to 60 feet in different parts. The chief seams in Flintshire are the Lower (4 feet),

[1] Chapter XIX, p. 554.

Main (7 feet 6 inches), Brassy, Hollin, and Four Feet coals. In Denbighshire the Main, Brassy, Twoyard, Powell, and Drowsalls are the principal seams worked. A small area of productive coal measures extends also into the Island of Anglesea, but these seams are not now worked to any considerable extent.

The coal is used mainly for manufacturing, gas, and household purposes, and hardly any at all is exported beyond the United Kingdom. The total output has increased slowly until quite recent years, as shown in the following table :—

PRODUCTION OF COAL IN NORTH WALES

	Output in Tons.	Number Employed
1870	2,329,030	
1880	2,429,315	
1890	2,975,646	
1895	2,848,072	12,236
1900	3,109,615	12,629
1905	2,901,998	12,743
1910	3,410,876	15,171
1911	3,443,078	15,424
1912	3,250,749	15,680
1913	3,505,724	15,948

The actual reserve of the coalfields in the three counties amounts to 2,564,714,618 metric tons.

IRELAND

A geological map shows that Lower Carboniferous strata are very widely distributed over the surface of Ireland, occupying, indeed, about half its area, whilst in Ulster they are overlain by newer rocks. Looking at the coalfields of Ireland from the broad point of view in their relation to Great Britain, we see that the north-east and south-west lines of folding, which determine the coalfields of Scotland, are continued across the North Channel into Ulster, producing a series of basins in which detached areas of coal may lie. On the other hand the coalfields in the south of Ireland are produced by east and west folds which are an extension across St. George's Channel of the great east and west system of folds stretching from Belgium through Kent and South Wales. Conformable with this correlation of earth folds is the character of the coals. In the north they are bituminous, but in the south they are semi-bituminous or anthracitic, probably with some true anthracite. The southern coals are, in fact, medium quality steam coals, requiring a strong draught to burn them, and not unlike steam coals of the same class which are extensively worked in South Wales and Kent.

In Ulster, coal mining has had a checkered history. The little field of Antrim is now worked out and only ironstone continues to be worked and shipped. The coalfield of County Tyrone lies immediately to the west of Lough Neagh. Very little of it is exposed, the measures dipping mainly under newer rocks. The industry has not been supported by sufficient capital, and accidents such as explosions and influxes of water have led to the closing of mines. Professor Hull remarks in his report to the

Royal Commission on the Coal Supplies (1904) :
" Want of enterprise and capital, especially in
Ulster, appears to prevent the opening up of a large
mineral field in this county, capable of supplying
fuel for the mills and factories of the surrounding
country."

The other northern coalfields are those of Con-
naught, occupying four detached basins in counties
Leitrim, Roscommon and Sligo. The coal is won
by very old-fashioned and costly methods, and only
about 10,000 tons per annum are produced in these
counties—scarcely a week's output of any large
Yorkshire or South Wales colliery.

Two basins occur in Queen's, Kilkenny and
Tipperary counties, just to the south-east of the
centre of Ireland, and these constitute the most
important coalfield of Ireland so far as output is
concerned, the average annual output being about
80,000 tons. The largest coalfield of Ireland is,
however, that of West Munster, in counties Kerry,
Limerick and Clare. Most of the coals are thin,
but there are two thick seams. They are scarcely
worked at all, however, being in an inaccessible
position as regards means of transport and industrial
markets. This field is probably incompletely explored, and eventually an export trade might be
developed.

The Irish coalfields are all comparatively poor,
both in regard to quality of coal, and to number
and thickness of workable seams. The exception is
the Tyrone field, which compares well with the

smaller fields of England in this respect. The total reserve of coal in all the Irish fields is only estimated at a little over 200,000,000 tons, or about ten months' consumption for the whole United Kingdom. This may be a very conservative estimate.

The statistics of the Irish coal trade do not indicate a healthy condition. The output has been falling off seriously in recent years, as the following figures of output for the whole of Ireland clearly show :—

COAL PRODUCTION OF IRELAND.

	Tons.	No. of Men Employed.
1870 . . .	141,470	
1880 . . .	133,702	
1890 . . .	102,267	
1895 . . .	125,586	
1900 . . .	124,700 .	997
1905 . . .	90,335 .	749
1910 . . .	79,802 .	725
1911 . . .	84,564 .	790
1912 . . .	90,307 .	862
1913 . . .	82,521 .	776

The output of Tyrone was reduced by 13,000 tons, and that of Queen's County, by 15,000 from 1900 to 1901. This was the conclusion of a period of boom in coal prices, and the set-back no doubt led to the closing of three or four collieries. The decrease from 1870 is, however, very striking ; and it is difficult to explain. It may be due partly to the decreasing population of Ireland, partly to the reduced freight rates for landing English, Welsh

and Scotch coal in Irish ports ; and partly, also, to
the rise in the rate of wages which has followed the
emigration, and the revival of agrarian life which
has resulted from the land-purchase, cottage-build-
ing and agricultural co-operation schemes promoted
by Government in recent years. It has often been
remarked how different might have been the history
of Ireland in the nineteenth century had she been
blessed with an abundance of good quality coal, like
England. The irony of Ireland's fate is that coal
measures as rich as those of Tyrone were probably
laid down over a wide extent of country, covering
many counties ; and they would have formed one
or more great coalfields had they been spared by
the denuding forces of nature. As it is, in the existing coalfields we usually have left only the lower
beds of the coal measures. Many valuable seams
have gone entirely, or have been left only as small
islands, which in many cases have been entirely
worked out already.

Yet the experience of Tyrone, where there is a
small but rich coalfield, and the experience of Spain,
Italy and Russia, not to speak of China, Japan,
Australia and Canada, in all of which countries most
valuable coalfields were for decades neglected, and
are so still to a large extent in spite of the presence
of a considerable population, shows clearly the
importance of capital and enterprise being available,
as well as coal and labour.[1]

Labour does not appear to be efficient in the Irish

[1] See Chapter XI.

coal trade. There were only 862 persons employed above and below ground in coal mines in Ireland in 1912, and 776 in 1913 (or less than in any one large colliery in Great Britain), and they produced only an average of 110 tons per person in that year, as compared with an average of 268 tons per person for the United Kingdom.

SCOTLAND

The Coalfields of Scotland

Several coalfields of considerable importance occur in Scotland in a wide zone extending from the Firth of Forth to the Firth of Clyde. This zone is about 95 miles long, and its width varies up to 30 miles. There are about a dozen distinct coalfields, the principal of which are the Midlothian, Fifeshire, Linlithgow and Lanarkshire, Clackmannan and Ayrshire fields. Geologically these are not all distinct fields, for the Fifeshire field is probably continuous with the Midlothian field under the Firth of Forth, and the Clackmannan similarly is continuous under the alluvial plains of the Forth valley with the great Lanarkshire–Linlithgow field, which also extends into the counties of Renfrew, Dumbarton and Stirling. These two and a third, the Ayrshire field, form the only three great coal basins of Scotland, the other fields being merely little basins contiguous to one or other of the three great basins.

Numerous coal-seams occur in the Lower as well as in the Upper Carboniferous Series, that is to say, in the beds (sandstones and shales) beneath the Millstone Grit, and equivalent to the Mountain Limestone of England and Wales, which is poorly developed in Scotland, though generally to be found. This appearance of coal seams in association with the Mountain Limestone appears strange to any one acquainted only with English geology; and we may note the interesting fact that it is characteristic of the northern province only. A gentle anticlinal fold crosses England nearly north-east to south-west from Ripon in Yorkshire to Clitheroe. To the north of this anticline, coal-seams are found in the Lower Carboniferous as in Northumberland, Cumberland and Scotland, and across the water in Ulster ; but south of this anticline, and also in the south of Ireland, coal is nowhere found in the Lower Carboniferous. This suggests the interesting reflection that the northern province was a shallow sea, occasionally subject to the conditions favouring the formation of coal-seams,[1] for long ages before the deep sea of the southern province became shallow and fertile for coal formation.

Dr. Walcot Gibson, in his book on the *Geology of Coal and Coal Mining*,[2] gives the following table of the sequence of the Carboniferous series in Scotland :—[3]

---

[1] See Chapter II.   [2] Edward Arnold, 1910.
[3] p. 221.

Coal Measures	Red Sandstone	600 ft.
	Upper Coal Series	300 ft.
	Lower Coal Series.	800 to 1,500 ft.

Millstone Grits : Roslin Sandstone or Moorstone Rock (almost barren of Coals) . . . 0 to 700 ft.

Carboniferous Limestone Series	Upper Series of Limestones . .
	Sandstone, shales and coals . .
	Lower Series of limestones . .

Calciferous Sandstone Series : Sandstones and conglomerates

Speaking generally the coals of the Lower Carboniferous are thin, and are more cut up by faults and igneous intrusions than those of the Upper Series. They are, however, well developed at Lochgelly in Fifeshire ; and between that district and Dunfermline are a number of thick and valuable seams in the limestone series.

Another peculiar feature of the Scotch coalfields is the extent to which the mining operations are hampered by intruded dykes or sills of igneous rocks. These have been forced in as molten lava from below, and form very hard walls or beds. When the molten rock fills a nearly vertical fissure, running across the strata, it is called a *dyke* ; when it has run along the bedding planes and formed a hard sheet parallel with them, it is called a *sill*. Dykes in particular are great obstacles to mining, as in working a longwall face the miners come upon an intensely hard, almost impenetrable wall, 20 or 30 feet thick, which has to be pierced in several places for roads to pass through, whilst new stalls must be opened out on the other side. In England and Wales these dykes are rare in the coal measures, being found only in South Durham, Northumberland and Derbyshire. In Scotland they are extensive in the Fife, Linlithgow and Ayr coalfields, and usually present, though less seriously, in the others.

The molten matter when intruded was generally white hot, and it is not surprising that the bituminous coals have been, as it were, coked by them, losing the volatile constituents. The structure of the coal is maintained, but it is found to be anthracite near the dyke, passing into semi-bituminous, hard steam coal further away. Anthracite is actually marketed from the neighbourhood of some of the very large dykes.

The *Lanark* coalfield in the restricted sense, that is to say, the part lying in County Lanark, consists almost wholly of the Upper Carboniferous measures. It is the centre of a great industrial district extending from Glasgow at its north-western end, up the valley of the Clyde, almost to the town of Lanark. This coalfield has been the industrial centre of Scotland for very many years, and practically the whole district is occupied by collieries, many of them in an advanced stage of development. The coalfield is, therefore, thoroughly well proved. The famous " black band " ironstone, exceedingly rich in iron, has been a great asset. The lower coals of the Carboniferous Limestone series rise to the surface on nearly all sides of the Lanarkshire basin, and to the north-west of Glasgow have been extensively worked and much exhausted, whilst in Renfrewshire they have been also much worked, though there the seams are thin and of poor quality, and continue so to the south-east to East Kilbride, and on towards the town of Lanark. On the east of the Lanarkshire basin from Bathgate to Wilsontown, however, the Limestone series of coals is well developed, the seams being here at their thickest. Only about 20 collieries work in the Limestone series, however, as against about 250 in the upper series, or coal measures proper.

The *Fife coalfield* is of rapidly growing importance, and is second only to the Lanarkshire field. It contains the whole carboniferous series of strata within its area, and the seams are of good thickness over large areas. The upper series (or coal measures proper) occur in the north end of a basin, along the Fife coast from Dysart to Largo. This basin is elongated from north to south, and is continuous under the Firth of Forth with the Lothian coalfield. The coal is now worked to a distance of two or three miles under the Firth of Forth, and is likely to be followed much further. On the Fife coast the seams successively crop out on the sea shore, and for two or three miles inland ; and it is especially this upper series of coals which has been worked for centuries. The carboniferous limestone series occupies a far larger area, stretching from Elie westwards, north of Largo, through Markinch and

Lochgelly, past Dunfermline to Cidross. It occupies a wide-stretching area of which Dunfermline is nearly the centre. The Limestone coals are comparatively thin near the coast at Kirkcaldy, but thicken as they are followed inland, attaining their maximum development at Lochgelly, where there are 16 seams, aggregating 57 to 70 feet in thickness. This coalfield of the Limestone series is from 15 to 18 miles long, from east to west, and 4 or 5 miles broad. West of Dunfermline the seams become thinner; but exploration by boring in recent years has shown many workable seams to be available right up to the border of the Clackmannan field, where the Limestone series dips west and north-west under the Millstone Grit series. This latter passes under the upper series of Clackmannan, which is geologically continuous with the Lanarkshire basin; so that the Fife and Lanark coalfields are continuous through the coalfield of the Limestone series.

The *Ayrshire coalfield* lies on the east shore of the Firth of Clyde; and is second only in area (330 square miles) to the Lanark-Linlithgow-Stirlingshire Field, and is third (after Fife) in its volume of production. Coal seams occur in both the Upper Series and Lower Carboniferous, and are generally bituminous with a high percentage of volatile matter. The seams are nowhere specially numerous, and are thick only quite locally. The quality is not specially good anywhere, and igneous intrusions cause much trouble. They have locally converted bituminous to steam and anthracite coals.

The *Lothian coalfield* is geologically the southern end of the Fife and Forth basin. The seams are nearly horizontal in the centre of the field, where the upper series is exposed ; but are highly inclined at the margins, particularly the west, where the Limestone series comes to the surface. The seams of the Limestone series aggregate 100 feet in thickness at the north, but deteriorate rapidly in thickness and quality southwards. They usually contain bands of cannel coal, one of them (in the Great seam) being fully two feet thick. This coalfield is probably the oldest worked in Scotland.

The relative outputs of the above-named coalfields may be gathered by grouping the county output figures as follows :—

COALFIELD PRODUCTION OF SCOTLAND, 1902 AND 1912

Coalfield.	Counties.	1902. Tons.	1912. Tons.
Lanark	Lanark . Dumbarton Renfrew Linlithgow Stirling Clackmannan	21,758,748	22,506,984
Fife	Fife Kinross	6,206,519	8,435,516[1]
Ayr	Ayr . . .	4,044,876	3,935,949
Lothian	Edinburgh Haddington Peebles	1,945,559	4,115,573[2]
Small detached *fields.*	Argyll . . .	154,565	524,607[3]
Brora (Jurassic)	Sutherland . .	5,042	

[1] Fife only: [2] Edinburgh and Haddington only.
[3] Argyll, Dumfries, Inverness, Kinross, Peebles, and Sutherland.

Character of Coal

The coals of Scotland are almost entirely of the bituminous kinds. They include good and medium quality steam coals, gas and house coals and coking coals. They are thus used for a great variety of manufacturing, railway and domestic purposes. A small quantity of dry steam coal and anthracite is mined from the neighbourhood of some of the great dykes and sills of Fifeshire and Ayr, and anthracitic coal is also found in a tiny coalfield in the north of the island of Arran, which may perhaps be regarded as a detached portion of the Ayrshire field. Cannel coal—a black, lustreless coal, breaking with conchoidal fracture, and resembling jet—is found in the Midlothian coalfield, as mentioned above. It is highly bituminous and is much used in gas manufacture.

Along the sea coast at Brora in Sutherland occurs a small coalfield of Jurassic age—the only coal of this age in the British Isles. There are two seams, the Main coal which has been worked for a long time, and the Parrot coal, lying below it. The latter is a fissile, shaly coal of little value, which is a pity, as it is in places as much as six feet thick.

Taken as a whole the Scotch coals command as good prices as those of the north of England, but, of course, do not approach those of South Wales. For example, in 1912, the bulk of the coal exported from ports on the east coast of Scotland was sold at 9s. to 12s. per ton, and from west coast ports (principally Glasgow) at from 10s. to 13s. From Durham ports the range of prices of the greater part is 8s. to 12s. ; from Humber ports 10s. to 14s. ; and from South Wales ports 14s. to 17s. for large and 6s. to 11s. for small. In Scotland and the north of England there is not so much separation of large from small.

History

The earliest records of the working of coal in Scotland relate to the Midlothian coalfield in the twelfth century, when the monks of Newbattle worked coal on the banks of the Esk, and to Linlithgow, where the monks of Holyrood had, about the same time, received a grant of land on which coal was dug. During the thirteenth and fourteenth centuries there are several records indicating the working of coal, probably merely on the outcrops by simple quarrying operations. These relate to Fifeshire as well as Lothian and Linlithgow. Early in the fifteenth century the " coaleries of Tranent " appear to be well known, and much coal was worked at Dysart and elsewhere in Fife. Wood was probably still fairly plentiful and generally used as fuel, or for making charcoal for iron smelting. Coal was, however, already used in smithies, and to some extent in cooking and probably for heating houses in towns and in the cottages of the poorer classes.

In the sixteenth century the coal trade became an established business, the export of coal having begun, mainly as ballast or return cargoes of Dutch and other foreign ships, which came with the finished manufactures of the Continent. In 1563 the ninth Parliament of Queen Mary passed an Act prohibiting the export of coals on account of " the dearth and scantiness of fewall " thereby caused ; and this was confirmed in 1579 with the addition of a reward for the apprehension of offenders. It is marvellous how Britain has progressed to greatness in spite of its Governments !

In 1592 it was necessary to pass an Act to prevent " the wicked crime of setting fire in coal benches, bi sum ungodly persones, upon privat revenge and despite " ; and in 1597 to take further stringent measures against the exportation of " grite burne coile," which means coal in large lumps, the small coal being left no doubt for the home market. In 1598 the Brora seam in Sutherland was worked for making salt.

Culross and its neighbourhood had already been producing coal for many years, and in the seventeenth century the mining operations here became so extensive, and working a mile under the sea was esteemed so wonderful, that the Culross mines became famous throughout the country. Much of the coal was used locally in the salt manufacture for the evaporation of sea water. About the year 1630 90 tons of salt were being made per week, using probably about 45,000 tons of coal per annum. From this date on, the coal trade continued to grow in importance on the Firth of Forth, which was then the only populous part of Scotland.

The development of the Ayrshire coalfield began in the seventeenth century, and in 1678 Sir Robert Cunningham introduced extensive improvements upon his colliery at Saltcoats. About 1690, Newcomen's steam engine was introduced in Scotland for pumping water from mines. During the eighteenth century the coal trade of the Firth of Forth continued to grow slowly, and the Customhouse returns of the Port of London show that from 1745 to 1765 about 3,000 to 6,000 tons annually were imported from Scottish ports. It was during the eighteenth century that the development of the Lanark coalfield began as the result of the growth of trade upon the Clyde, which received a great stimulus from the rising importance of the American colonies. Another cause of the growth of coal-mining near the Clyde was the presence of the rich " black-band " ironstone and the discovery of the use of coal for smelting iron, whilst the manufacturing industries which began to grow up in the latter part of the eighteenth century, especially after Watt's successful invention of the steam-engine, created an extensive demand for coal. The growth of iron and steelworks, machine-making, and the textile industries in Glasgow, Paisley, Hamilton and many surrounding towns, during the nineteenth century, caused a rapidly increasing demand for coal in Lanarkshire, and the output responded, since the steam-engine was available for keeping the mines dry. Thus, quite early in the nineteenth century, the Lanark coalfield caught up with and then com-

pletely outstripped the older coalfields of the Firth
of Forth.

The great development that has taken place in
the coal trade of Scotland may be seen in the following table giving the annual outputs at different
periods during the past 40 years.

GROWTH OF COAL PRODUCTION IN SCOTLAND.

Year.	Output in Tons.	Number of persons employed.
1870	14,934,500	
1875	18,591,500	
1880	18,294,800	
1885	21,288,500	
1890	24,278,500	
1895	28,792,700	88,371
1900	33,122,200	98,502
1905	35,839,300	108,132
1910	41,335,100	131,315
1911	41,718,200	131,314
1912	39,518,600	136,656
1913	42,456,500	147,549

CHAPTER VI

THE KENT COALFIELDS

THE story of the development of the Kentish coal field, and an account of the vast possibilities of its future, are of particular interest at the present time. The existence in that county of a great field with a vast store of mineral wealth has been amply proved ; but, as yet, it is in its earliest stage of development, for there are only four collieries from which coal is being actually raised and placed upon the market. Several other collieries are in course of being sunk and many more are projected ; new railways have been built and others commenced, and eventually some hundreds of miles of railways will be wanted to tap the coalfield. Great docks will also be needed for shipping coal abroad, and to London. A number of ironworks are sure to be started on a large scale and other new industries depending upon cheap coal will arise ; and all these surprising new industries will mean the growth in Kent of a hundred villages and many large towns. A strange transformation of the " Garden of England " !

But few Londoners realise yet that there is workable coal within fifty miles of London, and that it compares well in quality with many of the best coals in this country. The total thickness of workable coal is in many places remarkably large, and the collieries are being sunk and equipped for working upon a very large scale from the beginning of the development of the coalfield.

In this feature, as in so many others, the Kent coalfield is absolutely different from that of other fields in this country. That is because it is a concealed coalfield, in no part of which does the coal come to the surface. It is also different from any of the other concealed coalfields which have been proved and are not merely conjectural ; for these other coalfields are only continuations under newer strata of exposed and well-known fields. For example the concealed coalfield of South Yorkshire and Lincolnshire is a continuation of the Yorkshire and Derbyshire field. The procedure in sinking new pits there has been to follow the coal eastwards from its outcrop near the Pennine Range. Each new colliery sunk has been ventured further east-ward into the unknown but always only a little way beyond the area where coal was actually proved.

On the other hand, coal in Kent, near Dover, would be 200 miles from any other English coalfield ; and there was no reason to suppose coal to be there, further than the expert opinion of a number of mere geologists. Business men in London, many of the great financiers of the " City " scoffed at the idea, and " Coal in Kent " became the joke of the Stock Exchange. Contempt was mingled with pity for the poor fools who staked their money on the idea. They had not even the measure of sanity of those who hunted for Spanish treasure in the Cocos Islands !

And yet in 1890 coal was actually found at the Shakespeare Cliff, near Dover, and it was soon proved to be there in seams of workable thickness, though at considerable depth. Further borings revealed more coal, until now the area of the proved coalfield is over 200 square miles. But the unbelievers of the " City " were not to be outdone. There may be coal in Kent," they said, " but anyway it won't burn, so it's not worth bringing out of the pit."

New enterprises which require a vivid imagination, or a scientific training to initiate, require something of the same qualities to appreciate them. Most men are quite unable to discriminate between unaccustomed proposals having a reasonable probability of success, and the wild cat schemes of cranks or knaves which have no solid basis in science or experience. Unfortunately, the financiers of the City are prone to the same weakness and are shy of leaving the beaten track. The sceptics have at last been completely answered and silenced by the actual development of the coalfield—a triumph over a succession of disappointments and difficulties. Kent coal is now on the market, and is being eagerly purchased by local consumers to the amount of at least 4,000 tons per week at remunerative prices of 17s. per ton and upwards. The first cargo was shipped from Dover in August, 1913, and the coal has now every prospect of steady demand from many different quarters.

Geology of the Kent Coalfield

From the geological point of view the important feature of the Kent coalfield is that the Coal Measures nowhere appear at the surface, being completely concealed by a thick covering of Secondary and Tertiary strata. Over the greater part of the known coalfield the surface rock is chalk, overlain here and there by patches of brick earth. The chalk is in most of the borings from 700 to 900 feet thick, and beneath it comes the Gault, a stiff blue clay, 100 to 180 feet thick. Beneath it is the Lower Greensand, from 12 to 65 feet thick, and containing much water. Underlying this comes a few feet of the Wealden beds ; and these rest in some places on the top of the Coal Measures and in others [1] upon 200 to 300 feet of Oolite beds (Middle Jurassic). The Permian and Triassic strata are absent everywhere, so that, if they were ever deposited, they were denuded and wholly removed before the Middle or Upper Jurassic period. Probably the entire area was land during Permian and Triassic times, and much of it remained dry land [2] throughout the whole or most of the Jurassic period.

The Coal Measures lie as a platform with irregular surface on which the secondary rocks were uncon-

[1] Barfrestone, Fredville, Waldershare.
[2] North side of the coalfield.

formably deposited. In early Triassic times there was, doubtless, an extensive exposed coalfield just like the South Wales, Durham or Scotch fields now are. There was then elevated, almost mountainous, land in Essex, and the surface sloped southwards through Surrey and North Kent. The land surface then existing has been warped by subsequent earth movements, so that the slope from north to south has been lessened. In East Kent the surface of the Coal Measures is highest (at 860 to 950 feet below sea level) at the extreme east near Deal and Walmer, and to the north of the valley of the Stour. In the major part of the East Kent coalfield the top of the Coal Measures lies from 1000 to 1,100 feet below sea level ; but in the neighbourhood of Dover and generally towards the south it becomes somewhat deeper.[1] South of a line drawn from Dover in the direction of London the top of the Palaeozoic floor falls away rapidly to a much greater depth ; but the Coal Measures are not found there, Permian or Triassic strata lying on Devonian rocks.

The Coal Measures themselves are thick, as much as 2,730 feet of measures having been bored through at Oxney, near St. Margaret's, without reaching the base. According to the researches of Dr. E. A. N. Arber,[2] who has made a very detailed study of the fossil plant remains obtained in the various borings, both the Upper Coal Measures and the Lower Coal

[1] Dr. Aubrey Strahan, Presidential Address to the Geological Society : *Quarterly Journal*, Geol. Soc., Vol. 69 (1913), p. lxx.
[2] *Quarterly Journal*, Geol. Soc., Vol. 70 (1914), p. 54.

Measures, as known elsewhere, are entirely absent from Kent ; and we have here a thick development of the Middle Coal Measures, and of what are known as the Transition Measures, which lie between the Middle and Upper Coal Measures proper. In this respect the Kent coalfield is similar to that of Pas de Calais in the north of France. The fossil flora of the two fields also is nearly identical.

Taken as a whole the Kent field appears to be an elongated basin with its axis running from 30 degrees south of east to north of west, the beds dipping gently at about 3 degrees. It is bounded on the north by a fold, now fairly accurately located, and running from the mouth of the Stour to Herne Bay, and probably there are folds on the south and east sides, whilst there may be a great fault forming the western boundary, though here further exploration is necessary. The Coal Measures consist of sandstones, some of them massive, interstratified with shales, fire-clays and coal seams, and with one or two seams of ironstone. The measures are of greyish and bluish colour throughout like those of South Wales, and no igneous rocks at all have been met with.

From ten to fifteen workable coal seams were found in most of the deep bore-holes ; and, though pretty widely distributed in the vertical section, they tend to occur chiefly in the Transition series and in the lower part of the Middle Coal Measures.

The limits of the coalfield cannot yet be given with certainty. As stated above, it is improbable

that coal will be found north of a line from Herne
Bay to the mouth of the Stour, and the southern
boundary keeps to the north of Brabourne, where
no coal was found, and probably runs from Abbot's
Cliff towards Wye. Eastwards the coal doubtless
extends under the sea; but whether for more
than three or five miles it is impossible to say. It
is not thought that the basin is actually continuous
under the Straits of Dover with that of the Pas de
Calais, although it belongs to the same system.
Westwards the boundary is still quite uncertain,
though coal is proved practically to Canterbury.
There may be a fault cutting out the measures
west of Canterbury as some suppose; but if there
be a barren area here, the productive measures may
come in again still further west towards Maidstone.

Exploration of the Coalfield

The possibility of coal existing in Kent is by no
means a new idea. That distinguished geologist,
Sir Henry de la Bêche, threw out the suggestion that
coal measures probably existed there, as long ago
as 1846, after he had thoroughly studied the geology
of South Wales and of the south-west of England.
The first geologist seriously to take up the matter
was Godwin Austen who wrote an elaborate paper
for the Geological Society of London in 1855 giving
his reasons for supposing coal measures to exist
beneath the newer strata in the south-east of
England. The matter was carefully considered
by the Royal Commission on Coal in 1871; and for

them Sir Joseph Prestwich prepared an extensive report on the probability of finding coal under newer strata in the south of England. The general tenour of the report was that whilst coal measures might probably exist under a larger area in the south of England, there was doubt as to whether they would contain much or little coal, and a doubt as to whether it would not be of poor quality. It was also suggested that the difficulties of sinking to great depths through watery strata might prove insuperable wherever the Lower Greensand occurs amongst the strata overlying the coal. At the same time Sir Joseph recommended a few trial borings. It is interesting to find that he anticipated the trouble which has been actually caused by water in sinking through the Lower Greensand. Some boring was undertaken in the Weald of Sussex in 1876, but no coal was found, and the Kent Exploration Committee which was subsequently formed did not succeed in raising sufficient funds to begin systematic boring operations. The Committee's policy was, however, the right one, for they recommended boring in the valley of the Stour, on the ground that the Secondary rocks would thin out towards the north. Recent borings have shown this to be a correct anticipation.[1]

In 1882 the work at Dover in connection with the Channel Tunnel was suspended owing to the Government interdict, and it was decided to put down

[1] See series of articles on *The History of the Kent Coalfields* in the Joint Stock Companies Journal, Oct., 1913, to May, 1914.

boreholes in the disused vertical shafts sunk for the tunnel in order to try for coal.

The work of boring proceeded very slowly. The first coal seam was actually discovered in 1890, and as the boring was continued several other seams were successively proved. A company with very considerable capital, the Kent Collieries Corporation, was floated in 1897 to purchase all the rights of the syndicate which had proved coal, and to sink pits. There followed a most unfortunate period in the development of the coalfield, during which much money was spent but little progress made. A large amount of money was paid to the syndicate, and a good deal went in promotion expenses, whilst the management was at first arranged on too lavish a scale. Considerable difficulties were met with in sinking, owing to water and soft strata, and progress was stopped sometimes for lengthy periods. The adoption of the Kind-Chaudron system of sinking early in 1902 enabled better progress to be made, and the shaft struck the first seam of coal in September, 1903, at a depth of 1,190 feet. This seam was not of sufficient thickness and quality to be workable, however, and the company, though in serious financial difficulties, had to struggle on with sinking, which was done slowly. In February, 1905, the first workable seam was reached at the depth of 1,273 feet. At this date a reconstruction took place, bringing into existence the company which now owns the mine—Kent Collieries, Ltd.

Meanwhile the exploration of territory to the

north of Dover was being considered, and its organisation was actively undertaken by Mr. Arthur Burr,
whose foresight and untiring persistence in the face
of great obstacles has done so much for the development of the Kent Coalfield.   Mr. Burr's policy was
to acquire as nearly as possible a monopoly of the
mineral rights of a large area before proving it by
boring, so that the profits arising from organising
mining companies and the general development of
the coalfield should go, not to the landowners, but
to the parties who had done the work and risked
their capital.

In 1904, Kent Coal Concessions, Ltd., was registered, and sufficient capital was subscribed to enable
the purchase of the freehold of the mineral rights,
whilst in other cases mineral leases were taken at
6d. or 9d. per ton royalty and with a dead-rent rising
gradually to £2 per acre or thereabouts.   These
rates are low as compared with the royalties which
new colliery companies are obliged to pay in all
the principal coalfields of this country ; and the
Kent Coal Concessions, Ltd., has already been able
to dispose of some of its area at greatly enhanced
royalties.

A few words may be said here about the complicated system of finance which has characterised
the development of the Kent coalfield, and which
has suffered a good deal of criticism, most of it of
an ignorant kind.   It was force of circumstances
which made any other system of finance impossible.
A big programme had been outlined which, for suc-

cess, required the raising of very large capital, and the pursuit of a consistent policy over a number of years. Every other coalfield in Great Britain has developed first as a visible field. Mining has begun on a small scale, and wealth has grown in the hands of local families, who have ventured on larger and larger mines, opening up deeper and more remote measures, often in a concealed extension of the coalfield.

In Kent, however, there was no possibility of beginning mining on a small scale, it being a completely concealed field ; and it was further wisely decided that all mines sunk, after the Shakespeare Cliff venture, should be upon the largest scale, planned and equipped for an output of 5,000 tons or more per day. The sound policy of engrossing all the richest mineral lands necessarily meant locking up a great deal of the available resources, which had, therefore, to be spread over a wide field of uses, and to be raised by every legitimate means which ingenuity could devise. If Mr. Burr had been able to convert a multimillionaire to his own confidence in Kent coal, the history of the finance of development of the coalfield would probably have been very different. The financial magnates held severely aloof however, and it was necessary to raise funds by a multitude of small investments of the general public in a number of companies.

The Burr group of companies affords a splendid illustration of a principle generally applicable to the finance of risky undertakings which must necessarily take a number of years to come to maturity. The public invest in the parent company ; but as it is obliged to commence operations without having nearly sufficient capital to carry the whole scheme to completion, further issues of shares must be made from time to time. As no dividend can be paid for some years, the company loses any attractions it may have had for the public, even though the promoters may have anticipated a long wait. Issues of preference shares are then resorted to, and finally of debentures. Still more capital is needed, however, for certain purposes ; and these purposes can perhaps be defined as separate but correlated parts of the whole undertaking. Thus many industrial companies not in a flourishing way have formed subsidiary companies to erect plants for making use of their bye-products.

Kent Coal Concessions, Ltd., not having capital itself both to buy mineral areas and to sink a number of bore-holes, had to form a subsidiary company to do the boring,[1] the shareholders taking a big risk, but getting a substantial and immediate return if they were successful. Separate companies were formed to sink three or four large collieries, no one of them having adequate capital to carry the enterprise through. Hence were formed distinct auxiliary companies to finance sinking operations, for the equipment of the collieries with surface works, sidings, etc., for supplying electrical power,

[1] The Sondage Syndicate.

and for finance and promotion of colliery companies.[1]
The same parent company also formed the East
Kent Light Railway Company for the purpose of
building railways to connect the various collieries
with the South-Eastern & Chatham Railway and
with the harbour at Sandwich.

Kent Coal Concessions, Ltd., soon proved coal
four or five miles north of Dover, and in 1907
the formation of colliery companies began, two being
formed to sink at Tilmanstone and Guildford respectively, whilst in 1908 another colliery was commenced at Snowdown, adjoining the South-Eastern
main line near Shepherdswell. At the same time it
was desired to prove the extension of the coalfield
to the east, north and west of this proved area ;
and as Kent Coal Concessions, Ltd., could take on no
further obligations, three similar allied companies
were formed during 1907 and the next two years
to take options on mineral leases, or to buy the
rights, and to put down bore-holes.[2] Boring was
undertaken further and further afield, and proceeded
more rapidly and smoothly when a great German
firm undertook the work with far better equipment
than any English boring contractor possessed.

The success which Kent Coal Concessions, Ltd., had
achieved by 1910 in proving coal over a large area
led to the formation of other companies by quite

[1] Foncage, Mines Construction Co. Ltd., The Intermediate
Equipments Co. Ltd. ; East Kent Light and Power Co. Ltd. ;
East Kent Contract and Finance Co. Ltd.

[2] South-Eastern Coalfield Extensions Ltd., Extended Extensions
Ltd., Deal & Walmer Coalfield Ltd.

independent and competing parties, who negotiated options on mineral leases in areas not covered by the Burr Companies, and proceeded to bore. They have, generally speaking, been successful in proving coal, the most important being the Ebbsfleet Syndicate and the Anglo-Westphalian Syndicate, in which a group of German capitalists combined with English and Welsh colliery owners.

Character of Kent Coals

Vertical sections of the Coal Measures, as revealed by the deepest boreholes, show that there is a large number of seams, many of them being of considerable thickness. For instance, at Ripple 28 seams were found, with an aggregate thickness of 54 feet of coal, of which 14 seams are 18 inches or over in thickness and contain 49 feet of workable coal. In several other boreholes, from 40 to 50 feet of workable coal was found ; whilst it is probable that much the same result would have been obtained from all the borings had it been practicable to carry the earlier ones to the great depths achieved in later boreholes. The thickest seam of all is found low down in the Middle Coal Measure at a depth of about 3,000 feet.

Speaking generally, the upper seams are a rather soft gas coal ; whilst the lower seams are hard and firm house and steam coals, said to be not unlike those of South Wales in many respects. The seam which has been first developed in the collieries is known as the Beresford seam, and lies at a depth of about 1,500 feet. It is 5 feet 2 inches thick, which is a very satisfactory thickness for working. Its value would be very great were it not for two draw-backs ; the soft and friable character of the coal, and the existence of a stone-parting in the seam, the stone having to be picked out by hand. To effectively clean the coal modern washing plants have been installed at Tilmanstone and Snowdown, so that good quality large and small can be produced. The coal is so soft that at present only from 30 to 35 per cent. of large coal is obtained ; but this is likely to be improved later on when the workings are further advanced. The large coal sells at a good price for gas-making, for boiler-heating in electric power stations and laundries, factories, etc., whilst it is used as a stand-by for railway purposes. As a house coal it is economical, if not attractive, for it burns slowly, giving a dull hot fire, very similar to the South Wales house coal. The small coal is a very satisfactory and economical fuel for steam raising if used with mechanical feed and strong draft, and it is also used for other industrial pur-poses and will eventually be largely used for coking and for making briquettes, both of which it makes of good quality.

The deeper seams were shown by the bore-hole samples to be, with one exception, hard coals of excellent quality. The next important seam below the Beresford, after passing three rather thin seams,[1]

[1] Thickness ; 1ft. 10in., 2ft. 9in., and 2ft. 6in., which, however, will all be worth working eventually.

is the 4-foot 7-inch seam, so denoted from its thickness, as proved by the Barfrestone boring. The
Snowdown Colliery Company has just sunk to this
seam, and has found it to be only 3 feet in thickness
there, but of excellent quality, giving 70 per cent.
of large coal, so that it will pay well to work. It is
better adapted for a house coal than the Beresford
seam ; and in view of the very large local market
for house coal in Dover and the Kentish watering-
places, not to speak of London, this seam is of great
importance.

Below this Snowdown seam, if we follow the
Barfrestone boring, we pass two thin seams, and
come to one of 6 feet 9 inches at a depth of 2,762
feet. This seam contains two stone partings ; but
the coal is reported by the analysts to be of the
finest class, being of "exceptional value for steam
and heating purposes." After passing a divided
seam at 2,877 feet, we come to another 4 feet 7 inch
seam at a depth of 2,944 feet, reported by the
analysts to be an exceptionally good steam and
house coal, with a small amount of residual ash.
Going 250 feet or so deeper we come upon three
thin seams, which are of no value at this depth, and
then reach at 3,260 feet a four-foot seam of medium
quality steam coal. Below this, at 3,318 feet, is a
magnificent seam of high quality navigation steam
coal, 9 feet 6 inches thick. It has a high evaporative
power, and is low in ash and sulphur, and so should
prove of great commercial value, and will almost
certainly be worked at no very distant date.

Reserves of Coal

The Royal Commission of 1904 did not attempt any estimate of the coal reserves in Kent, as they considered that too little was then known to form any idea of the extent of the coalfield. In *The World's Coal Resources* Mr. Aubrey Strahan made, in 1912, a tentative estimate, based upon an area on land of 150 square miles, and an undersea extension of 56 square miles. He assumed only an average thickness of 10 feet of workable coal seams ; and calculated the reserve on this basis to be 2,000,000,000 tons.

There is no doubt that Mr. Strahan's is a gross under-estimate, due probably to the fact that the results of the deeper borings which passed through many workable seams below 2,000 feet had not then been published. Taking as evidence only borings which have penetrated to a depth of more than 2,500 feet, it is now well established that over an area of at least 150 square miles there is a thickness of from 30 to 40 feet of coal in seams of 18 inches and upwards in thickness, not exceeding 4,000 feet in depth.

Mr. Strahan took the northern limit of the coalfield a little north of Canterbury and Sandwich ; but it is now known to extend some miles further north and north-west. It is safe, therefore, to take the area of the coalfield, including an extension undersea to four miles from the coast, at 250 square miles, and to assume that there is an average of 25

feet thickness of workable coal throughout. Reckoning upon the usual basis of 1,500 tons of coal per
foot of thickness per acre, the total reserve is found
to be 6,000,000,000 tons.

This may be taken, I believe, as a conservative
estimate ; and it may be worth while to consider
the *probable* as apart from the *proved reserve*. For
this purpose I extend the western boundary to ten
miles west of Canterbury, and allow an extension of
one mile in each of the other directions, including
undersea. The additional area is 150 square miles,
representing 3,600,000,000 tons, and making the
grand total of proved and probable reserves
9,600,000,000 tons.

Future of Kent Coalfield

The development of the coalfield is proceeding
apace. Three collieries, namely, those at Tilmanstone, Snowdown and Shakespeare Cliff, have reached
the producing stage ; and there are at least four other
collieries, at Guildford, Wingham, Stonehall and
Chislet, in process of sinking. In addition to the
foregoing, some five or six areas have been leased
to newly-formed colliery companies, or promotion
syndicates. I am including not only collieries of
the Burr group, but all of which any information
is available. Before long there are likely to be
twenty colliery areas taken up, covering well over
100 square miles of the coalfields. As it takes
from seven to ten years from the inception of a
colliery undertaking to bring it to a volume of

production approaching its intended normal, it will be some time before any large output is raised annually in Kent.

The following are the Home Office figures for output in Kent, together with an estimate of probable output in the present year, which I have made from the information available to date.

Output in Kent

	Tons.
1912	1,099
1913	59,203
1914[1] (estimate)	110,000

It is quite possible that by the year 1925 the annual output will have reached, or surpassed, 10,000,000 tons per annum.

The coalfield is served by the South-Eastern and Chatham Railway, there being the Ramsgate and Eltham Valley lines, both radiating from Canterbury in addition to the main line from Dover to Ramsgate. There was, however, a large central area of the coalfield destitute of railway connections, and it was therefore necessary for the Kent Coal Concessions to proceed to build several railways to enable the coalfield to be developed.

The East Kent Light Railway Company was formed for the purpose, and obtained powers from the Light Railway Commissioners to construct

[1] The output of the Tilmanstone Colliery during the first ten months of 1914 was 76,000 tons. For six weeks in April and May the mine was flooded and no coal was raised. The average output per week during October and November was 2,000 tons.

over 50 miles of railway, some 30 miles of which are built, or in course of construction. The most important line runs from Shepherdswell, on the South-Eastern, past Tilmanstone to Sandwich. It is interesting to note that the intention is to construct a railway beforehand to every point fixed upon for sinking a colliery, as this so much reduces the cost of hauling to the site the necessary machinery and materials.

The Kent coalfield has a highly important location with reference to trade routes and centres of population, so that there is an enormous demand assured from the existing centres of consumption. Coal will be shipped to London, possibly from docks constructed in the estuary of the Swale, near Faversham. Electric power may be generated in the coalfield to be transmitted in bulk to London. A great export trade will be possible from Dover, or from a new harbour constructed at Sandwich, because the south-east coast of Kent is nearer to many important continental ports than any of the Northern, Yorkshire or South Wales ports of shipment. The Straits of Dover, too, is one of the greatest highways of the world's shipping, and a great trade in supplying ships' bunkers is almost certain to arise. Further, there are in Kent rich and extensive ironstone beds, not only in the secondary rocks, but also in the coal measures ; so that a great iron and steel industry may quite possibly be established, besides general manufactures depending on the proximity of coal and of great markets for their produce.

In the old days when iron was smelted with charcoal and the county was covered with extensive forests, Kent was the principal district of iron manufacture. Now that coal is discovered and worked, there seems nothing more certain in civilisation than that Kent will again become a centre of industry, but on a far greater scale. Great towns and numerous mining villages will arise ; and it is greatly to be hoped that an intelligent supervision may be exercised over all urban developments. To the great credit of Mr. Burr I am glad to be able to say that he is an enthusiastic and practical housing reformer ; and Kent Coal Concessions has itself on his initiative built three garden villages for housing the miners. Land being comparatively cheap, there are large gardens and spacious roads, and the villages are in every way models which ought to be followed by all colliery companies or local authorities which build. There need be no Black Country in Kent.

CHAPTER VII

Prospecting for Coal

BEFORE undertaking the huge risk and expenditure likely to be entailed in the opening of new mines, it is necessary to receive some satisfactory assurance that coal is obtainable and that it can be worked with a reasonable prospect of financial success. Inspection of the strata at the various outcrops will soon indicate to the geologist whether any coal seams exist in the locality, but a superficial examination does not indicate whether such coal seams can be worked at a profit. The promoters of colliery undertakings are not likely, therefore, to lay out large amounts of capital in opening new mines merely upon the assurance of geological experts that in certain places undeveloped coal resources are waiting to be tapped. Before any such action is taken a number of important facts must be ascertained. Amongst these may be mentioned the number and probable thickness of the various seams, the quality of the coal contained in them, the character as to hardness, etc., of the overlying strata and the amount of water present. The presence or absence of faults or other irregularities

176

and the area of unworked coal must also be inquired into, whilst the rent and royalty payments required, the price of labour, the availability of adequate transport facilities, etc., are further important conditions to be ascertained.

Most of the data first mentioned are now obtained by putting down bore holes through the strata. If the surrounding coalfield has been proved by other collieries, boring is undertaken only as an additional precaution ; if, however, the district is unproved, boring is of exceeding importance, as it provides the engineers with reliable information, not otherwise obtainable, concerning the nature of the strata in the areas where sinking is proposed. Sometimes also in areas where the upper seams only have been worked and it is contemplated sinking to lower measures, the driving of bore holes from the bottoms of existing mines is undertaken with the same object.

Boring not only determines the depth, quality, and the approximate thickness of the various strata, but may also provide data from which the direction and the amount of inclination of the seams can be fairly accurately ascertained. As a rule if the fullest information is desired more than one bore is necessary.

Boring by Percussion

There are two methods of boring—the Percussive and the Rotary. The former method is now gradually being superseded. It consists in chipping

away particles of rock by the dropping of a sharp
chisel fastened to the ends of rods fitted into one
another by means of screw joints. At the spot
selected for the bore-hole an iron or wooden tube
is driven vertically into the ground, and is kept in
position by a framework erected on the surface.
This tube serves as a guide to secure that the hole
does not deviate from the perpendicular. Above
the hole is erected a derrick to which is attached a
pulley and windlass which are used for raising and
lowering the boring gear. Other alternative
arrangements are also provided. The boring
operation consists in lifting and dropping the rod
and chisel ; at each such movement a slight turn is
given to the gear so that a perfectly round hole is
obtained. As the hole becomes deeper extra lengths
of rod are added at the top. To remove the débris
it is necessary to withdraw the rod and replace the
chisel by a hollow instrument fitted with an inward
opening valve called a *sludger*, into which, when the
upward and downward operation is repeated, the
particles of stone and dirt are forced but cannot fall
out again. By examining the contents of the
sludger and also by observing the character of the
successive percussions, the borers are able to
ascertain fairly accurately the nature and thickness
of the strata through which they are passing.

The process of boring is a lengthy one. When a
considerable depth has been attained the greater
part of the time is absorbed in lifting and lowering
the rods for the purpose of changing the chisels and

removing the débris. The great weight of the rods also sometimes results in fractures. When these occur specially devised instruments are required to grip and extract the lower portions left jammed at the bottom of the hole. Where the strata are of a soft or loose nature a tubular lining of steel or wrought iron has to be inserted to prevent the sides of the hole from falling in.

As a rule when very deep holes are being bored by the percussive method, mechanical power is used for raising and lowering the rods, though formerly the windlass was turned by hand. The rods are sometimes replaced by ropes, as in the Mather and Platt system, the up and down movement being produced by attaching the ropes to a pulley which is raised and lowered by the piston of the vertical cylinder of a stationary engine. In deep holes, to minimise the possibility of rod breakages, sliding or " free-falling " cutters are sometimes used. The lower section of such a rod is of such form that it can easily be released from the main portion and allowed to fall independently. Improvements have also been introduced by which the débris can be more easily removed. One plan is to force water at a high pressure down hollow boring rods and back to the surface through the angular space between the rod and the sides of the hole.

Rotary Boring

Some of the more recent machines of the percussive type give very efficient results ; but the superior advantages of the rotary system must result in the eventual abandonment of all percussive arrangements. The special advantage of the rotary system is that it enables the strata to be extracted in the form of an unbroken *core*. The hole is made by the turning of a ring of some material harder than the rocks to be bored through. The rapid rotation of this material wears away the rock, making an annular cutting through it, while the central core rises up the hollow tube and is in due course brought to the surface and broken up into short lengths. There are two principal systems of rotary boring—the Diamond System and the Davis-Calyx System.

In the Diamond system the boring piece consists of a steel crown in one rim of which black diamonds are set. This crown is screwed on to a cylindrical core box from which extends to the surface a hollow wrought-iron tube attached above to a rod about 3 inches in diameter. This rod is connected with an engine which imparts to it the rotary motion necessary for obtaining a properly rounded hole. As the crown revolves and the hard core is cut, a current of water forced at high pressure down the rod forces the fine débris upwards between the sides of the hole and the cylindrical core box where it is collected in an annular sediment tube. Diamonds are very expensive—the price is from £6 to £10 per carat—and are therefore used only in very hard strata ; in soft strata properly tempered steel cutters give equally good results.

The Davis-Calyx system differs from the Diamond system only to a small extent. Instead of diamonds a cylindrical steel shell is used, the lower end of which is cut into a jagged, saw-like edge. It, also, is provided with a calyx or annular cup in which is collected the débris or sediment forced upwards between the rod and the walls of the hole by the introduction of a current of water at high pressure into the boring rod.

Position of Shafts

When all the required knowledge regarding the strata, etc., has been obtained and the prospect of success assured to the promoters of the proposed undertaking, preparations are made for sinking the shafts in order to work the coal. The choice of position is an important matter, and depends upon a variety of circumstances both underground and on the surface. The old plan was to sink the shaft so as to reach the seams at their lowest points. This was because traffic and water would run to the pit bottom by force of gravitation, and haulage and pumping operations would be facilitated. This is not nearly such an important consideration now, as haulage and pumping operations have undergone great improvements in recent years. Nevertheless, other things being equal, the old practice is still a good one to follow. In a flat seam the best position to sink is in the centre of a taking. The presence of faults, soft strata, and other conditions below ground, has to be borne in

mind, however, as the injudicious selection of the
site may lead to much unnecessary expenditure.
It is surface conditions which most often determine
the precise location of a new shaft. Easy access to
railways at favourable gradients, the existence of
ample space for rubbish tips, buildings, screens,
sidings, etc., facilities for obtaining or removing
water,—these are some of the points which the
engineer must consider with great care. The conditions of mining leases, etc., must also be borne
in mind, as it is possible that wayleaves charged at
one point may be much greater than if the sinking
is made a short distance away.

Kinds of Collieries

Mines are usually classified by their outlets
into (1) levels, (2) slants or inclined planes, and (3)
pits. Large collieries are not now usually worked
by levels. Slants are, however, still often used,
especially in South Wales, where the seams can
still be worked extensively from the outcrop. By
far the larger number of mines, however, are worked
by means of vertical shafts usually called pits.
In form, levels and slants are rectangular or arched ;
pits are rectangular, polygonal, oval or circular.
In England and Wales pits are usually circular
in form, but in Scotland rectangular shafts lined with
timber are still often sunk.

Pit shafts are increasing in dimensions. This
is because the royalties or takings are larger now
than formerly and the output from each mine

must be proportionately greater. The diameters of shafts vary from twelve to twenty-five feet, according to the probable duration of the mine, the anticipated output, and the size and number of the trams to be employed. From 16 to 20 feet is the usual size of pits with moderately large outputs.

Sinking Shafts

When the form and size of the shaft have been determined sinking operations are commenced. Gangs of men called " sinkers " set to work with picks and shovels to remove the soft earth, rocks, sand and other soft material that overlie the harder measures below. When a depth of ten or twelve feet has been cleared the bottom is accurately levelled and a temporary timber framework is erected to support the sides until a hard layer of rock or *stonehead* is reached. A circular *crib* or *curb* made of hard wood is first laid ; on this *punch props* or vertical timbers from 3 feet to 5 feet long are erected to support a second crib or curb above which a further series of punch props are raised, and so on. Behind the curbs are fixed *packing deals* which are secured tightly by inserting wedges behind. The entire framework is bound together by nailing strips of plank known as *stringing deals* to the cribs. It is now usual to substitute for curbs iron binding rings made in segments, and for the stringing deals and punch props strips of flat iron, bent and shaped like the letter " S " and called *hangers*. When the depth has become too great to enable the débris to be shovelled out a bucket or *kibble* attached to a rope passing over a pulley suspended from an overhead derrick is used.

Walling Pit Shafts

Having reached the *stonehead*, the sinkers prepare very carefully an absolutely level ledge of solid rock to serve as a bed for a walling curb made of oak timbers or cast iron, the segments of which are bolted together. Care is taken to ensure that the curb is perfectly circular, and when the correct position has been determined it is wedged firmly in its place and is sometimes grouted in cement. Upon this bed the permanent brick lining of the pit is built. The usual plan is to make the wall of uniform thickness, and to ram the interstices behind with ashes. As the wall rises the temporary timber lining is, section by section, removed.

During the building operations the bricklayers stand on a scaffold or platform suspended with chains either from hooks driven into the pit sides or more usually from a central wire rope connected with an engine at the surface.

In order to give additional security the scaffold is sometimes fitted with a number of flat bolts which, when pushed out radially, rest on one of the brick courses. Another method is to connect the platform by means of hooks and links to chains which are hung down the sides of the shaft and are secured at the surface. In most large sinking pits patent scaffolds are now commonly used. One of the best of these is known as the Galloway scaffold. It consists of two platforms about 8 or 10 feet apart, the upper one forming a protection for the masons while at work. The platforms are provided with openings through which the kibbles can easily pass. The great advantages of the Galloway and similar scaffolds is that the sinkers need not suspend their operations while the bricklayers are at work above them.

Later Stages of Sinking Process

Sinking operations now proceed below the first walling curb. For a couple of feet the sinking is continued narrow so that the diameter of the excavated portion is the same as that of the walled upper portion. This is done so as not to weaken or undermine the brickwork. The work of excavation now becomes more difficult. Where the rock cannot be dislodged by ordinary means blasting is resorted to. Holes are driven into the floor by means of long chisels, or by drills worked by compressed air or electricity. Some very elaborate machinery has been devised for this purpose. Into these holes charges of powerful explosives such as gunpowder, dynamite, or blasting gelatine are placed. These are sometimes fired by means of ordinary powder fuses. The uncertainty of firing speed, the danger from *hanging* or *miss-fires*, and the dense smoke produced, however, are great disadvantages ; and consequently firing is now usually done from the surface by passing a current of electricity through a cable to detonators fixed to the individual charges.

Temporary supports for the sides are erected as before until a second suitable stratum for the laying of a walling curb is reached. The second section is then bricked up until the bottom of the first foundation ledge is reached. This ledge is then either removed piecemeal and bricks inserted, or if the rock is hard enough it becomes part of the shaft lining.

Water and Air in Sinking Pits

In sinking pits through strata containing large quantities of water provision has to be made for allowing the water to pass through the wall and for its collection. This is usually done by leaving spaces in the walling through which the water may flow and pass down inside to be collected at intervals by *water rings* or *garlands*. These are usually ordinary cast-iron walling curbs provided with a flange at the outside edge so as to form a groove or gutter around the shaft. To these are connected pipes which convey the water to the *sump* or pool at the bottom of the pit from which it may be pumped to the surface.

The water which collects during sinking operations is removed by means of water barrels or pumps. One of the best form of water barrel is the Galloway Pneumatic Water Barrel. In most modern mines, however, elaborate electric or steam sinking pumps are used, these being hung in the shaft.

As the shaft increases in depth arrangements have to be made for its proper ventilation ; this being extremely important as soon as the coal measures are reached, especially if the strata are charged with gas. It is done by fastening a line of sheet-iron pipes about 18 inches in diameter to the side of the shaft, down which the air is forced, the shaft itself acting as an upcast. Occasionally the air is drawn up through the pipe by means of fans, the shaft in this case acting as a downcast.

Various means are used for lighting the sinking pits. If there is no danger from gas naked lights are used ; otherwise the sinkers carry safety lamps. Sometimes also electric lights are hung in the shafts.

Coffering and Tubbing

The method of brick lining above described is customarily employed under ordinary circumstances. Where, however, the strata are so heavily laden with water as to impair the safety of the shafts special methods of lining must be used. A method known as *coffering* is commonly used where the flow of water is not excessive. This consists of several concentric rings of brickwork, the intervening spaces being filled with cement. The thickness of this masonry lining is sometimes as much as 3 feet. All the bricks are laid in good cement. By this means perfectly watertight joints are obtained. The system is only used in the waterbearing strata. When impervious layers of rock are again reached the ordinary brick lining is used.

Where, however, the rush of water is very great coffering is not sufficient to restrict the flow and

pressure, and metal *tubbing* has to be inserted.
Sinking is carried on for a few yards beyond the
water-bearing strata, the sides being temporarily
supported as before. A smooth and level curb bed
of solid stone is prepared in the non-water-bearing
rock and a special kind of *tubbing curb* is laid, the
open space behind being packed tightly with dry
timber secured with wedges. Tested tubbing plates
are of cast iron about $3\frac{1}{2}$ feet long and $2\frac{1}{2}$ feet wide,
and of an appropriate thickness, provided with
flanges so arranged as to fit the plates closely together. These are put in position above the crib,
and are fastened together tightly by driving wedges
between each segment and the sides of the shaft.
Any intervening spaces are filled with concrete or
well-rammed earth, clay or ashes, and any accumulation of water behind the tubbing is allowed to run off
through holes in the tubbing segments.

Special Methods of Sinking

The difficulties sometimes arising from rushes of
water, beds of quicksand, or loose ground, are such
as to render impracticable the methods of sinking
briefly described above, and special methods of
sinking have to be adopted.

Where the sinking is commenced in loose ground
which is not of great depth, the difficulty is sometimes overcome by driving piles. By this method
the thickness of the stratum must first be ascertained, as each round of piles diminishes the diameter
of the shaft. The number of sets of piles required

is carefully calculated so that the diameter of the innermost ring at the bottom of the loose strata is equal to the required diameter of the shaft. The piles are driven down vertically, and as the sand, gravel, etc., within this ring is removed oak curbs are inserted. The piles overlap one another, and the sides are thus stepped downwards to the stonehead.

The use of piles for sinking is costly ; and often the results are unsatisfactory. Cylinders or drums of various materials are, therefore, sometimes used. One form of cylinder is constructed of timber and bricks. The site of the shaft is marked out, a curb of wood about 18 inches broad and 6 inches thick is laid perfectly level upon the site, and a section of brickwork erected above. When a height of 3 feet has been attained another curb is laid, and the building process repeated. The whole structure is bound together by means of iron tie-rods fastened from curb to curb. As the workmen inside the ring dig out the sand the weight of the brickwork causes it to sink gradually. Other sections of brickwork are now added until the quicksand stratum has been passed through.

The great difficulty of maintaining the vertical position of the brick drum is a serious objection to its use, and iron cylinders are now much more frequently used. These cylinders are made up of thick segments strengthened with ribs inside, of the exact curvature of the shaft, the sides being perfectly smooth to allow of their easy passage through the sand. The lowest segments are made watertight by the insertion of sheet lead. The rings of segment are put together on the top, and as the cylinder descends other segments are added. The drum is kept vertical by means of guide frames at the surface or by suspending it from a derrick. It is forced down by hydraulic jacks, or by imposing heavy weights on timbers laid across. Sometimes the sand is dug out by means of mechanical excavators. The liability of cast iron to fracture is a great drawback, and often the cylinder is made of wrought iron or steel plates riveted together and lined on the outside with brickwork.

An ingenious method of sinking through quicksand is that known as the *Haase Process*. This consists of driving a ring of vertical wrought-iron tubes into the sand, and closing the spaces between the tubes with webs of angle-iron. A hollow boring rod is placed in each tube and a current of water introduced under pressure forces the sand to the surface, so that the ring of tubes continually sinks.

The Poetsch Method of Sinking

A very successful though expensive method is the *Poetsch Freezing Method*. In this plan a number of bore holes are driven around the circumference of the proposed shaft and two tubes inserted, one being inside the other. The bottom end of the inner tube is open. Brine at a low temperature is forced down the central rod and back to the surface through the annular space. The great cold freezes the damp earth around the bore holes and a solid wall of frozen ground encircles the shaft. The freezing process is kept up until the excavation has been effected and the permanent brick lining has been constructed. The Gobert Method is a modification of that of Poetsch, the only difference being that liquefied ammonia or carbonic acid gas is used instead of chilled brine. Another method, similar to that used in excavating the London "Tube" and tunnels, consists in sinking a cast iron cylinder through the sand in the ordinary way, the cylinder being closed at the top and compressed air forced in to keep back the water. The disadvantage is that the men cannot work with comfort in the shafts owing to the air pressure, and that short shifts and high wages are necessary.

The Kind-Chaudron Method

One of the best-known methods for sinking shafts through very wet strata is that known as the Kind-Chaudron Method. It consists in boring out the shaft with *trepans*, sludging away the débris and lowering into the shaft a cast-iron lining. The boring tool or *trepan* is raised and lowered by a beam engine and rotated by men at the pit-head. After each stage the trepan is removed and the sludger introduced. The tubbing is cast in rings without vertical joints in a foundry adjoining the shaft; and the rings are bolted to one another at the surface and lowered into the pit when the impervious layer has been reached. The joints of the tubbing are made water-tight, and between the tubbing and shaft a lining of concrete or hydraulic cement is imposed.

The Cement Method

An exceedingly interesting method of overcoming difficulty in watery strata is that which was devised by M. E. Remaux, of the Lens Mining Company and successfully carried out in connection with one of the collieries of that body. The method aims at solidifying strata by the injection of cement. A ring of bore holes was put down some yards away from the shaft, to which pumps were attached. Cement was poured through pipes into the sinking pit. The pumps were then set in motion, and as the water passed through the strata in the direction of the boreholes, the cement was conveyed with it into the apertures of the ground, where it set, and so made the strata impermeable. The cost of applying this method at the colliery mentioned was over seventy pounds per yard of depth.

Cost of Sinking

The sinking of shafts is generally entrusted to contractors, who are paid a fixed sum per yard. Where, however, the conditions are abnormal, as, for example, when sinking through sand or water-bearing strata, the day work system is used. The cost of sinking new mines may be gathered from the following estimate given by Professor Galloway.[1] The shaft in this instance was 1,800 feet deep, 20 feet in diameter, the volume of water being between 3,000 and 4,000 gallons per hour. The progress

[1] *Lectures on Mining.* Shaft Sinking, p. 37.

made with the sinking was 25 feet per week. The shaft was sunk in 1892–3, when wages were 30 per cent. above the standard.

Wages and Salaries	.	.	.	.	£14,400
Materials	.	.	.	.	16,548
Total	.	.	.	.	£30,948

Average cost per foot, £17 3s. 11d.

The cost of sinking and lining a pair of shafts during 1898–99 at Hylton have been published, and are worth quoting as the conditions were normal :—[1]

Cost of Shafts at Hylton Colliery

West Pit, 20 feet diameter, 1,440 feet deep—

	£	s.	d.
Sinking (labour and explosives) .	5,048	14	10
Making beds and laying cribs . .	273	0	0
Extra water consideration . .	330	0	0
Out shifts	376	7	10
Shaft walling (labour) . . .	1,767	4	6
Hanging-on excavations and arching	1,030	17	0
Total cost of Labour . .	£8,826	4	2

Materials

	£	s.	d.
Plate Bricks in shaft, 1,235,000 at 30s. per thousand	1,852	10	0
Plate bricks, arching, 300,000 at 30s. per thousand	450	0	0
Cement, Lime and Sand 3 to 1 .	1,151	5	0
Metal ring and wall cribs . .	552	0	0
Temporary cribs and backing deals .	175	0	0
Materials	£4,180	15	0

Total cost : £13,006 19s. 2d.
Equal to : £9 0s. 8d. per foot.

[1] *Practical Coal Mining*, Vol. 1, p. 151.

South Pit.—Depth 1,578 feet ; diameter, 15 feet.

Labour	£	s.	d.
Sinking (labour and explosives) .	4,480	8	0
Making beds and laying cribs . .	228	7	0
Out shifts	283	9	7
Shaft walling (labour) . . .	1,458	18	11
Temporary hanging-on . . .	215	7	6
Labour	£6,666	11	0

Materials	£	s.	d.
Plate bricks in shaft, 840,000 at 30s. per thousand	1,260	0	0
Plate bricks for arching, 55,000 at 30s. per thousand . . .	82	10	0
Cement, Lime and Sand 3 to 1 .	671	5	0
Metal ring and walling cribs . .	520	0	0
Temporary cribs and packing deals .	130	0	0
Materials	£2,663	15	0

Total cost : £9,330 6s.
Equal to : £5 18s. 3d. per foot.

Some engineers arrive at the approximate cost of sinking by taking about 7s. per foot depth for sinking and walling for each foot of diameter, and from £5 to £10 for each walling curb, making an additional allowance for the removal of water, etc.

Pit Bottom

As the pit approaches completion, its diameter is enlarged at the bottom. This is necessary as

considerable room is required for the efficient handling of the traffic. A common method is to widen
the shaft in a bell-shaped manner ; this method is
technically known as *bell-mouthing*. The widen-
ing out has, of course, to be very carefully and very
gradually done, the usual rule being to slope the
walls outwards at a gradient of 1 in 18. Sometimes,
however, the necessary space is obtained by continuing the shaft to the bottom, and then constructing
around it a wide roadway. The arched spaces and
roadways at the pit bottoms are made very wide—
sometimes 30 or 40 feet ; and sidings are provided
for full and empty trams. Upon the character of
these tramway arrangements depends very largely
the economical working of the colliery. It is essential
that the traffic operations should be carried on
without unnecessary delay, and that the unloading
of empty trams and the loading of full trams should
be continuous. In some mines wagon roads on
opposite sides of the shaft are on the same level,
and as the empties are passed out of the cage at one
end the full trams are pushed in at the other ;
this is a preferable method to that in which both
loading and unloading take place on the same side.
When double decked cages are employed it is
common to have two landings corresponding in
level with the upper and lower decks. The trams
from one district are loaded into the upper deck, and
those from another district into the lower deck.
In opening out the roads leading from the pit bottom
into the mine interior great care must be taken that

the gradients are in favour of the full trams ; the usual dip is from 1 to 60.

Shaft Pillars

The working of coal at moderate depths often results in subsidences, and it is necessary for this reason to leave unworked a certain area of coal in the vicinity of the shaft for the purpose of supporting the various erections on the surface, and more important still, for protecting the shaft lining itself. The amount of subsidence varies, on the average, from 50 to 70 per cent. of the thickness of the seam or seams which are being worked, and if the strata are disturbed in the vicinity of the shafts there is risk of the brick or iron lining being seriously weakened and injured. The dimensions of the shaft pillar are determined by a variety of conditions, such as the depth and thickness of the seams, their inclination, the nature of the coal, the strength and character of the overlying and underlying strata, and the presence or absence of faults, irregularities, or disturbances. Considerable diversity of opinion exists as to the proper extent of shaft pillars, and several rules have been suggested for the guidance of engineers. The conditions in individual mines, however, vary so much that no hard and fast principle can be laid down. In flat seams containing coal of fairly hard quality underlying a strong roof it is usual to leave rectangular pillars, the sides of which are equal in length to not less than two-thirds of the full depth of the shaft ; in many cases, however, the sides of these pillars are greater than the depth of the shaft. If the coal seam is highly inclined, the shaft pillar is so arranged as to leave a greater width and area of unworked coal on the higher or *rise* side of the shaft than on the lower or dip side.

In addition to the pillars left for support of the shafts, it is necessary in some districts to leave unworked areas of coal for the support of valuable property at the surface. If this were not done large numbers of buildings would sustain considerable injury on account of the subsidence of the underlying strata. In the older coalfields, for example Northumberland and Durham, considerable areas of coal beneath the towns are thus left unworked. The inadequacy of such pillars in some localities has been productive of considerable loss, not only to the surface property-owners, but also to the community at large. Often cottages and other structures have been damaged so much that they have had to be demolished. Some idea of the amount of subsidence may be gathered from the fact disclosed in the Rhondda Valley in the course of the year 1910 by a re-levelling of 176 bench-marks by the Ordnance Survey Department. As compared with the records of 1898, only twelve years previously, the whole district had sunk in many localities over 5 feet, the maximum difference being 7·95 feet. This is in a mining area of comparatively recent date ; it is probable that in the older coalfields the results are much more striking.

If the subsidence took place regularly, probably little harm would be done. Usually, however, there are considerable variations between the amounts of subsidence at different points, and the irregularity is responsible for considerable damage to canals, railways, sewers, water and gas mains, etc. In the Rhondda Valley considerable nuisance has been caused by the leakages in the sewage system arising from broken pipes, and the local authority also sustains a great loss annually through the shattering of gas mains. In addition roads are constantly needing repair and re-levelling.

Main Roads

The directions of the main haulage roads of the mines depend upon the dip of the seams, the shape and area of the royalty to be worked, the position of known faults, the system of working, the method of haulage, and a variety of similar conditions. The number of the main roads depends upon the anticipated output of the mine. These roads are so arranged as to serve the largest possible area, with the least amount of waste of energy. As a rule main roads should be quite straight; and they should dip towards the pit bottom so as to drain off the water and facilitate the haulage of the full loads. It is also desirable that main roads should be solidly and permanently constructed, as carelessly alid roads often impede the passage of the traffic and cost too much in upkeep. The usual gradient recommended is a little more than $\frac{1}{4}$ of an inch to

the yard, or 1 in 130. It is customary to leave barriers
of unworked coal on both sides of the roads for
protection.   These are usually from 40 to 60 yards
wide.

These barriers, while they bear the bulk of the
weight of the overlying strata, are not alone sufficient, and the roofs and sides of the roadways have
to be supported. Sometimes brick arches are
built ; stone or cement walls carrying steel cross
bars are also often used.   The most usual method of
supporting main haulage roads, as well as ordinary
gateways, as subsidiary roads are called, is by
erecting a framework of timber.   Larch or fir are
the varieties of timber most frequently used because
they grow straight and are capable of bearing immense weight without breaking.   Pine and oak are
also utilised for certain purposes.   In main roads
two upright *props* or *posts* are erected at the sides.
On top of these, and stretching across the roadways,
is laid a *bar* or *lid* also of timber.   Sometimes if the
sides are firm the props are dispensed with, and the
bar is fitted into niches cut into the rock or coal.
The posts are erected at intervals, the frequency of
which varies according to the character of the roof
and sides.   If the rock or earth is loose and broken,
lagging, or strips of timber, are laid longitudinally
over the cross bars, and just under the roof, and
also behind the posts at the side.   The timbering of
roadways is a matter of very great importance as
falls of the roof and sides are responsible for a very
large proportion of mine accidents.   The work is

usually entrusted to timbermen, paid, not as in the
case of the collier on the basis of the output, but on
day rates, so that they have no incentive to hurry
over their tasks.

Ventilation

In opening up a mine, the engineer has to bear in
mind the necessity of providing for an adequate
supply of air for the workmen. It is for this reason
that each mine has to be provided with two shafts,
one called the *downcast*, through which fresh air
is admitted to the mine, and the other the *upcast*,
through which the foul air returns to the surface.
Similarly it is necessary to drive the underground
roads in pairs. The road along which the fresh
air passes from the downcast shaft is called the
intake, and the roadway along which the impure
air is conveyed to the upcast shaft, is called the
return. Between the intake and the return, cross
cuts or *splits* are made at various points which are
opened only for special reasons ; and *stoppings*
are erected in the intake roadway so that some of the
air is diverted along a side passage to some of the
working places. At points where roads have to
be closed to the air but open for traffic, doors are
erected in pairs. The interval between the two doors
is just long enough to admit the train of wagons,
and one door is closed before the other is opened.
The opening and closing of these doors is entrusted
to boys. Regulating doors also are used ; these are
provided with an opening, the area of which can be

varied by means of a sliding panel. Sometimes it is necessary to convey one current of air over another. To prevent the mixing of the two currents, part of the roof above the roadway is removed to a sufficient height ; brick walls are built in the road-way on top of which are laid planks made air-tight by nailing strips of wood over them. A less common but more efficient method of constructing an air crossing is by arching the roadway and carrying the cross current through a narrow drift above it. In mines worked on the pillar and stall method, described in the next chapter, and also in dead-ends, such as the ends of headings where no second road-way is available, it is customary to run lengths of air pipe of wood or iron, and these act as substitutes for intakes. Sometimes also the road is divided longitudinally into an intake and return by means of a partition made by stretching brattice cloth over a wooden frame-work. The divisions are of unequal sectional area,—the smaller one being used as the intake, and the larger one as the return.

The air is set in motion by means of a fan erected, as a rule, on the surface of the mine. One kind of fan is used to force air downwards through the down-cast shaft ; more usually, however, the fan is placed at the top of the upcast shaft and is used to exhaust the mine of foul air, thus causing a current of fresh air to proceed downwards through the down-cast shaft.

CHAPTER VIII

METHODS OF MINING COAL

A NUMBER of important considerations have to be borne in mind when determining the method to be employed in winning coal. The two chief methods are known as *Longwall* and *Pillar & Stall*, but various modifications of these are in use, some of them being of such a mixed and complicated character as to render their classification under either of the heads quite impossible.

The longwall method aims at removing the whole of a seam in one continuous operation, and allowing the overlying strata to settle down gradually on the rock and débris packed by the miners in the space from which the coal has been removed. This space is called the *goaf* or *gob*. In the Pillar and Stall system the coal seam is cut up into a number of pillars of square or rectangular form, varying in size according to the depth and thickness of the seam, the nature of the coal, etc., which pillars are removed in a second operation.

Choice of Method

The chief factors determining the choice of a method are the locality of a mine, the market for which it caters, the nature and thickness of the seam, and the character of the over—and under—lying strata. Local custom counts for much in working a colliery, and in districts where a certain method has been in operation for a considerable period the introduction of a different method may entail great expense and difficulty on account of the inability of the miners to adapt themselves to the change. The purpose for which the coal is required is also an important point. If the coal is wanted large, the longwall method is usually best ; if the whole of the coal is to be crushed and the slack can be utilised to economic advantage the pillar and stall method serves equally well, and in some circumstances may prove cheaper. Thick seams, again, and seams containing much refuse can be worked satisfactorily by pillar and stall, whereas the longwall system would be best where seams are very hard and roofs and floors are soft. These and a multiplicity of other factors have to be well considered before opening-out operations are commenced, and no hard and fast rules can be formulated regarding the method to be adopted in any mine. The longwall method is fast displacing the pillar and stall method in all the coalfields ; but there are a few collieries in which it cannot advantageously be adopted.

The Longwall Method

The longwall system provides for the removal of the coal at one working. In some of the coalfields the working face advances in nearly a straight

line, often three-fourths of a mile long ; in others
it is gradually stepped so as to form stalls of
from 30 to 60 feet in length. The direction of the
working face is determined by the *cleat*, or jointing
of the coal, the inclination of the seam, etc. As a
rule, if it is desired to obtain a high proportion of
round coal, the advance should proceed at right
angles or across the cleavage ; if the character of
the output is immaterial it should be parallel to
the line of cleat. Main roads or headways are
driven away from the shaft bottom into the coal ;
and as the faces advance, other roads or *gates* are
made at right angles connecting the headings
with the roadways made along the face for the
conveyance of the trams. The latter roads are
kept open by means of *pack* walls and timber.
When the face has proceeded for a considerable
distance from the headings, in order to lessen the
expense of road maintenance, a new diagonal
cross gate is made joining the various gates to the
main gate, the old roads through the *goaf* or *gob*
being then allowed to fall. In order to keep the
various gates open the roofs are ripped, and the
rock and débris thus obtained is constructed into
a wall or *gate-end pack* on either side. Often, also,
in order that the subsidence may be regular and
steady, an intermediate pack wall is built between
the gate-end packs.

Two Systems of Longwall

There are two systems of longwall in operation— *longwall advancing* and *longwall retreating*. In the former all the coal is worked away from the shaft bottom towards the boundary ; in the latter from the boundary towards the shaft. Longwall retreating is usually regarding as the more advantageous method, where the conditions for its application are satisfactory. The main roads are driven to the boundaries of the royalty or to the boundaries of the district, and the coal is worked backwards. The advantages of the method are that when the coal has been removed rails and timber can be withdrawn and the trouble and expense of maintaining roadways through goaves is rendered unnecessary. In gaseous mines, and in seams liable to spontaneous combustion the method is specially useful, as the goaves in which quantities of gas are apt to accumulate do not intervene between the shafts and the working places. The disadvantage of the method is that a huge amount of capital must be expended and considerable time lost before any considerable output can be obtained and the shareholders receive any return on their investments.

The longwall face where the men are employed requires to be very carefully protected, as it is only a few yards at most away from the goaf where the roof has been allowed to come down. The protection provided need only be of a temporary nature. For this purpose timber is generally used. The usual way is to set two parallel rows of props ; and as the face advances and another row is set, the row farthest from the face is withdrawn. The latter process is necessary in order that the subsidence in the goaf may be even and unbroken, and no extra weight may be thrown on the face or on the timber supporting it. When the roof is very bad, or pressure is expected to be heavy, special kinds of supports are used.

The Pillar and Stall Method

The pillar and stall method was in vogue in the earliest days of coal-mining and has prevailed in the coalfields of Northumberland and Durham to the present time. This method consists of two operations.

(a) The coal is cut up into rectangular blocks or *pillars* by means of roads driven from the shaft towards the boundary and crossed at right angles by other roads at distances of 60 to 200 yards apart, and known as *walls* or *endings*, Between 15 and 25 per cent, of the coal is worked in this operation.

(b) From the endings other roads called *bords* are driven out and stalls are opened from these at intervals of 10 to 30 yards apart, and the pillars gradually removed. The process of *robbing the pillar* is the most dangerous part of the miner's work, and taxes his skill to the utmost.

" Robbing the Pillar "

In olden days the pillars were left unworked to prevent subsidences. The desire to win as much of the coal as possible evolved a system of wide roads and narrow pillars. This system proved uneconomical and unsafe in working ; and often owing to the crushing of the pillars or the rising of the roadways whole mines were destroyed. These phenomena are known as *thrust* and *creep* respectively. Taught by experience of the dangers of small pillars the mine-owners increased more and more the pillar dimensions, and adopted the system of removing them altogether in a second operation. At first they divided the entire taking into pillars, and began their removal, often years afterwards. The disadvantages of this delay were numerous. The coal being subject to pressure, and in the faces to exposure, suffered a considerable deterioration ; ventilation often became very difficult, road maintenance was heavy, the effects of local explosions became more widely distributed, and the danger from *creep* increased. Most collieries employing the pillar and stall method are now divided into districts of moderate size divided from one another by wide ribs or barriers of coal ; and in these districts the " robbing " process follows very closely after the conversion of the seams into pillars.

Several plans are in operation for the actual getting of the pillar, or *working in the broken*, as it is sometimes called, many of them being closely

allied to longwall. In some mines the pillars are
removed by dividing them into two, four, or more
smaller pillars called *juds* by intersecting roads, and
gradually working, the débris and timber being used
to secure the safety of the miners. In others,
slices called *skirtings*, are taken off the ends of the
pillars, and *jerkins* or *lifts* off their lengths. The
last slice of lift is removed in small sections by
driving short lifts at right angles. When the pillars
have been worked and the roadways hitherto maintained for access to the coal are no longer needed,
the timber and rails are removed and the roof is
allowed to fall.

Advantages of Longwall System

The growing popularity of the longwall system is
a recognition of its superiority over pillar and stall
working ; and it seems likely that in the future,
except where exceptional circumstances prevail,
all new collieries will be worked according to this
method. The mine-owners seek to obtain as large
a quantity of well-conditioned coal as possible,
consistent with safe and economical working. To
enable this object to be attained the longwall
method is recommended on the following grounds :—

(1) Practically all the coal is removed, and none
    is lost or abandoned as frequently happens in
    the pillar and stall method.

(2) For a given length of coal face a larger
    number of men can be employed than in
    any other system.

(3) The workings are less liable to sudden and unequal subsidences, the ventilation arrangements
are simpler, and there is consequently less
liability to accidents.

(4) There being fewer roads to be maintained the
cost of working is much lower.

(5) A larger proportion of *round* coal in comparison to *slack* is obtained, which renders the
output more valuable.

Coal-Getting Operation

The coal-getting operation needs great skill,
and every stall has to be in charge of an experienced

FIG. 10.

miner. The coal is first holed or undercut with a
special tool known as a pick or mandril. That is,
a groove one to two feet in depth is cut either in
the lowest part of the coal or in the underlying
fire-clay. The mass of coal is supported during the
operation by sprags or props. When the holing
process is completed several sprags are withdrawn

o

and if the immense downward pressure of the overlying strata proves insufficient to break down this
coal, wedges are driven in at the top of the seam
or explosives are used in non-gassy mines.   The
broken   coal   is   then   filled   into   *trams*,   *tubs*
or *corves* which run on rails close up to the working face, and is conveyed to the pit shaft.   Where
explosives have to be used shot-holes are bored either
by percussion or by a rotary hand-boring machine,
and into each one is placed an explosive charge
which is fired sometimes by powder fuses, but more
usually by means of electricity.

The use of explosives in mines is very strictly
regulated, and shot firing is now only used in places
where there is not much inflammable coal-dust,
gas or fire-damp.   Home Office rules prescribe the
conditions under which alone explosives may be
used, and the only explosives allowed are those
included on a list of " permitted explosives."   The
use of electricity for firing shots has such great
advantages that in most cases it has superseded the
old powder-fuse method.   It does away with the
use of naked lights, dangers and delays from miss-
fires are avoided, several shots can be fired simultaneously, and the operation can be performed
safely from a distance.

An enormous increase has been recorded during
recent years in the number of machines employed
for cutting coal, which is due partly to a scarcity
of skilled labour, and partly to a desire for the more
economical working of the coal.   The number of

machines of different kinds employed in Great Britain during 1912 and the number of tons got by each kind, are given in the following table published by the Home Office :—

COAL-CUTTING MACHINES IN MINES.

Machines Worked by	1912.	1913.
Electricity . .	1,134	1,305
Compressed Air .	1,310	1,587
	2,444	2,892
Output of Coal (Tons)	20,025,964	24,367,726

KINDS OF MACHINES, 1912.

	Driven by		Total.
	Electricity.	Compressed Air.	
Disc . .	623	480	1,103
Bar . . .	345	123	468
Chain . .	148	34	182
Percussive .	14	666	680
Rotary Heading .	4	7	11
	1,134	1,310	2,444

Most of the machines are in use in Yorkshire and the Midlands, where the conditions are such as to make them specially advantageous.

Advantages of Coal-Cutting Machines

The advantages of coal-cutting machines are well stated in the general report of the Royal Commission on Coal Supplies, 1905. The following are extracts from the report mentioned :—

" Briefly stated, the chief advantages of coal-cutting machines are, according to the evidence :— (1) That an increased percentage of large coal is obtained, and the coal got is in a firmer and better condition. (2) A more regular line of face is obtained, which facilitates ventilation and leads to a more regular and systematic timbering, and the weight being more regular and uniform the roof can be more easily kept up. The greater rapidity of working also tends to keep down the cost of repairs, and causes less damage to overlying seams and the surface, the subsidence being more even. (3) The regular and systematic working tends to increase the safety of the workmen. (4) Seams, which either because of their thinness or hardness, or both, could not be worked at all, or could only be worked at a profit in good times, can be worked profitably by machines. (5) Holing is less frequently done in the coal and, when it is, there is much less small made than in the case of holing by hand. (6) The output is increased, and is more regular, and the work is more easily superintended. Fewer explosives are used for getting down the coal ; in some cases none. Generally machine work is less costly than hand work, especially in thin

ASHINGTON COLLIERY, NORTHUMBERLAND.

ELECTRIC COAL CUTTER.

ASHINGTON COLLIERY.

seams. According to one witness the saving is much greater in the narrow work or headings than in the longwall faces. From the point of view of the men the work is safer and easier, and the wages are better. The importance of lightening the labour of the men will probably be more appreciated as the working places become deeper and the temperature becomes higher."

"Conditions Adverse to the Use of Coal-Cutting Machines

" There are, however, certain conditions under which machines cannot at present be worked to advantage ; viz.:—(1) where the roof or floor is bad, (2) where there are numerous faults or dykes, or (3) where the seams are highly inclined. So, too, in the case of very soft coal there is the danger of falls from the face and damage to the machines."

Methods of Haulage

When the coal has been extracted it is removed in trams, tubs, or corves to the pit bottom. The process of haulage is often a complicated one, and entails the use of mechanical arrangements of considerable complexity. The ancient method of bringing out the coal from the stalls to the pit mouth was by means of baskets carried on the backs of women and children. The pits were not provided in those days with winding apparatus and the women and children had to climb to the surface by means of ladders. Later, small sledges were introduced and eventually tubs or trams pushed along wooden rails came into use.

The history of the development of underground haulage is one of remarkable interest, and there is a great contrast between the old-fashioned methods referred to, and the elaborate systems now in vogue. In some large collieries to-day, all the haulage work is done through the medium of stationary engines fixed underground or on the surface, or of locomotives propelled by compresed air or by electricity. A well-known method of underground conveyance was that formerly used near Manchester. In this case the Duke of Bridgewater's canal, constructed in the latter part of the eighteenth century, ran through a tunnel nearly four miles long. The coal in the vicinity of this tunnel was mined and removed to numerous barges for conveyance out of the mine. These boats were 56 feet long, 5 feet wide, and 3 feet deep and carried loads of 10 tons each. They were propelled by the workmen lying down on the coal and pressing their feet against projecting portions of the roof.

In modern times coal is conveyed from the workings to the shafts in two operations. The trams are conveyed from the various stalls to collecting sidings or *passbyes*, which should be in the immediate vicinity of the working places. This is done sometimes by young men called *putters* or trammers. These workmen grasp the trams firmly with both hands, and also use their heads to assist in pushing them along. Tramming is a costly method of haulage, and for this reason great care is taken to limit as far as possible the distance along which haulage by manual labour is necessary. In many mines ponies are employed to collect the wagons from the various workings ; and in the most up-to-date mines small engines driven by electricity or compressed air are used. The conveyance of the coal from stalls to sidings is known as *secondary haulage*. In modern mines the trains of wagons are conveyed from the sidings to the pit bottom mechanically. There are three systems of haulage in common use ; these are known as (1) Direct Haulage by Single Rope, (2) Endless Rope or Chain, (3) Main and Tail Rope Haulage.

The single rope or direct method of haulage can be used only when the gradient of the road is steep enough to allow the empty trams to run to the bottom and drag the rope with them. This rope passes over a drum provided with an efficient brake, the drum being capable of being thrown out of gear with the engine. This is the method in common use in South Wales for winding coal from inclined planes or slants. The advantages of the system are that only one line of rails and one length of rope is required, and that, consequently, the roads may be narrow and less costly to construct and maintain. A modification of this method is sometimes effected by having two drums, and hauling one train of trams up, whilst the other runs down. Where the direct system of haulage is in use, to prevent accidents the trams are fastened together by means of a bridle chain, and also iron drags pointed at their extremities and attached behind the last tram, are in common use. The object of the drag is to throw trams off the road if the rope or shackles should break.

In the endless rope system a double road is necessary. The rope passes round a driving pulley, either at the surface or at the pit bottom, along the main road or roads, and around a sheave or pulley at the opposite terminus. This endless rope travels continously in one direction. The trams are connected either singly or in sets with a rope which may pass either above or beneath ; in some cases a chain is used instead of a rope. The advantages of this method are : that on account of the slow motion it is safer than the others ; that the supply of trams is constant ; and that the weight of the trams going downhill assists the raising of those going up.

The main and tail system is used when the road is uneven or not sufficiently steep to enable the empty trams to run back into the workings and take the rope with them. The system requires two drums and two ropes, one in front of and one behind each train of trams. One drum carries the rope which pulls the train of full trams from the workings to the shafts ; this is called the main rope and is equal in length to the whole distance of haulage. The second drum carries the tail rope, which must be twice the distance of haulage and is fastened at its loose end to the last wagon in the train. Each of the drums can be put in gear with the engine or run free, as may be necessary, and each is provided with a strong brake. When the train of loaded wagons is brought from the sidings to the main haulage road, the main rope is attached to the front wagon and the tail rope to the hindermost wagon. The engine is then started, the drum beside it, carrying the tail rope, is put out of gear; and as the wagons are drawn towards the pit bottom, the tail rope uncoils itself from the loose drum, and is extended along the whole length of the line, both downwards and back behind the wagons. The full trams are now removed and replaced by empty trams, and the process is reversed.

In many mines in the United States, compressed air locomotives are in use, and are found very convenient as a method of haulage where the gradients are not very steep. In this country also compressed air locomotives are used in a very few mines. Electric locomotives taking their current from the overhead trollies in exactly the same way as street cars, are also frequently used both on the Continent and in America. They have a complete air-tight steel shell, every part being covered. The advantages of locomotives for underground haulage are that the cost of maintenance where sparking could possibly occur, is less, delays on account of derailed trams are fewer, and accidents can be prevented when the trams can be stopped without delay.

When rope systems of haulage are in use it is

necessary to provide means for communicating messages from one end of the road to the other. Electric signalling is by far the most efficient method ; two galvanised iron wires are fixed on small insulators, and run along the side or roof of the roadway ; these are connected to a bell and battery in the ordinary way. When it is desired to transmit a signal to the engine-house, the rider or person in charge of the train of trams rubs the two wires together, so causing the bell to ring. There is in every mine a well understood code of signals, and the number of rings given conveys to the engineman the instructions of the person in charge of the underground traffic.

CHAPTER IX

Objects of Preparation

IN former times coal was usually sold as *through and through*, that is to say just as it came from below ground, when the hewer merely picked out what were obvious pieces of stone so far as he could see with his dim sight. The small coal, if it contained much dirt, that is, stony matter, was generally not sent out, but thrown into the *goaf*, or if that were dangerous owing to the coal being liable to spontaneous combustion, the small coal was wound from the mine and dumped on a rubbish heap, where it was generally set on fire and would burn for years. During recent years, however, much attention has been paid to the importance of utilising to the greatest advantage the whole of the coal output from mines ; and an elaborate system of treating the coal after it is brought to the surface has been evolved. The objects of this are three-fold. (1) In most modern collieries the mineral is carefully sorted into various qualities and sizes, which are in accordance with the varying uses made of it by different buyers ; and in a large number of collieries the coals are washed, (2) for one reason to

219

remove all dirt and refuse, and, (3) for another, to produce dustless coal. The higher prices obtained for such prepared coal much more than repay the expenditure so incurred.

Weighing and Screening

When the loaded trams or tubs of coal are discharged from the cages in which they have been brought to the surface they are conveyed on rails to a weighing station, where the weights are carefully recorded on separate sheets or books by an official acting on behalf of the management and also by a checkweigher appointed to safeguard the interests of the men. The identity of the coal hewer is indicated to the weighers by numbers chalked upon the trams. Some of the weighing machines indicate the weights by pointers on dials; in others automatic recorders are provided. In the South Wales coal-field, where separate records of the weights of small coal are required, special weighers and checkweighers, the latter being locally described as " billychecks," are often employed. The small-coal weighing machine usually takes the form of a spring balance, from which is suspended the receptacle into which the small coal falls after it has passed through the screen. This receptacle is emptied by a lever arrangement.

When the coal trams leave the weighing machine they are run on to a tippler to be emptied. Tipplers are of two kinds, known as end-tipplers and side-tipplers. The former are arranged to empty **the** trams over the end, and the latter over the side. The revolving side-tippler is the device most generally used. It consists of a cage or framework, the ends of which are formed by two cast iron-rings, capable of revolving on rollers upon a horizontal axis. The tram enters the cage, a lever is pulled, the cage is turned upside down in a sideways direction, and its contents shot out upon the screen. When the revolution is completed, and the tram is again in an upright position, the frame stops automatically and the tram is pushed out at the opposite end to be replaced by another full tram.

An end-tippler consists of an iron platform suspended on pivots with their axes parallel to the ends of the tram, and capable of revolving through the whole or part of a circle. The weight of the tram overbalances the platform and the coal is discharged over the end. This form of tippler is now gradually being discarded in favour of some variety of side-tippler, as the latter distributes the coal over a larger area and so facilitates screening ; and no time is wasted in withdrawing an empty tram to make room for a loaded one.

When the trams have been emptied they are returned along a different route to the mouth of the shaft, usually with the aid of machinery.

When the coal leaves the tippler it passes through a long and complicated process before it is ready to be consigned to the purchasers. In no branch of mining engineering has greater progress been recorded than in the grading and cleaning of the coal,

and the complicated screening arrangements of
to-day show a remarkable advance on the old fixed-
bar screen which was commonly used a few years
ago. The old type of screen consisted of a series
of steel bars fixed at a gradient of 1 in 2 or 1 in 3,
with spaces of desired width between them. The
screens were usually constructed in sections or
combs, the widths of the spaces in the various
sections increasing from above downwards. Such
screens would enable the coal to be graded roughly
into different sizes, but not in so careful and clean a
manner as is now deemed desirable. They have
for this reason been superseded in important collieries by much more effective arrangements by which
all dirt is removed, a more perfect classification is
achieved, and damage from breakage of the large
coal is reduced to a minimum.

These modern devices are of varied types. Some
consist of parallel bars, alternate bars being arranged
in two sets which are capable of a vibrating motion
longitudinally and vertically in opposite directions.
In a second type the coal passes over a series of
rollers carrying triangular shaped plates which, as
they revolve, carry the largest coals forward, allowing the smaller portions to drop down between the
plates and the rollers. A popular type of screen is
known as the *jigger*, which in its numerous modifications is in use at probably the majority of modern
collieries. This screen is so called because of the
shaking or jigging motion imparted to it by the
revolutions of a number of eccentric wheels fixed at

various points. The screening takes place over wire
masks or perforated iron sheets, the to and fro motion
shakes out the small, and at the same time carries
forward the larger lumps and discharges them on a
picking band or *belt*, which conveys them to the
wagon allotted to receive them. Somewhat similar to the jigger is the *rocking screen*, which consists of a number of short narrow sections which are
hinged together and are made to rock in opposite
directions by the revolutions of an eccentric. In
some systems the backward and forward movements
of the jigger are combined with the up and down
movements of the rocker. The arrangement of
such a screen is very simple. The upper portion is
attached to a shaft passing through a pair of eccentric wheels, while the lower portion is suspended from
a crank which in some cases is fixed, and in others is
capable of an up and down motion. *Gyrating
screens*, of which there are several makes, are constructed to impart motion in a nearly horizontal plane
to the screening surfaces so as to avoid breakage
of the coal where it is soft, and are usually employed
for screening coals of small size only. *Revolving
drum-shaped screens* are largely used for anthracite,
and for grading coals which pass through washing
plants. The screening surfaces are made of wire or
of perforated plates arranged as concentric cylinders
round a revolving shaft. Such drum-shaped screens
are from 20 to 30 feet long, and from 6 to 8 feet wide.
The coal is fed into the inner drum, which has the
largest holes, the small falls through into the second

screen, and is again separated, the smallest particles being collected in the space between the largest drum and an outside iron casing. When the coal has travelled through the whole length of the screen, it is discharged into different wagons. The proper discharge is facilitated by making the inner screens project at the discharging end over the outer cones, and by arranging shoots in the proper positions to receive the various sizes of coal.

Hand-Picking of Coal

The grading of coals alone is not sufficient for modern marketing purposes. The presence of shale, stone, pyrites, dirt, etc., in the coal tends to depreciate its value. Great care is taken to ensure the loading of " clean " coal only into the trams by the miners, but this precaution is usually not sufficient, and in all up-to-date collieries arrangements have to be made on the surface for the removal of all impurities. In the case of the large coal, hand-picking is the usual method ; where the coal is small or very dirty washing is more thorough and economical.

Where the coal is hand-picked an endless *travelling band* of overlapping iron or steel plates from 30 to 70 feet long and 3 or 4 feet broad, and enclosed on either side, is arranged on sets of links which revolve longitudinally over rollers. The coal is spread out on the band, and as it revolves slowly (the ordinary rate is about 50 feet per minute) boys or youths standing at intervals along the sides pick out any pieces of rubbish that may be present. It is essential for effective picking that the coal should be well and evenly distributed so that the pickers can see any impurities without the slightest difficulty. This is secured by arrangements called *distributors* attached to the screens. In cases where coals of several different sizes or qualities have to be treated, separate picking bands must be provided. The rubbish picked off the bands is dropped into shoots ranged at frequent intervals along the platform on which the pickers stand and which deposit it on a travelling conveyor, from which it is discharged into a tram or overhead corve for dumping on the rubbish heap. At some collieries revolving *picking tables* are used. These are concentric in shape and revolve on rollers, and the pickers are stationed both in the inner and outer circumference. These have proved quite satisfactory, and in some respects are more advantageous than travelling bands. There are some modern types of screens so constructed as to permit of picking and screening being carried on at the same time.

Coal Washing Machines

The washing of coal is undertaken for two reasons. In cases where there is a large proportion of small or dirty coal, handpicking is ineffective or too expensive for removing all the fragments and chips of stone ; and it is because of the possibility of separating them in running water owing to the difference of specific gravity that in many modern

P

collieries costly and elaborate washing machines have been installed for removing fragments. The use of such plant for the purification of coal intended for conversion into coke is long-standing ; the application of washing to coal intended for sale has, however, been of comparatively recent date ; and the economy effected by its use has been very substantial where the coal contains much " dirt," *i.e.*, stone, and there is a plentiful supply of water. Besides the separation of " dirt " washing is valuable as a means of cleaning the coal from coal-dust. Dusty coal is liable to spontaneous combustion when stored in a closed place, as in a ship's bunkers ; and the washing practically eliminates the risk of fire. The coal is also cleaner to handle and a trifle freer burning, which are considerations on ocean liners.

The separation of coal from its impurities in washing machines is based upon a difference of specific gravity. If materials of varying gravities are allowed to sink in still water it will be found that they will not all reach the bottom at the same time ; the heaviest bodies will reach there first, and the others will arrive later in order of their specific gravity. Similarly, in running water the lighter matter is carried forward faster than the heavier, which is the principle so widely used in washing alluvial sand or crushed quartz for gold.

In most types of washing machines the coal is separated from the impurities directly by the action of water only. In others a foreign substance, usually felspar, of intermediate specific gravity to the coal and the rubbish, is introduced, the water is agitated and the various substances arrange themselves in layers, the coal, having the lowest specific gravity, being at the top, from which it is washed into troughs for removal.

Washing machines of the first type are known as *troughs*. In its original form the trough washer consisted of a wooden trough about 2 feet wide, 1 foot deep, and from 50 to 100 feet long. At intervals *riffles* or *dams* three or four inches high are fastened to the bottom of the trough, which is slightly inclined. The water and small coal are fed into the top of the trough, and as the stream proceeds down the slope, men stationed at intervals along the sides keep up a stirring process with shovels. The heavier fragments sink and are kept back by the dam, while the particles of coal are conveyed by the stream over the end of the trough to a *collector*. Washers of this kind are too crude and too expensive for general use, and are fast being superseded by more elaborate and much more effective machines.

An improved trough is the *Merton Washer*, which is similar to an endless picking band. The bed of the trough travels upwards at the rate of 8 or 10 feet per minute, and the slack and the water are fed into it at the top. The heavy particles of dirt are rolled along the bottom, where they are caught by *stoppings*, and conveyed upwards to the top to be tipped into a refuse wagon, while the washed small coal is carried downwards by the current of water into perforated hoppers, which allow the dirty water to escape. In the *Elliot Washer* the trough is stationary, but is provided with scrapers attached to chains travelling uphill along the trough bottom. They are usually constructed in sets of three, each size of small coal being fed into a separate trough by means of a revolving screen.

Satisfactory though some of the latter types of trough washers undoubtedly are, they are not nearly so frequently used as the numerous rotary and jigger washers now on the market. The rotary machine depends upon the principle that if coal and shale are stirred together in water, the shale, on account of its higher specific gravity, will sink to the bottom before the coal. A well-known and exceedingly ingenious example of this type is the *Robinson washer*. In an iron pan, shaped like an inverted cone, a number of vertical iron stirrers are caused to revolve. The slack is fed into the pan from above, and a strong current of water rushes in from below and forces the clean coal over the rim of the cone, whilst the shale and dirt gradually find their way to the bottom to be removed through sliding doors.

Jigger washers, of which there are numerous types, require the coal to be carefully graded in size before treatment. They are much used in metalliferous mines. In its simplest form the jigger consists of a gully-shaped iron box divided into compartments connected at the bottom. A piston or plunger in one compartment moves regularly up and down as a pump. A horizontal strainer with fine perforations is arranged in the larger compart-

INTERIOR OF THE COAL WASHERY AT THE BARGOED PIT OF THE POWELL DUFFRYN STEAM COAL CO., LTD.

COAL WASHERY AT THE BARGOED COLLIERY OF THE
POWELL DUFFRYN STEAM COAL CO., LTD.

ment above which the coal is fed. The water is
introduced at the bottom of the machine. The
action of the piston causes the water to rise and fall,
thus imparting to the material on the sieve an upwards and downwards motion. The lighter particles
are lifted higher and fall more slowly, so that whilst
still suspended in the water the coal and shale are
separated into an upper and lower layer, each of
which is washed out through an opening specially
prepared for it.

Felspar washers are used for treating very fine or
slack coal, which may contain as much as 20 to 30
per cent. of stone chips and dust, rendering the coal
practically valueless. The felspar washers are of
the jigger type ; but the perforations in the sieve are
large enough to allow particles to pass through, and a
layer about 3 inches thick of felspar is imposed upon
it before the machine is set in motion. The jigging
motion results as before in the mixing of the different substances, but now the small and heavy particles
of shale fall through the meshes, and the lighter
particles of coal rise above the felspar and are
washed out. The dirt below the sieve is evacuated
through adjustable outlet valves.

In the machines hitherto described it is an essential
that the coal should be carefully graded before the
washing process is begun. This entails the use of
a separate machine for each size of coal. There are
washers, however, which proceed on the principle
that classification should follow washing, and
others which perform the two operations at the same

time ; but they are rather too complicated to be described here. Washing without previous sizing was, indeed, the plan first tried when washing was first introduced ; but it was abandoned owing to waste in loss of coal. There came a stage in which sizing was, perhaps, carried too far ; and the tendency in recent plants is not to size closer than is necessary or useful for market purposes.

In the matter of preparing coal for market we come far behind the continental countries, and are consequently heavily handicapped in markets where we have to face the competition of Germany, which has adopted highly scientific methods of treating coal before offering it for sale. It has been clearly established that not only is the value of coal greatly enhanced by the adoption of washing and grading processes ; but also that much small coal formerly unsaleable at a profit, and therefore not worth bringing to the surface, can now be brought out and sold to great advantage after washing. British consuls have complained that we do not lay ourselves out to suit the particular requirements of our customers abroad ; and it seems inevitable, if our lead as a coal-exporting nation is to be maintained, that we adopt improved methods of treatment much more generously than has hitherto been customary. The importance of the thorough grading of coals was emphasized by many witnesses who gave evidence before the Royal Commission on Coal Supplies, and the Commission adopted the following recommendation :—

" We desire to urge as strongly as we may the importance of cleaning, sizing, and sorting coal for the market. The more consumers realise the advantages that accrue from the use of coal selected to suit their special requirements and appliances, the more they will expect and demand uniformity of quality and size. Uniformity is important ; and there is no question that a consumer is willing to pay more, if he can rely upon always getting what his experience has proved to be the best suited for his purpose."

CHAPTER X

Utilisation of Small Coal

IN the previous chapter some account was given of the great advances which have been made during recent years in the reduction of waste in the disposal of the output of coal mines, and in the economical utilisation of much mineral that was formerly rejected as rubbish. This development has proved very profitable, and in many of the coalfields, especially in the South Wales anthracite district, huge quantities of small coal that have been lying waste in ugly tips, in some cases for nearly half a century, have recently been disposed of at satisfactory prices. It is not improbable, also, that in the future means will be found for utilising commercially the huge mountains of more or less carbonaceous shale that now usually disfigure the landscape in mining districts. Our concern in this chapter, however, is with the processes now in general use for manufacturing from the least valuable portion of the coal gotten from mines, other products which command a much higher market price. The dust and small of many kinds of coal are of little

232

commercial value. When, however, these are subjected to a coking process, often to obtain coal-tar products as well as coke, or are combined with other constituents to make patent fuel briquettes, their value is greatly enhanced ; and from the sale of the manufactured products many collieries derive a profitable revenue.

Caking and Non-Caking Coals

Coke is a form of fuel made by heating small coal in an airtight chamber. It is in considerable demand for the manufacture of iron and steel. For this purpose it is necessary that all the volatile constituents of the coal should be removed, leaving behind a high proportion of carbon. Viewed from the point of view of the coke manufacturer, coals belong to two classes : those which when burned or heated in a retort congeal into a paste-like mass, very dissimilar from coal in appearance, but possessed of great calorific power ; and those which do not combine well together when heat is applied, but merely form some kind of cinders, retaining more or less their original form. The former are known as *caking coals*, and are used very largely for the manufacture of coke. Coals which do not run together in the manner described are *non-caking*, and are useless for coke-making purposes. The coking properties of coal cannot be discovered by chemically analysing its constituents, as there appear to be certain physical conditions, not yet clearly ascertained, which determine the quality of

the product of the coking process.   The coal must
be tried in the retort.

The following analyses of coal previous to coking,
and of the resultant product, illustrate the changes
that take place :—

	Coal per cent.	Coke per cent.
Carbon	60 to 85	85 to 90
Volatile substances . .	20 to 30	$2\frac{1}{2}$
Ash	3 to 5	5 to 10
Sulphur	1 to 5	·05 to 2
Water	3 to 6	1 to 5

The greatest change is manifested in the reduction
of the volatile or gaseous constituents : very small
quantities remain after the coal has been converted
into coke.   These substances are in many recent
installations carefully collected during the distillation, and put to commercial uses.   In some cases the
gases are wholly used to heat boilers, or are burnt in
flues for the purpose of coking the coal.   More
usually, however, they are treated in order to separate out valuable bye-products.

It is essential for the production of coke of the
best quality that the coal from which it is made
should be as free from dirt as possible ; and most
coke ovens, therefore, have *washeries* attached to
them where the small coal is subjected to a thorough
process of cleansing in the manner already described
in a previous chapter before the actual coking process
is commenced.   When this has been done the
" slack " is finely ground, and is then ready to be
charged into the coking ovens.   Several varieties

of installations are in existence, but these resolve themselves into two main classes : those which admit air during the process of manufacture, and those which do not. By far the most common example of the former type is that known as the *Beehive*, so-called from its shape. In ovens of this form the volatile matters are not fully utilised, and no arrangements exist for the recovery of bye-products. The quantity of coke produced, also, is considerably less than that obtained from more recent types, and the quality is said to be not so good.

Coking Ovens

An ordinary Beehive oven is a dome-shaped chamber, usually from 8 to 14 feet in diameter, and about 8 feet in height ; and they are built in blocks of from twenty to sixty. The structure is of firebrick cemented together with fireclay mortar. Surrounding each oven are non-conducting masses of clay, ashes, sand, or other material, the function of which is to retain in the ovens as much of the heat as possible. The outer shell of the battery or series of ovens is constructed of ordinary masonry clamped together with iron rods to resist the expansion produced by the heat, which is generated in small furnaces at either end. The ovens are generally ranged in two rows back to back. Each chamber is capable of taking from ten to twelve tons of *smudge*, or small coal, which is fed into it through a circular *charging hole*, about 12 inches in diameter at the apex of the dome. The coal is conveyed to the charging holes from the drying *bunkers,* or chambers of the washery in iron wagons running on rails laid on the flat surface which forms the roof of the ovens. These wagons are known as *hoppers,* and are of funnel-like construction. When a wagon is exactly over a charging hole the iron lid of the hole is removed, a slide in the bottom of the wagon is drawn out, and the contents fall into the coking chamber. When the chamber is full the surface of the coal is levelled by means of a specially designed appliance, and the lid of the charging hole is carefully replaced. The coal is soon fired by the heat from the adjacent ovens. Each oven is provided at the front with a door, constructed usually of brickwork in an iron frame, and suspended from a chain passing over a pulley to the free end of which counter-weights are attached, or some mechanical power is applied. An adjustable ventilator is fitted to the door to allow of the ingress of air to facilitate the combustion of the coal. Each oven is also fitted near the roof with a flue to carry off waste gases, which are conveyed under boilers and used to generate steam. The coal remains in the oven for periods which vary according to the size and type of the installation from 24 to 48 hours. The ovens are loaded and drawn alternately, so that there is no suspension in the process of manufacture, and the ovens never become cool. When an oven is ready to be discharged, the door is raised and a current of cold water is directed by means of a hose on the heated mass within. Labourers armed

with long iron scrapers now set to work and draw
out the coke. The cost of hand-drawing varies
from 1s. 6d. to 2s. 6d. per ton of coke ; and, as this
is considered too heavy, at many collieries mechanical
extractors are used. These are engines running
on rails laid on the platform in front of the ovens,
and carrying a hook-shaped shovel attached to a
long iron arm, so arranged that it can swing around
in the oven at all angles. As the coke is drawn out
the big lumps are broken up, the coke is further
cooled with a spray of water, and it is then loaded
into railway wagons standing in the siding laid
alongside the platform adjacent to the ovens.

Beehive ovens are not nearly so efficient as later
types, as no provision is made for the recovery of
such bye-products as tar, ammonia, and benzol.
Some other kinds of ovens also have the same defect.
Most of the installations provided during recent
years, however, are of the *retort* type ; that is, all
air is excluded, and the ovens are fitted with arrangements by which the volatile matter can be carefully
treated with a view to the recovery of its principal
constituents. Retort ovens are rectangular arched
chambers about 30 feet long, 6 feet high, and 2 feet
wide, separated from one another by thin firebrick
walls. Between and under the ovens are flues from
which heat passes into the oven interiors. The
ovens are arranged in batteries of from 25 to 60, and
each one is provided back and front with vertical
sliding doors. Charging holes are sometimes constructed in the roofs, and also ascension outlet pipes,

fitted with valves, through which are drawn off by means of exhaust pumps the gases generated in the ovens. These gases are conveyed through iron pipes to cooling and condensing chambers. The first stage in the cooling process results in the condensation of the tar. The uncondensed gases are next conducted into chambers fitted with arrangements for breaking up drops of falling water into fine spray which absorbs much of the ammonia contained in the vapour. Benzol is extracted by passing the gas over heavy coal-tar oils. The process of recovering bye-products is a highly complicated one, and has not even yet attained the stage of perfection. In all installations a considerable amount of gas is incapable with present methods of exhaustive treatment. When as large a proportion of the products named as possible has been recovered the incondensible surplus, instead of being allowed to run to waste, is mixed with hot air and conveyed to the oven flues, where it is burned to generate the heat required for coking. The products of combustion, and the hot gases passing out from the flues are conducted under boilers and used to raise steam before they are finally ejected through the chimney outlet provided for them. In some districts where the gaseous product is very large, much of the colliery machinery is driven by gas engines ; and in others so great is the quantity available that it is fired in order to illuminate the surface at night. In some few cases it is sold to a gas company or an electric generating company.

Charging and Discharging Coke Ovens

As in the Beehive system the ovens are loaded and discharged alternately ; and in this case also the " smudge " in a newly-charged oven is fired from the heat of the adjoining chamber. In some installations the charging is done from above. This method, however, is gradually being abandoned on account of its great disadvantage as compared with the " compressor " method. The latter process consists in stamping damp smudge into solid blocks, which are loaded laterally into the chambers by means of mechanical appliances. In the best systems a combined stamping, charging, and discharging machine, driven by steam or electric power, runs on four rails in front of the battery of ovens. The smudge is fed through a hopper or funnel from above into a stamping chamber with movable sides and bottom in which a ram works vertically. This chamber is of slightly smaller dimensions than the oven. When the immense pressure applied has consolidated the coal, the adjustable sides are relaxed and the block of coal is ready to be charged into the oven. The machine is now brought up so that the charge of coal is exactly opposite the oven. By means of a rack and pinion arrangement the iron bottom is projected forward into the oven carrying the block of coal with it. The sliding door is now dropped nearly to the bottom, and the machine is reversed, drawing out the iron plate while the coal remains within to be burned. Compressing the coal is said to result in a larger quantity of coke of a better and more solid quality than that produced by the old method. When the coking process is concluded, the back and front sliding doors are opened, the machine is brought into position at the back. The power is applied, and an iron ram extends forward, pushing the contents of the oven before it on to the platform in front, where it is cooled, and afterwards loaded into the railway trucks.

Financial Advantage of Coke Making

About one hundred tons of small coal are necessary to produce sixty tons of coke ; but the prices obtainable for the latter are much higher than those paid for the former, so that the installation of coke ovens generally proves very profitable to the colliery owners. The prices f.o.b. at Cardiff obtained for the small of the best Rhondda bituminous coals in 1910–1911 ranged from 7s. to 8s. 3d. per ton for No. 2 quality, and from 9s. 6d. to 10s. 6d. for the No. 3 quality. In the same year the prices of coke were approximately 12s. 6d., 17s., 19s. 6d., and 25s. per ton for the four different qualities marketed at Cardiff. If 9s. per ton is assumed as the average price of small coal, and 19s. per ton as the average price of coke, we shall be able to ascertain roughly the difference in value to the mine-owners of the raw material and the manufactured product. Thus :

60 tons of coke at 19s. ..	..	.. £57	0	0
100 tons of small coal at 9s.	..	.. 45	0	0
		£12	0	0

That is, on each 100 tons of small coal coked the owners receive £12 more than they would receive if they put the small coal on the market in the ordinary way. Of course the cost of the installation and of its working have to be paid out of this profit. On the other hand, we must add also the income derived from the sale of tar, ammonia, and benzol, and in a few cases of surplus gas. Each 100 tons of coal yields about three tons of tar, the market price of which is about 17s. per ton, and about 1½ tons of sulphate of ammonia, which is marketed at £12 per ton.

The financial advantages of coking coal are becoming yearly more widely realised, and large numbers of coking plants are being installed in all the chief coalfields. Most of the newer installations are equipped with facilities for the recovery of the tar and ammoniacal liquor driven off from the coal in the process of coking. There is considerable room for development in this direction, however, as less than 20 per cent. of our total output of coke is obtained from recovery ovens.

The Royal Commission on Coal Supplies inquired very carefully into the possibilities of utilising commercially the gaseous constituents of coal, which, in this country, are too often allowed to run to waste during the process of coking ; and their remarks upon this subject are of great interest. The following extract is taken from the Final Report of the Commission :—

" Bye-product recovery ovens are, however,

costly both in capital expenditure, and in running
repairs and renewals. Whether it will pay to work
the bye-product recovery process depends first on the
nitrogen in the coal recoverable in the form of
ammonia, and secondly, on the volatile matter in
the coal recoverable as tar and oils. According to
evidence, 16 or 17 per cent. of volatile matter
in the coal probably represents the paying minimum
worth recovery. It has been suggested that a
largely increased supply would decrease the value
of the bye-products ; but, even if this should occur as
regards some of them, there seems no reason to anticipate a serious fall in the price of sulphate of
ammonia. Mr. Darby, in his evidence, has given
particulars of the comparative advantages and
costs of the different types of ovens, and in the
result it would appear that in the case of rich coal
the extra initial outlay upon the bye-product
recovery is soon reimbursed. The richer the coal the
greater the advantage of using these ovens. . . ."

Utilisation of Volatile Products

" The production of coke, as it is extensively carried
on in this country, without full utilisation of the
volatile products, is condemned by all the witnesses.
In the best modern practice these products are either
burnt in flues round the ovens, or are separated by
cooling into liquids and gases, the latter of which are
used for heating the ovens themselves. The surplus
gas can be used for the production of power under
steam boilers, or with greater advantage in gas

engines. Coke oven gas is a rich gas approximating to illuminating, and far richer than producer, gas."

Importance of Coking

" The importance of the extended adoption of coking cannot be exaggerated. It is one of the methods by which small coal can be rendered marketable, and in some districts it has reduced the waste by furnishing the collieries with an outlet for the small coal, without which outlet it is doubtful whether they could have been carried on."

British Coke Production

The following table gives the output of coke in the United Kingdom during the years 1910, 1911, and 1912. The number of coke ovens in use during the latter year was 21,076, 70 per cent. of which were of the Beehive class :—

COKE PRODUCTION IN THE UNITED KINGDOM [1]

	1910. (Tons.)	1911. (Tons.)	1912. (Tons.)
South Wales & Mon. . .	1,512,255	1,450,293	1,378,192
North Wales .	28,525	28,643	30,061
Rest of England	16,343,684	15,940,969	15,420,858
Scotland .	1,283,592	1,357,029	1,343,462
Ireland . .	153,831	170,358	168,121
Isle of Man .	9,574	9,987	9,374
	19,331,461	18,957,279	18,350,068

[1] These figures include Gas Coke made at municipal and privately owned gasworks. The amounts as produced in each year were as follows :—1910, 144,638 tons ; 1911, 151,547 tons ; 1912, 150,329 tons.

Foreign Production of Coke

Germany is far ahead of the United Kingdom in respect both of the output of coke and of the recovery of bye-products. The great majority of German installations are equipped with recovery plants, and the results attained in that country have been remarkably good. The total output of coke in the principal coal-mining countries is given below :—

PRODUCTION OF COKE IN CERTAIN FOREIGN COUNTRIES IN 1911, 1912, and 1913 IN METRIC TONS.[1]

Country.	1911.	1912.	1913.
Germany	25,405,000	29,141,000	32,168,000
Belgium	3,161,000	3,187,000	—
France	2,487,000	2,603,000	2,635,000
United States	31,225,000	37,243,000	42,101,000
Austria	2,058,000	2,308,000	2,584,000
United Kingdom	19,641,000	19,261,000	18,644,000

Patent Fuel Briquettes

Another important method of utilising small coal to the greatest commercial advantage is by mixing it with some cementing material, and compressing it into rectangular blocks. This method is capable of being used with coal that lacks suitability for manufacture into coke, and the high prices obtained for the blocks of fuel, or *briquettes*, as they are called, make it exceedingly profitable to pass the small coal

[1] Figures from *South Wales Coal Annual*.

through a patent fuel plant. In this country, however, the colliery proprietors do not usually own their own installations, but sell their coal to different companies who have extensive works, usually near the seaboard. Much the greater part of the fuel manufactured in the United Kingdom is produced in South Wales at Swansea, Cardiff, and other ports in the vicinity of the docks. The advantage of this situation becomes apparent when it is stated that most of the output of patent fuel works is exported for consumption abroad. It means that loading the blocks into railway trucks before shipment can be altogether avoided. The small coal must be loaded into trucks as it falls through the screens at the colliery ; it is, therefore, taken direct to the dockside to be made into blocks, or to works situated on a canal at no great distance from the port of shipment, whither it is brought in barges right alongside the steamers, which can load their blocks from the water side whilst loading coal or other cargo from the wharf. Another reason for locating patent fuel works at the docks is that a good deal of small coal is made by the jolting of the trucks in the transit of the large coal from pit to dock ; and it is removed by tipping the coal over a screen into the ship. The cost of taking this small coal out of the dock area, and back again for shipment as blocks, is so great as to make it worth while to pay a high rent for space to manufacture at the dockside.

The manufacturing process is one of great interest.

When the small coal arrives at the works it is tipped from the wagons, and elevated direct to the screens and graded into different sizes. These different sizes are next carefully washed in order to remove the dirt, slag, and other impurities. The water from the coal is now drained off, and the coal is conveyed into drying ovens, where it comes into contact with hot gases, which effectually remove all the remaining moisture from the coal. Next, the coal is taken to a *disintegrator*, or crushing machine, where it is crushed into minute particles and thoroughly mixed with molten pitch. One of the important points to be observed in the fuel manufacture is the proper adjustment of the proportions of coal and pitch. Dry kinds of coal require larger amounts of pitch than most bituminous coals; in Cardiff, for example, from 8 to 8½ per cent. of pitch to coal is a sufficient proportion, whereas in Swansea 10 per cent. has to be used. When the mixture has been ground sufficiently fine, it undergoes further treatment, and is then fed automatically into the moulds of a lever or hydraulic press, where it is subjected to a pressure of about two tons per square inch. The press, also stamps upon each block the trade mark and name of the manufacturer. The blocks are then forced out of the moulds and passed on to a travelling belt, which conveys them to the stacks or the weighing machines. The latest patent is capable of turning out 500 tons in 20 hours, or 4,480 briquettes, each weighing 12½ lbs., per hour.

The blocks are of varying weights and dimensions. The largest manufactured in South Wales is 12 inches long, 8½ inches wide, and 5½ inches thick, and weighs 26 pounds ; the dimensions of the smallest are 9 inches by 5 inches by 3 inches, and the weight 6½ pounds. The bulk of the fuel is made in large briquettes ; in only two out of thirteen South Wales brands does the weight fall below 10 pounds. On the Continent, however, where patent fuel is largely used for domestic purposes, very small briquettes weighing only a few ounces are manufactured, and are found very convenient. The quality of foreign briquettes is rather better than that of the English varieties, probably because the small coal is crushed finer and greater pressure is applied.

Use of Anthracite for Briquettes

Many experiments have been made during recent years in the improvement of manufacturing processes, and the utilisation of coals which have hitherto proved unsuitable. The reason why South Wales has become the chief, almost the only, centre of patent fuel manufacture in the United Kingdom is partly because so much small coal is on the market owing to the general practice of screening to obtain large coal, and because the special suitability of Welsh steam coal for generating steam applies also to the fuel blocks manufactured from it.

Considerable attention has recently been paid to the effective utilisation of small anthracite coal for fuel manufacture. This coal, however, although an excellent heat-producing agent, does not ignite easily, and will not, as is the case with other coals, burn easily, except in lumps. A considerable amount of anthracite small is used in the patent fuel factories at Swansea, mixed with the small of steam and bituminous coals. The reason why anthracite cannot be used more generally is that it has no caking or adhesive tendency during burning, so that the pitch burns first, leaving the coal in a disintegrated form in the fire. A cementing material is needed which, while being equally strong and weather-resisting, does not consume so rapidly as pitch. A number of substitutes for pitch have been tried, but none have proved perfectly satisfactory; and fuel manufacturers are eagerly awaiting an invention which will provide a satisfactory matrix for anthracite blocks, and will place at their disposal an equally efficient but cheaper matrix than pitch for blocks made of steam or bituminous small. One of the best of the substitutes is a mixture of starch and lime. It binds well, generates considerable heat, is comparatively smokeless, and is less costly than pitch. Its greatest disadvantage lies in the fact that it produces much ash. The growing demand for briquettes makes it probable that a satisfactory material will be devised at no distant date. The latest invention, which has yet, however, to prove itself a commercial success, is a method of making briquettes, or large blocks, out of coal alone, without any cementing material whatever. The principle is to use a mixture containing a large proportion of good coking coal. This is strongly heated in retorts and automatically fed into powerful presses whilst in a soft, half-coked condition. The plant is, unfortunately, very costly.

Advantages of Fuel Briquettes

The future of the patent fuel manufacture is assured. " Preserved coal," as the briquettes might very truly be called, possesses merits which enable it to command a ready and increasing sale in places where coal can be conveyed only at great trouble and expense. Coal cannot be delivered at foreign destinations without considerable breakage ; both in trucks and in ships the vibration and jolting result in the making of much small which depreciates more or less the value of the consignment. In addition there is the danger arising from spontaneous combustion on board the vessels ; whilst the storage of coal in the open leads to considerable deterioration and loss of calorific power. From all these disadvantages patent fuel is free, for the blocks, on account of the regularity of their shape, can be packed tightly into trucks and the holds of vessels so that they do not move about, and are delivered to their destination practically without breakage. The blocks also hardly deteriorate at all with exposure, and can be stored in the open in all climates without sustaining any loss of calorific value. Some years ago, for example, a block of Cardiff patent fuel was recovered from a ship that was wrecked in Balaclava Bay in the fifties. In

spite of forty years' immersion it still retained its
original properties. Another point in favour of
patent fuel is that the space required to store the
blocks is only about 32 cubic feet per ton, as compared with about 44 cubic feet for a ton of large
coal. This is a very important factor, for a smaller
storage space per ton means larger cargoes, heavier
truck loads, and bigger margins of profit.

Production of Patent Fuel

As has been mentioned already, the bulk of patent
fuel manufactured in this country is exported abroad,
mainly to France, Spain, Italy, Algeria, and Mexico,
and the South-American republics. The following
table gives the production of patent fuel in the
United Kingdom in 1910, 1911, and 1912 :—

PRODUCTION OF PATENT FUEL

County.	1910. Tons.	1911. Tons.	1912. Tons.
Glamorgan .	1,354,969	1,554,449	1,507,220
Monmouth and Gloucester .	155,249	112,404	137,083
Nottingham, Sussex, Somerset, Stafford	14,937	17,051	16,095
Derby, Devon, Essex, Hants	10,554	9,919	9,521
Yorkshire .	5,921	6,543	6,464
Lancashire .	2,002	2,329	2,028
Scotland .	42,075	57,021	60,454
Ireland . .	21,959	19,417	17,004
Total for United Kingdom .	1,607,666	1,779,133	1,755,869

From this table it will be seen how important is the industry in Glamorgan ; indeed, this county produces about 85 per cent. of the output of the whole country. The total export during the same year from Cardiff, Swansea, and Port Talbot was 1,319,603 tons, leaving only about 35,000 tons for consumption at home. The fuel is used very largely for driving locomotives, and also for generating power in warships, manufactories, machinery works, electric lighting and traction stations, etc. Britain has made much progress in the manufacture of recent years ; but it is far behind some other nations in the bulk of its output, as the following statistics show.

PRODUCTION OF PATENT FUEL IN CERTAIN FOREIGN COUNTRIES

	1911. Metric Tons.	1912. Metric Tons.	1913. Metric Tons.
Germany	21,828,000	24,392,000	27,242,000
Belgium	2,518,000	2,441,000	—
France	2,520,000	2,650,000	2,685,000
United Kingdom	1,808,000	1,784,000	—
Austria	349,000	400,000	438,000

It is probable that the statistics for 1913 will show a distinct advance in British manufacture and consumption, as coal users in this country are beginning to realise the advantages of the blocks, many having used them for the first time during the national coal strike.

Financial Advantage of Fuel Manufacture

The economic advantage of converting small coal into fuel blocks may be gathered from the following particulars regarding the sale prices of small coal and of blocks. In 1910 the prices of small steam coal f.o.b. at Cardiff were : Ordinary 7s. 4d., Seconds 8s. Best 8s. 8d. Taking 8s. as the average, the price obtainable for 100 tons would be £40. A batch of 100 tons of fuel would be made up of 92 tons of small coal and 8 tons of pitch ; and the average price of the latter in 1910 was about 35s. per ton. The total cost of materials in 100 tons of fuel, therefore, works out at £50 16s. The average selling price of patent fuel at Cardiff in 1910 was 15s. 3d., and the revenue from 100 tons would amount to £76 5s. This leaves a margin of £25 9s. The cost of making fuel, inclusive of labour, interest on capital, depreciation, etc., is said to vary between 1s. 6d. and 3s. per ton, or on an output of 100 tons about £11 5s. By deducting the cost of manufacture from the gross profit of £25 9s., we have a net profit on 100 tons of fuel of £14 4s. This is, of course, only a rough estimate, and must not be taken as strictly accurate. It serves, however, to show that the process of converting small coal into fuel can yield a profit of at least 2s. per ton ; and is thus a highly profitable one.

In its final Reports the Royal Commission on Coal Supplies summarises its conclusions and recommendations regarding the manufacture of briquettes in the following terms :—

" Hitherto this industry has been mainly confined to South Wales where the small coal made in the screening and in the transit of the best steam coal is mixed with 8 to 10 per cent. of pitch and converted into briquettes. Large quantities of similar small steam coal are exported to the Continent for the same purpose. Of the value of these briquettes as a fuel there is no doubt, and they are extensively purchased by the Royal Navy as a reserve stock in hot climates, where they are said to deteriorate less than Welsh coal.   In England and Scotland briquettes are seldom made, probably because there is a good market for small coal.   There is, however, every reason to anticipate that in the future they will be more largely used for steam and domestic purposes, and there appears to be a good field for the discovery of a suitable binding material, pitch, which is the chief binder used at present, being rather too smoky for domestic purposes, and also high in price.

" The evidence points to the conclusion that a suitable briquette plant, if well managed, should pay in connection with a colliery ; at present the briquette factories in this country are mostly situated at, or near, docks. Suggestions have been made that partial distillation, in addition to washing and cleaning, would give a much wider choice of material for the manufacture of first-class briquettes of good calorific value out of inferior coal."

The possibilities of increasing the value of coal by manufacturing it into patent fuel and coke, and by extracting its valuable oil constituents, are likely

to be more and more appreciated as the years roll
on.   The day may not be far distant, indeed, when
coal will be regarded not as a fuel in itself, but as
the raw material from which different kinds of fuel
may be manufactured.   The aim of the coal-owner
to-day is to obtain the coal in large lumps ; and it
is the general practice to pay the miners a lower rate
of wage for small coal.   In South Wales, indeed, the
miners receive an inclusive payment based only on
the large coal, so as to give them an incentive to
produce large.   If, however, coal comes to be regarded merely as material for the manufacture of
coke, patent fuel, etc., the form in which the coal
is mined will cease to matter, as the total output of
the mine will be crushed at the surface before it is
conveyed to the fuel factory or the coke oven.   The
growth in popularity of internal combustion engines,
also, must result in an increased demand for suitable
oils ; and there is a possibility that the need can be
supplied by extracting from coal in liquid form
much of the power that makes it now so valuable a
prime mover of industry.

Both the gaseous and liquid products of distillation, and even coke, the solid product, can be used
in internal combustion engines ; so that the intermediate stage of raising steam now necessary in
converting the energy of coal into motive power,
may, in time, become obsolete.

In this and the preceding chapter reference has
been made to the great importance of the preparation of coal for sale and for use, so as to reduce

waste to a minimum. A very considerable proportion of the coal seams of many of our mines is now wasted because of the difficulty of extracting the contents in the form of large lumps. The perfecting of methods by which small coal can be commercially utilised will enable the bulk of the coal now wasted to be successfully worked and marketed, thus husbanding the national resources.

The installation of screening and washing machinery at the collieries, and the treatment of small coal in coking ovens and fuel factories, must result in the greater prosperity of mining undertakings. This, at any rate, is the conclusion of the Royal Commission :—

" The evidence shows that seams which cannot be worked at a profit will in the future be rendered profitable by washing, sorting, coking, and briquetting the coal, or converting it into gas, and that no small coal need be left in the mine. It has been proved that large quantities of the best Welsh steam coal are left underground in the form of ' small ' solely because under present conditions it does not pay to bring it out. It appears that much of this ' small,' although it is frequently dirty, is of similar quality to that now made into briquettes in South Wales, and we look to washing and briquetting as one of the available methods by which such coal can be brought out and sold to advantage."

In the ten years which have elapsed since this evidence was given the price of small coal has risen,

and little is now left underground in South Wales
except in a few places where not enough water is
available for a washing plant.   In other coalfields,
however, there is much room for extending the
economical use of small coal.

CHAPTER XI

The Demand for Coal

I PROPOSE in this chapter to give some explanation of the economics of the mining industry, and of the coal market; or, more correctly—to indicate the special application to the coal trade of the general economic laws which characterise all industry and commerce. Foremost among them is the Law of Supply and Demand, which is not quite so simple in conception as is generally supposed. The point of central interest—the pivot of economic forces— is the price of coal; the price existing at any moment being that which creates an equilibrium between the conditions of demand and the conditions of supply.

Demand means not only the quantity of coal per annum which a given community buys at a given price; but at the same time the quantities which the community would buy if the price were higher, and if it were lower. Thus the Argentine Republic would be ready to buy, perhaps, 4,000,000 tons per annum of Welsh coal when the price is 30s. per ton, or 3,000,000 at 35s. per ton, or alternatively 6,000,000 tons if the price were only 20s. per ton.

There cannot be more than one price at one moment
in the same market ; and assuming a given condition of demand the market price is determined by
the supply.   Putting it in another way—if we could
vary the price at will, as by Act of Parliament, the
above quantities of coal would be bought and consumed annually by the people of the Argentine
Republic whenever the corresponding prices pre

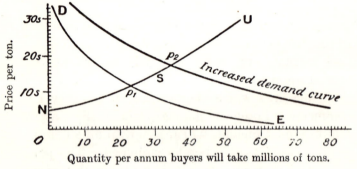

Quantity per annum buyers will take millions of tons.

FIG. 11.

vailed in the market.   The price might be fixed too
high, so that the supply forthcoming would be
greater than the quantity demanded at the price,
and then stocks would accumulate in the merchants'
hands.   If fixed too low, stocks would be rapidly
exhausted and many would-be buyers would have
to go without.

The condition of demand is a state of people's
wants, and involves the various different quantities
they would buy at different prices.   There is *no
change of demand* simply because more coal is bought
at a lower price.

Assume, however, that people's want of coal actually changes : more factories are built, and railways are extended—the great wheat farms of the Argentine are perhaps adopting steam ploughs and steam tractors, and coal is beginning to replace wood as the domestic fuel, owing to exhaustion of forests in certain regions. All these are causes which would make people *want more coal at the same price*, whether that price were high or low. This is a true increase of demand (or *condition of increased demand*, it is perhaps best to say) as distinct from a mere change in the *quantity demanded*, which may depend solely on price, and not on a change of the wants or habits of the people.[1]

It is important to note that the condition of demand in any market at any time does not depend only on what uses people have for coal, and how much they want it, but also upon their ability to pay for it. A family living in straitened circumstances may burn but one ton of coal throughout the winter ; but if their income suddenly becomes doubled, they will probably burn at least half a ton of coal more each year, though they remain in the same house. This *ability to pay* is responsible, with other causes, for a good deal of the increase of

[1] The conditions of demand and supply at any moment can be represented by curves as in Fig. 11, where heights above the base line represent price, and distances from left to right the quantities per annum which buyers will take or sellers supply in a particular market. The demand curve DE crosses the supply curve NSU at P and the height of P1 above the base line indicates the market price at the time. If the condition of demand increases as shown the price will rise to P2, though the supply conditions are unaltered.

the price of coal in recent years, which we all deplore.

One rule worth remembering is that fluctuations of the price of coal taking place in periods ranging from a few weeks to a few years are generally due to changes of demand, rather than of supply, excepting such as are due to strikes. This is obvious when we consider that the price of coal is a result of the balance of a certain condition of demand with a certain condition of supply ; and that, apart from strikes, there is no way in which the condition of supply of coal can easily change, as we shall see by a fuller consideration of the economics of mining.

Conditions of Supply

Supply must be considered in the same way as demand, not merely as the particular quantity which would be supplied at any particular price ; but as a series of different quantities of coal which coal-owners would be ready to put on the market at a series of different prices. Every mine manager knows that there are certain portions of the mine, perhaps a whole seam, which do not pay to work when the price of coal falls below a certain figure. If we imagine the price of coal to fall to successively lower levels there would be one part after another of the mine closed down, until it might be that at a certain low price the whole mine would become unprofitable. On the other hand, if coal rose to successively higher prices, new parts of the mine would be opened up, the output of the mine as a

whole would increase ; and seams too thin to be
worked at lower prices would be opened up. Thus,
there is a considerable range of total output from
existing mines with changes of prices. If a marked
change of price occurs in a short period of time there
is no perceptible change in the condition of supply ;
and a curve such as N C in Fig. 12 illustrates this
case well, showing the way in which the price must

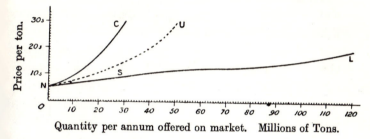

Fig. 12.

rise in order that the output from existing mines
shall increase.

In a longer period of time, however, it is, of course,
possible to sink new mines, opening up measures
which have previously been considered too expensive
to work in proportion to the value of the coal likely
to be won. To sink and properly equip a new pit
will occupy from four to five years. We find, however, that after a considerable rise in the price of coal
due to a trade boom, such as that now (1914) proceeding, the sinking of a large number of pits is
commenced with a view to taking advantage of the
prevailing high price ; and until these new pits are

in a position to supply coal in marketable quantities
there is no real increase of the supply, but only a
prospective increase. When these new pits are in
full working order there will be much more coal put
on the market at given prices than there was before,
and particularly at the higher prices ; so that the
curve of supply will take the form shown by the
second curve, N S U.

There are a great many causes which may influence
the position of the curve of supply, which is the
graphic illustration of what I have called the condition of supply. For example, an increase of the
cost of working, due to a rise of wages, or an increase
in the price of pit wood, or legislation requiring
more costly appliances for safety, all tend to reduce
the condition of supply, because a higher price is
needed to induce colliery owners to put the same
quantity of coal upon the market.

Balance of Demand and Supply

We are now in a position to understand that the
price of coal is determined by a balance of the conditions of demand and supply—always assuming that
there is free competition amongst buyers and sellers,
and no combination to withhold orders or to restrict
output. The balance comes about automatically
through the necessity felt by both producers and
consumers not to allow stock to accumulate at any
given time. There is only one price at which the
quantity demanded in a given market is equal to
the quantity which will be supplied at the same price

in other words, the price must at any time adjust
itself so that the quantity of coal raised may be
sold, and at the same time buyers may get all that
they can take at that price.   If they are not getting
all that they want, they will offer slightly higher
prices until the enhanced price calls forth a quantity
equal to their wants.   This postulates no change
in the condition of supply.

A change in the condition of supply is a secondary
matter following usually as a result of a change of
demand.   Thus, if the demand increases rapidly for
some reason, so that under the prevailing condition
of supply the price rises higher in order to make the
quantity supplied equal to the quantity demanded,
we have a condition of supply which is unstable ;
because, after a period during which new mines
can be sunk, the quantity demanded can be supplied
at a lower price.   In Fig. 12 are drawn two curves of
supply, the steepest (N C) showing how the price
must rise to increase, in a short time, the quantity
supplied, and the flattest curve (N L) showing the line
upon which the prices would be required ultimately
to call forth supplies of the stated quantities,
assuming the necessary period of several years to
elapse for new mines to be opened and developed.

The sum total of the various factors which go to
determine the condition of supply at any time are
very numerous and complex ; but most of them
are summed up in that convenient though rather
vague term, *cost of production*.   In the long period
the cost of production tends to change owing to the

investment of fresh capital in sinking new pits ; to the invention of new machinery, or adoption of new methods of working ; and to changes in the supply of labour due to the growth, or comparative dwindling, of the mining population. One of the most important factors in the increase of supply over a long period is the number of persons ready to take the risk of mining enterprise, and the amount of free capital they command.

Opening the Mine

The manner in which the price of coal is determined, so far as it depends on supply, can be understood only by studying the economics of mining operations. Every industrial enterprise receives its initiation from some man of wide knowledge and ability who plans out the whole undertaking and satisfies himself that the conditions are present for making the venture a commercial success. Such a man, unless possessed of a large capital, must be a man of the highest character and credit, possessing the confidence of business men and of the investing public.

The distribution of the risk of an industrial undertaking is most varied in different cases. Sometimes the promoter simply borrows capital at a fixed rate of interest upon the best security he can offer. He is then taking practically all the risk himself. In many cases other persons having capital to invest associate themselves with him, and share the risk of the success or failure of the undertaking by becoming shareholders with him in a partnership or joint stock company.   As the risk of loss is often considerable, it is now customary to adopt the legal form of association provided by the limited liability company, the advantage of which is that the shareholder cannot be called upon to pay the debts of the company beyond the sum which he agreed to subscribe to the share capital.   Many profitable undertakings beneficial to the community have been delayed many years in their initiation, because no one could be found to undertake the risk of loss of capital necessarily involved in starting them.   It is not generally understood how important is this function of risk-taking, and how much the community benefits thereby.   The Italians are very largely the labourers of the United States, and the contrast of wealth between the United States and Italy is partly the result of the fact that in America there are many men who will take big risks with their money whilst in Italy there are comparatively few.

Risk-taking is, however, but one of the five functions which are exercised in the production of wealth.   Let us discover the others by following out the commencement of a coal mine.

When plans for the undertaking have been fully prepared the next step is to raise such *capital* as may be required.   The sinking of a shaft, the development of the underground roads and ventilation passages, and the equipment with most expensive hauling and winding machinery, not to speak of a multitude of accessories, all require the provision of a large capital.

The owners of all the capital invested in the company as shares or debentures, whether it belongs to the promoters themselves or to members of the investing public, are exercising the economic function of the capitalist—which is to provide the capital for the necessary expenditure in opening the mine and properly equipping it with machinery. The ordinary shareholders are, of course, taking a considerable proportion of the risk, whilst the debenture holders are taking little risk, and practically their only function is that of lending capital.

There is some risk of confusion of thought between the use of the word *capital* in the sense of the economist, and the sense in which it is used by the business man. To the latter any sum of money, either in the form of money, or in a form immediately convertible into money, is termed capital ; because it is capable of being applied by the purchase of the proper goods to the commencement of any commercial or industrial enterprise. To the economist, on the other hand, money is but one form of capital, the most important being *fixed capital*, which is the term applied to the machinery, buildings and other works employed in producing anything. What happens is, that free capital becomes *fixed*, as it were, by being used for paying wages of workmen for their support, and in purchasing materials, the price of which also goes largely in paying wages. Such *fixed capital* can be used in the industry so long as it resists wear and tear, and so long as it retains its commercial value.

The last observation raises a very important consideration : that in many industries capital is lost, not by the machinery or buildings becoming worn out, but by their becoming commercially useless. This may take place either by the demand for the commodity produced falling off so that, with a lower price, it no longer pays to make it ; or it may result from the invention and extensive use of improved machinery or appliances which will do the same work at less cost. There is not much fear that the demand for coal will fall off, as it is shown in the chapter on oil fuel that oil can never replace coal to any great extent ; but there is some risk of loss of capital from the second cause. New mines are constantly being sunk, and they are so planned as to reduce the cost of working much below that of the older collieries ; besides being equipped with far more powerful and economical winding, pumping, and ventilating machinery. The large scale upon which they are worked is also sufficient in itself to give them advantage over their smaller and older rivals. It may, therefore, well happen that many mines upon which a great amount of capital was expended years ago may become unprofitable, and have to be closed down before the coal is worked out, the whole of the capital invested in them being thus lost.

Land is, of course, a necessary factor of production ; for, apart from the right to work coal, many acres of

the surface are required for the various pithead
works, sidings, offices, etc. Wayleaves for laying
railway and tram lines across land of other owners
have also to be negotiated. When the necessary
land is leased, the landowner becomes, along with
the capitalist and the enterpriser, one of the agents
of production ; but if land be bought there is no
economic difference—the firm or company merely
becomes its own landowner.

The function of *management* is extremely important ; and men have to be found who will make all
the arrangements, and exercise a continuous supervision. The function is exercised by an organised
body of salaried servants : the directors or partners
who are in supreme control, the colliery agent, the
engineer, the pit-managers and undermanagers,
and the various grades of overmen, firemen and
clerks.

Closely following on the organisation of the
management comes the engagement of workmen,
or manual labourers in their different grades, for
the manifold tasks which have to be performed.
Enginemen, stokers, pumpmen, banksmen, carpenters, surface labourers, repairers, and as development of the mine proceeds, an increasing number of
hewers, hauliers, etc.

Five *factors of production* have now been noticed,
namely :—enterprise (the taking of risk), capital,
land, management and labour ; and all of these five
must of necessity be associated in every industrial
undertaking, though two or more of their functions

may be undertaken or controlled by one man, or one body of persons.

In the case of coal mining the securing of the mineral lease, or purchase of the mineral rights of a certain area, or of certain seams, must be regarded as equivalent to the operation of purchasing raw material, which is characteristic of most industries. In working under a mining lease the colliery proprietor pays the royalty owner 1s. per ton, or some other figure, as each ton is raised from the pit, so that he is able to buy the raw coal in the ground, and sell the worked coal at the same time, and so avoid locking up idle capital. He is really buying raw coal from time to time under a long contract.

The sinking of shafts and opening up of new mines often takes several years ; and it is frequently forgotten how serious is the loss of interest arising from the slowness with which this work is sometimes carried out. Of course it is inevitable that a good deal of interest should be lost during the period of construction of any great enterprise ; but it is often profitable to use a somewhat more expensive method in cost per unit of work done if the more expensive method is at the same time much more rapid. This can be easily proved from a simple example, taking the case of sinking a shaft to a depth of 700 yards. Let us suppose that the sinking costs £20 per yard, if the whole depth is accomplished in 140 weeks ; and £20 10s. per yard if accomplished in 70 weeks. Assuming that 6 per cent. per annum had to be paid upon the money borrowed to carry on the sinking, or that at least that rate would be earned by the capital invested when the mine is completed (so that at least 6 per cent. would be paid as dividend) and reckoning at compound interest on the outlay, we find that by the method costing £20 per yard, the total cost will be £15,107 and by the method costing £20 10s. per yard, the total cost will be only £14,936.   The higher the rate of interest, the greater is the advantage in using the more rapid method.

The Pit at Work

We may now examine a coal mine at work, and analyse the revenue and expenditure of a normal company or firm, with a view to understanding the economic laws which determine the rates of rent, wages, interest, profits, etc.

The revenue of the colliery is derived almost exclusively from the sale of coal of various sorts and qualities, which may have been washed, sorted or screened ; and from the sale of bye-products, coke and anything else which the colliery manufactures. Obviously the amount of the revenue depends upon the price at which the coal and other products are sold, and upon the volume of business, or quantity sold per annum.   A falling off in price may be made up by an increase of quantity sold ; and the revenue is greatest when high prices coincide with a large volume of business.

The expenditure of our representative company includes :—(1) Royalties, usually so much per ton of coal extracted, which is like the purchase of a raw material, except that the price is fixed by the lease for a long term of years ; (2) The salaries of managers, overmen, firemen, clerks, etc., that is to say, of all those engaged in the management of the mine ; (3) Wages paid to the hewers, hauliers, and all other manual workers below and above ground, reckoning as part of the wages the value of any coal, house accommodation, or other commodity, given as part payment to the workmen. (4) There is rent paid for the land on the surface, and for wayleaves above and below ground, and (5) the interest paid on mortgages or mortgage debentures or to preference shareholders. The remainder is (6) what may be broadly called profit. This use of the term profit is not quite the same as that of the business man, because the interest or dividend paid is generally included as profit. From the scientific point of view, however, it is desirable to distinguish any payment at a fixed rate for the use of capital, whether it be on the basis of a loan, or by cumulative preference shares, from any fluctuating residue which may remain after meeting the whole cost of production and all fixed charges (including interest). It is this residue, which usually goes to the ordinary shareholders and holders of deferred shares (if any), which is the remuneration of risk-taking, or the result of special advantages, and is properly called profit in the scientific sense.

Profits

In every industry profits tend to adjust themselves to such a level as will, over an average of several years, equal that which enterprisers controlling capital could obtain in other industries. Let us see what this means in relation to the coal trade.

Industry is created by the men who are possessed of capital, and are willing to risk it by investing it more or less under their control in a manner (usually in ordinary shares) in which they take the bulk of the risk. A few wealthy men may have between them one million pounds, which they are willing to risk in starting some industrial enterprise. They are quite able to equip a big cotton mill, build a huge motor works, construct a light railway, or use their money and their business skill in a hundred different enterprises. They may decide to use their money and their skill in opening a large coal mine, partly because they happen to be well acquainted with the business and technique of coal mining ; but they will not do so unless they are satisfied that the opening of a coal mine in the best locality available to them is the most profitable use of their capital and their business knowledge combined which they are able to make. It might be true that one million pounds expended in rubber planting in Africa, or mining for radium, would bring them great profit, if they had the necessary technical knowledge to venture upon one of such undertakings ; but the decisive consideration with them is to make what appears to them the most profitable use of their money, mental labour and experience with the least degree of risk of loss, and the risk is least by venturing in an industry they thoroughly understand.

Profits may easily remain high in an industry, the technique of which is little understood by men controlling large amounts of capital ; but sooner or later, capitalistic-enterprisers recognise the opportunity, and enter the industry. The competition of new factories or mines soon becomes felt ; and there is a reduction of the market price at which the products can be sold. This reduction of selling prices continues until the industry is no longer especially attractive for the investment of more capital. Profits are then said to have reached their normal level, and the normal price that has been reached tends to persist so long as there is no change in any of the factors determining it. It is easy to think of many examples during recent years ; bicycles were very profitable to manufacture as soon as they came into general use, but now there are so many makers that the profits are no greater than in older industries such as cotton spinning or woollen manufacture. Incandescent mantles, metal filament electric lamps, the manufacture of chocolates and cocoa, are a few amongst the industries which are going through the process of reducing profits towards the normal level.

In coal mining the problem is slightly different from that of other industries, mainly because of the great bulkiness or weight of coal in relation to its value. This leads to the special profits due to the advantages of locality or special privileges, such as wayleaves, being much larger than is common in other industries. From the scientific point of view, we may say that the profits of coal mining, which are usually lumped together into a single dividend, consist of incomes derived from at least four different sources, being returns from different economic services or advantages :—(1) Interest on capital represented by ordinary shares ; (2) the ordinary average return for risking capital in the coal mining industry ; (3) the profits of special advantages in locality, contracts, etc. ; (4) the profits of unusually skilful management, being the excess of the produce of careful management above the salaries paid to secure such management.

The first two items together may yield dividends of 7 or 8 per cent. per annum, taking the average of 10 years. If more than that is paid, it may be assumed that there is some special advantage, such as proximity to port of shipment, or low royalty paid for a cheaply workable seam, or that there is a special profit arising from good management. One other cause of high dividends must be mentioned, however, though it is more a matter of accountancy than of business or economics : I refer to the practice of investing surplus profits which are not paid as dividends in the extension of the mine or its equipment so that the real capital invested comes largely to exceed the issued share capital.

Demand for the Factors of Production

The most interesting, as well as the most difficult part of economic theory is that which explains how wages, rents, interest, and profits are obtained. Each of these is a payment at a certain price for service rendered ; and such price is established by the conditions of the demand and supply for the factor of production in question. It makes no difference that the service is a continued one, and that the prices and rates of payment are so much per week or so much per annum. It is to be observed, however, that the price of such continued service is often fixed by a long contract which may even run to a great number of years—as for example a 99 years' lease. Employment of labour is based upon contracts varying from one hour up to several years in the case of salaries to officials. In the case of skilled workmen Trades Union agreements between owners and men often run for as long as five or seven years. It is most important to remember that, in every case where long contracts are adopted, it is a balance of supply and demand at the time the bargain is struck which determines the remuneration whether of labour or of capital during the whole period of the contract. The price determined will probably have been influenced by the fact that it referred to a long contract, and that the parties made certain estimates of the trend of the market. Often, however, such expectations are falsified ; and one of the parties stands to gain by the divergence of the current market price from the contract price.

We are now concerned with the various economic forces which determine the prevailing or market price at which contracts for labour, loan of capital, or use of land are entered into from time to time. There are two familiar, but most important, principles which are the immediate, as opposed to the ultimate, causes of the price paid for any service. The enterpriser says to himself : "How much is this labour or this land worth to me ? " And also he asks himself, "How much can I get it for ? " For, if the maximum limit of what it is worth to him is greater than the price which he has to pay for the service in question, he can make a profit by employing that service. "How much is it worth ? " is a question of demand ; "What will it cost ? " is a question of supply.

In the same market at the same time there cannot be more than one price for the same service ; consequently all employers must pay the same rate of wages for the same labour of equal skill ; and all must pay the same interest for the use of capital. This assumes, of course, a rather ideal condition of things : namely, that each purchaser is in touch with every possible seller, a condition which is scarcely ever realised in practice, though the Stock Exchange and Money Market approach it very nearly. In the labour market the organisation of Trade Unions and their standard wage rates have practically brought about the same result ;

but in the hiring of managers and other special services there are often opportunities of picking up bargains at a lower price than other people would have paid if they had known of the offer.

At any moment the manager of every colliery has before him the question of how many men he can advantageously employ ; and in general it is true that the higher the rate of wages which has to be paid to induce men to work the smaller is the number that can be profitably employed. This is evident when men are employed by time ; but is only true in an indirect manner when they are employed by piece work. For simplicity's sake we may at first assume a workman to be paid by time, and we can see afterwards how to translate our theory into terms of piece work. So complicated are all the economic adjustments of modern industry that the economist must of necessity proceed to eliminate many of the actual circumstances of the case whilst arriving at any conclusions. Once eliminated, however, and the conclusion reached, he must take care to replace the eliminated conditions before he applies his conclusions to give any advice on the affairs of practical life. It was by failing to do this that several economists of the nineteenth century made serious mistakes, and came to be distrusted by practical men.

In this discussion we must at first eliminate the differences of skill and endurance between the different men, and also assume the men to be willing workers who maintain a normal output with the present amount of supervision, these being the conditions necessary for payment by time. There are two reasons why the number of men that a colliery can employ depends upon the price at which their labour can be obtained ; and both combine in producing the same result : that the larger the number of men employed the less is the rate of wages that can be paid. This is a statement which is, generally speaking, true for all industries, and is called the *Law of Demand for Labour*. Let us see why it is true.

We may assume that we are dealing with the simplest class of labour, that of the hewers ; that they are paid a day wage, and that they are all of equal skill and can produce in a working place of average productiveness, three tons of coal per day. Different working places differ a good deal in productiveness so that in some the men could produce five tons a day with the same exertion and the same skill and energy as would be required to produce one ton per day from another place, as is fully explained in the chapter dealing with abnormal places (Chap. XIX.). Sometimes a whole district of the colliery is less productive than another district. The result is that if the men are paid 7s. 6d. per day each, it costs 2s. 6d. per ton on the average to produce the coal ; in some places it will cost 7s. 6d. per ton, whilst in the most productive places it will cost only 1s. 6d. per ton.

Suppose that the coal is being sold on the average at 15s. per ton. The difference between what is

paid to the hewer and the selling price represents
what remains to pay all the other expenses of the
colliery, such as haulage, repairs and maintenance of
the pits, winding of coal up the shaft, ventilation,
and salaries of managers, foremen, etc. Some
interesting figures have been published of the
percentage of the different classes of expenditure
of a colliery ; and though they vary considerably
the following may probably be taken as typical :—

TABLE showing approximately the principal items
in the cost of production of coal at pit's mouth
at two Welsh steam coal collieries during years of
high and of low wages.

	Year of low wages, 1897. Per cent.	Year of high wages, 1900. Per cent.
Labour	73.86	77·36
Stores and Materials ..	15·54	13.94
Royalties	6·64	5·69
Rents, Rates, etc. ..	2·65	1·92
Incidentals [1]	1·31	1·09
	100·00	100·00

NOTE.—The items of cost in this table do not include management,
but in the case of one of the two colliery companies represented
in this table the management expenses amounted in 1900 to about
4 per cent. of the total cost of production. The items given
include, however, interest on debentures.

The larger is the less productive area of the coal
mine which is worked the higher is the average cost

[1] Includes charges in respect of Workmen's Compensation Act
and workmen's coal.

of hewer's labour in working the coal. But this is not exactly how the colliery manager looks at the matter. He says : " I am getting a good volume of production from certain districts and places now being worked. Will it pay the company to work also such and such a district ? or such and such a place ? " He considers the wage he will have to pay each additional man he will have to engage, and the additional output which these men will produce. It is a fact, therefore, that in many mines there are working places, sometimes a whole district, not profitable enough to be worked at all.

If, however, labour could be obtained at a lower rate, say 7s. per day, several working places, perhaps a whole district, would become just profitable enough to work, assuming the price of coal remained unchanged. When the current rate of wage fell to 7s. per day, therefore, there would be a demand by this colliery company for the employment of more men. The same would be true of almost all other colliery companies ; so that if, in a certain coalfield, 100,000 men could be profitably employed when the rate of wages was 7s. 6d. per day, a total of 110,000 might be employed if the rate of wages fell to 7s. per day.

It may be thought, at first sight, that what a colliery owner would do would be to fill up the working places which could profitably be worked with men at 7s. 6d. per day and then to advertise that he could employ men but offer only 7s. per day ; and if he could get them, of course it would be worth his while to do this. Here, however, he would be acting contrary to the most fundamental of Trade Union rules : that all men doing the same work in the same locality must be paid the same rate of wages. Therefore, if any hewers are to be employed at 7s., it can only be by the wages of the whole number employed being reduced to this figure. The employer, of course, saves sixpence per day on the wages of all the men who could have been profitably employed at 7s. 6d. per day, so this rule may, at first sight, appear disadvantageous to the workmen. In fact, however, experience has shown that the rule tends to keep up the general level of wages. The employer is not likely to employ any of the men at 7s. 6d. if he can get others to replace them at 7s., unless they are specially skilled. Attempts are sometimes made to put men on to the less productive work at lower rates of wages ; but this always tends to undermine the wages of equally skilled men in more productive work so that it is strenuously resisted by Trade Unions. Many good examples could be quoted from the building trades, printing, etc.

We may take it then that the lower the rate of wages per day falls the larger is the number of places in a colliery that could be worked, and the larger the number of men that would be employed ; which is one reason for the truth of the Law of Demand for Labour. A colliery is really very similar to agricultural land in regard to variation of productiveness, and the law of diminishing returns applies to the different parts of a coal seam just as it does to land of differing fertility, which is taken into cultivation whenever the cost of working it decreases sufficiently in relation to the money value of the produce.

We now come to the other causes creating the law of demand, namely, the price at which coal can be sold. The consumer's demand for coal, as already explained, is such that the lower the price at which it can be bought the larger the amount that will be purchased. The converse, looking at it from the colliery owner's point of view, is naturally also true; namely, that the larger the amount of coal that he attempts to sell, the lower will be the price that he has, generally speaking, to accept for it. Of course, individual colliery owners may make lucky contracts for sale which enable them to dispose of all their possible output at a high and remunerative price; but it is certainly true of nearly all colliery proprietors that, apart from any improvement of trade, they must accept somewhat lower prices for an increase of output. Consequently we see that the law of demand for labour also follows the law of the consumer's demand for coal. If more men are put on to work they produce more coal, which must be sold at a lower price, so that whatever they produce in coal is less productive in money.

We have assumed so far that the price of coal remains fixed. Let us see what interesting results follow when it varies. Suppose the price of coal rises, then working places which were not quite productive enough to be worth working at the former price become profitable to work. Thus it often happens that when trade improves, and the price of coal rises, not only are a few abandoned working places taken into use again but a whole district of a colliery may be re-opened. Indeed, it has sometimes happened that when the price of coal fell very low the whole of a colliery fell below the margin of profitable working, but it was re-opened again when the price of coal rose.

Supply of Labour

We have now examined pretty thoroughly the demand for labour and may turn to consider what is meant by the supply of labour and how it may vary, still dealing with the unreal conditions of assuming men all to be of equal skill and to be paid a time wage. The supply of labour simply means the different numbers of men who would be available to work at a given place, at a given time, assuming alternatively a series of different rates of wages. Thus, if a man has opened a colliery in a somewhat isolated position and there is not lodging accommodation, men will not be keen to go there. He might find that by offering 7s. a day he could only get ten men, whereas by offering 7s. 6d. per day 30 men would be available. At 8s. per day he could have 150 men, at 8s. 6d. per day 500, at 9s. per day 1000 men, and so on. According to the equipment of his colliery and the productiveness of his coal seams, and according to the prevailing price of coal, he will have a certain demand for labour, and there must be a certain rate of wage which will just supply the number of men whom he could employ at that wage. Take any wage you like and increase the figure and you get more men available, but at the same time you diminish the number of men which the colliery can profitably employ. Hence at some figure there will be an exact balance between demand and supply.

The number of men who would offer themselves for work at a given colliery at different rates of wages depends upon a great variety of circumstances. In the first place there is a demand for the same kind of labour at other collieries in the neighbourhood. If for any reason their demand increases, so that they can offer higher wages, the supply at our particular colliery falls off; and a higher wage rate must be offered in order to keep a sufficient number of men. In so far as labour is mobile and can move easily from one colliery to another, an increase of demand at a few points in a coalfield—as where new large collieries are being developed—draws upon the existing supply of labour for the whole coalfield and leaves the other part with a reduced local supply.

There is, however, another way in which economists are accustomed to explain the balance of supply and demand, which leaves out of account the intermediate process of adjustment. The total demand for labour on the coalfield is the series of numbers of men who would be employed at various

rates of wages ; and the total supply is the series
of numbers of men who would be available at any
given time at different rates of wages. Obviously
if the development of a few collieries increases their
demand, it also increases the demand for labour on
the coalfield as a whole and the balance with supply
can only take place at a higher wage.

When mention was made in the foregoing para-
graphs of changes of supply and demand, the
intention was to regard these as taking place within
a short period of time, say a few months or a year
or two.   During such a period the resultant change
of the market price of labour (assuming it is settled
entirely without Trade Union or Conciliation Board
machinery) would be considerable in relation to the
number of additional men employed or dismissed.
In other words the supply of labour is inelastic.
If we take into consideration, however, a longer
period of time, there is opportunity for the supply
of labour gradually to adjust itself to the demand.
In this country, generally speaking, we are most
accustomed to an increase of demand, sometimes
very rapid, as when new colliery developments are
accompanying trade expansion.   There must then
be a rise of the rate of wages, which draws a few
more men in to the pits who perhaps have had more
or less experience of colliery work, but have not
been working underground, as until the wages were
raised they could do better in some other employment.   In course of time persons are attracted from
all manner of other employments, and a larger

number of boys are sent by their parents to work in the pits, because of the high wages which, it is learned, are now being paid in the coal trade. In South Wales, for example, large numbers of agricultural labourers from South and Mid Wales, also from Somerset and Devon, have flocked to the mines during recent years. Large numbers of men of the most miscellaneous previous employments, mostly from Bristol, Gloucester, and the Midland towns, have drifted to South Wales, as a district of high wages, and have gone underground either immediately or after trying two or three other trades. In the new pits, and in older pits where the price-lists are low relatively to the conditions of work, one finds a most extraordinary assortment of men of different occupations. Working with a few men who have been miners from their boyhood there is, perhaps, one who has just been a baker's van-driver, and one who was a grocer's assistant. Beside him is a late traveller for a firm of paint manufacturers ; and asking another you will find he is a market gardener who has fallen on evil days. Yet again, there is a man brought up in a saddlery business of his father's, which after many years' struggle came to an end ; and working with him is an intelligent young labourer who came from a Devonshire farm to the tin mines of Cornwall and then thought he would try his luck in South Wales ; a postman from Pembrokeshire, and a railway signalman dismissed from service owing to a mistake which derailed a few trucks. All these,

and many like them, have found in South Wales
their El Dorado with steady work at high wages, if
none too pleasant and somewhat dangerous. The
same kind of thing is going on in South Yorkshire,
and in every growing coalfield. It takes two years
for a man to learn the hewing of coal sufficiently
to be given the charge of a working place. In
five years any intelligent man can become a
thoroughly skilled hewer, knowing all the peculiar
ways of the coal and the roof and able to get his coal
down and into the trams in the quickest possible way.

If the demand, after increasing, remained steady,
the supply would gradually increase over a long
period, and there would be a permanent readjustment though at a somewhat higher wage rate than
before the increase of demand. On the other hand
there may be a falling off of demand for labour in
the coalfield, due either to a long continued depression of trade or to exhaustion of some of the best
seams of coal in certain of the collieries, or possibly
to the growth of German or American competition,
if the coalfield depends mainly upon export. When
a demand falls off, men are dismissed from the
collieries and these reduce the wages they pay. The
fact that there are a number of unemployed miners
about makes it possible, by reason of the competition, to reduce wages considerably, and then a readjustment takes place over a long period of time,
men gradually leaving the coal trade and finding
other occupations, whilst fewer boys are brought up
in the trade.

Piece-Work and Differences of Skill

In the above examination of the economic forces determining the wages in the coalfield it has been assumed throughout that the workmen were being paid time wages by the day. We must now see what differences are introduced by the actual method of payment of the hewers, and a few other grades of workers, by piece-work.

Payment by piece-work is adopted mainly on account of the difficulties of supervision in a coal-mine, and the differences of skill of different hewers; and also payment by the piece is always a stimulus to workers to keep up the volume of production. Most good workers prefer, indeed, to feel that they are being paid in proportion to the amount of work accomplished.*

In the mind of the workman the important point is what price per ton he is paid for cutting coal and whether the working conditions are favourable or not. This is the equivalent to him of a certain day-wage rate, if we look at the matter from the point of view of the foregoing analysis. He says to himself, " Where I am working now on a tonnage rate of 1s. 10d. plus percentage, it is an easy seam to work and there is good clearance of trams; and over the last three months I seem to have averaged about 7s. 6d. per shift. Is it worth while throwing

---

[1] Some of the reasons for payment by piece-work are more fully dealt with in the early part of Chapter XIV on " Methods of Paying Wages."

this up, and trying the pit over in the next valley where I hear they have a number of places open for experienced hewers ? The cutting price there is 2s. 1d., but I am told there is a lot of dirt in the coal. It won't be worth while moving unless I can get a good place and so be sure of 8s. 3d. per shift at least." Another man may be quite content to work for 7s. a day, whilst yet another is accustomed to make 9s. per day.

As is explained in Chapter XIV, there is a great deal of other work than hewing coal to be done in the pit ; and whereas in the North of England there is a deal of division of labour and the hewer does practically nothing else but hew coal, in South Wales he does a good deal of so-called " dead " work, such as putting up timbers to support the roof, laying rails to bring his trams to the face, ripping the roof or building a wall where necessary. Such work is also paid by the piece, and the rates paid for that will combine with the cutting price of coal in determining what each man can earn.

By paying a piece rate with no minimum wage, the coalowner transferred to the miner most of the risks and loss of working in unremunerative places and parts of the mine. Hence it was not a question of the colliery owner deciding that certain places were not worth working because with the existing price of coal and daily wage the amount of coal produced gave no profit. As long as men could be found to take the places which yielded little coal they would be worked, because

the cost of working to the colliery was the same per
ton whether the men got much or little coal.   It
was, therefore, a question of the supply of labour.
If many men were seeking work then they would be
put into poor places and left to earn what they could,
even if only 15s. or 20s. a week.   If the supply of
labour decreased there would be no men willing to
take such places, and the colliery manager had perforce to leave them unworked.   As explained in
a subsequent chapter, colliery proprietors would
sometimes give a " consideration " or allowance
extra to the piece rate to men working in abnormally
poor places.   In most cases this was only given when
it was profitable to work the coal at a somewhat
higher cost than the piece rate only ; or when some
special purpose connected with the haulage or
ventilation of the colliery rendered it desirable to
have that particular part worked out, even though
the result in coal production might be poor.

Under this system of payment by piece rate, the
local demand and supply are adjusted by the level
of prices relative to work necessary as determined
in the colliery price list ; and any temporary fluctuation of local supply or demand is made up by
altering the additional allowances.

The general adjustment of supply and demand
throughout the coalfield is obtained by altering the
percentage addition to the standard piece and time
rates.   This affects all the workmen in the mines
throughout the coalfield simultaneously, as is
explained in Chapters XIV and XV.

CHAPTER XII

Coal Contracts

No less complicated and technical than the business of mining coal and converting it on the pit top into its most marketable form is the system of arranging for its sale and transport. The method usually followed is that of sale by contract, the coalowner or merchant binding himself to sell and the buyer binding himself to accept a quantity previously agreed upon at an agreed price, the delivery to be spread over a specified period. These periods vary in length from two to three months upwards, but twelve months is the usual period in the case of the larger contracts entered into by the principal colliery companies, though in exceptional instances periods as long as three years have been covered. The period agreed upon for the contract depends, however, to some extent upon the state of the market during the months of September to November, when most of the contracts are made. Obviously, when trade is depressed and the price of coal is apparently at its lowest, buyers press for a long contract, which the sellers invariably refuse, believing that the chances of a rise in the market are all in their favour. On the other hand, when there has been a great rise of price, buyers are naturally fearful of committing themselves for long. At the very bottom and the very top of the market contracts are, therefore, usually short, that is from three to six months—at the former time from the exigencies of the producer, at the top of the market from the caution of the buyer. Contracts run into longer periods when, immediately after the period of depression, demand and prices have improved and trade conditions at home and throughout the world's markets indicate a steady average price level.

Contracts are, as a rule, made only after much bargaining ; and the process entails great business skill and acumen on the part of the parties concerned. So many factors enter into the situation that the commercial agents have to be possessed of very extensive knowledge regarding not only the commodity with which they are dealing, but also of various industrial and political conditions. Whether the price is a profitable one is not always the only consideration for colliery companies. If the market is slack, rather than entail the loss incurred by closing a mine, the coalowners may, in order to guarantee a minimum amount of work, accept a price which is only just sufficient to meet the cost of production. In some cases indeed, contracts have been arranged at prices which show an actual loss, though a loss less serious than if the mines were closed.

Big industries, such as collieries, cannot be worked advantageously on day-to-day sales, which would probably mean irregular working, as coal cannot be stored without great expense. What the colliery proprietor wants is to deliver his coal regularly to the docks or elsewhere in a steady stream of so many tons per day, and to have all of this sold beforehand under contracts at remunerative prices. If the colliery relied entirely upon day-to-day sales the labour of selling would not only be greatly increased, but he would be obliged at frequent times when ships are scarce, perhaps through storms, to sell his coal at a ridiculously low figure to any one who could store it, or else stop his colliery. In contracts, the purchaser is usually made responsible for finding the ships to take the coal in equal monthly instalments of the whole contract.

Newspaper Quotations

The proportion which the coal sold under contract bears to the total output of a colliery varies in different coalfields, and even in the same coalfield, though it usually exceeds 50 per cent. The lowest percentage prevails in the Midland and Yorkshire coalfields, where much of the business consists of small orders for industrial purposes and domestic consumption. In South Wales, on the other hand, where the mines cater very largely for a foreign demand, the proportion sold on contract ranges from 60 to 75 per cent., and, at times, even 85 per cent., of the total output. The margin of coal

not sold under contract is disposed of in lots of
200 or 300 tons up to 4,000 or 5,000 tons for prompt
delivery to meet current requirements, such as
coals for ships' bunkers, special or mixed cargoes,
or making up shortage in contract delivery by other
collieries due to a local strike or other cause.

The prices quoted from day to day in the columns
of newspapers are those at which the " free " coal is
sold. It is not usual for " day to day " or " newspaper " prices to agree exactly with those at which
the larger proportion of coal is being sold under
contract. Sometimes, if the purchasers have mis-
calculated market tendencies, and a sudden fall in
prices occurs, the newspaper prices are below those
on the basis of which contracts have been made.
This is not usual, however, for as a rule current prices
for spot, or prompt, delivery of small or large parcels
are higher than contract prices. This circumstance
explains why it often happens that the quarterly
averages of prices for all coal exported from Newcastle, or from the Bristol Channel ports, differ very
much from the current market prices during the
same periods. Let us take, for example, the case
of the best Welsh Admiralty steam coals, which in
bulk exceed half of the total coal exports of the
Bristol Channel ports. During 1907 the current
market prices varied as widely as from 16s. to 20s. 6d.
per ton f.o.b. but, owing to contracts having been
made at lower prices in the autumn of 1906, the
average selling-price during 1907 of all coal shipped
did not exceed 14s. 6d. Further, it ought to be

stated that newspaper quotations, useful as they are, can give only an approximate representation of the current prices at which parcels of coal are being sold, as there is no system of notification of prices at which dealings are made, and the newspaper reporter relies on such information as he can pick up " on 'Change."

Conditions of Sale

Having considered the factors determining the duration of contracts, and the prices at which the coals are sold, let us now discuss the general conditions on which contracts are made. As a rule coals intended for home consumption are sold at pit-mouth prices. The cost of transportation to the point of consumption is then borne by the purchaser. The terms regulating coal for shipment, however, are different. Coal intended for use as bunkers is usually sold f.o.b. (free on board) ; that is, it is delivered free of charge past the tip chute and into the ship's hull. The " trimming " of the coal, that is to say, distributing it evenly from the chute over the ship's bottom, is charged to the vessel. When the coal is bought for export by foreign buyers the price is usually quoted either c.i.f., or f.o.b. By c.i.f. is meant that cost, insurance and freight are covered by the contract price, and the coal is, therefore, delivered at the seller's cost to the port of discharge, the unloading being paid for by the buyer. When sold f.o.t. (i.e., free on truck) the coal is delivered at the cost of the seller on the trucks at the port of destination. On what basis the price is arranged is generally determined by the business of the buyer.   If a foreign railway or manufacturing company buys directly from this country, its manager does not want to be bothered with all the technical business of arranging the freight, insurance and discharge of the vessel, and so he buys f.o.t., or at least c.i.f., which is popular with some of the foreign railways. On the other hand foreign buyers are often coal merchants, and they make it their business to know as much about freights as the English exporter, and generally prefer buying f.o.b., as do also the Admiralty and the principal steamship lines. The greater portion of all Welsh coal shipped is sold on the f.o.b. basis, as nearly all the larger customers prefer it. The Germans have lately been very severe competitors of British coal merchants in the c.i.f. business in all markets where German coal is suitable ; and with a splendid selling organisation they are likely to secure a great deal of our trade.[1]

The following are facsimiles of Sale Notes (f.o.b. and c.i.f.) ordinarily used :[2]

1. *Form of F.O.B. Sale Note*

" We confirm sale to you of about .... tons of............ coals..................screened ready to load in........................ Dock (Cardiff, Barry, Newport) about ....

---

[1] This chapter was written before the European War. It is possible that German competition will now be much less severe for some years.

[2] For a form of contract for the Sale of Welsh Coal see Appendix 4.

Price.........................per ton of 20 cwt. Payment in cash at 30 days from date of shipment, less 2½ per cent. discount, or by your acceptance of our draft of 30 days' date, less discount as above. Trimming to us. Buyers to pay export tax (if any), and wharfage as customary. Subject to usual exceptions for strikes, lock-outs, accidents, etc., including force majeure, preventing or delaying deliveries, and such deliveries to be proportionately extended as customary. None of above coal to be sold by purchasers to any other person or persons in Great Britain under a penalty of five shillings per ton on any quantity so sold or shipped, as liquidated damages. Sellers have the right of suspending delivery or cancelling sale, so long as payment for any delivery is in arrear, or in any case of purchasers being bankrupt or making any acknowledgment that they are unable to pay their debts in full."

2. Form of C.I.F. Sale Note

Cardiff,

....

" We confirm sale to ....

    Quantity ....

    Quality   ....

    Price   ....

per ton of 20 cwts. C.I.F. ....i.e., free on board

........................., but including freight, wharfage, export tax (if any) and insurance to ....

    Payment :—The coals, coke and fuel, to be paid for on bill of lading, quantity as ascertained by dock or railway, or fuel company, at port of loading, without deduction ....

    Shipment :—By ....

    Delivery :—To be taken at the rate of ....tons per day as per charter.

    This sale is made upon the conditions of the Chamber of Shipping Welsh Charter, 1896

    Colliery certificates to be conclusive proof of quality, quantity, and description of coal, coke, or fuel, buyers having the right of inspecting cargo during shipment.

In case of mixed cargoes, cost of mats, separation and levelling to be paid for by buyers.

Insurance effected free from war risk. If buyers require war risks covered, same to be at their expense, and they to advise sellers accordingly.

Sellers not to be held responsible for any loss, damage, or delay to cargoes caused by strikes, lock-outs, disputes, force majeure, epidemics, combination of officers, engineers, crew, dock labourers, stevedores, lightermen, or any hands or agencies, connected with the loading or working of the ship, or supply of the coals, coke or fuel.

This contract is subject to the usual exceptions of strikes, lock-outs, riots, accidents, epidemics, dismissal and usual or unusual stoppages of all descriptions, at colliery or collieries, factory or factories, from which the above coals, coke, or fuel, are to be drawn, and on or at railways or docks, or of vessel's crew, or dock hands, including force majeure, preventing or delaying production, deliveries or chartering, and such time of deliveries to be proportionately extended as customary, except in case of a general strike of associated collieries in South Wales, when sellers shall have the option of cancelling this contract. In case of European war or epidemic, at port or ports of shipment and of delivery, or imposition of quarantine on vessels from such port or ports, sellers shall have the option of cancelling this contract.

It is also agreed that sellers shall have the right of suspending deliveries or of cancelling balance of contract, if payment for any delivery is in arrear, or in case of purchasers being bankrupt or making any acknowledgment that they are unable to pay their debts in full. Any vessel being chartered and advised, and which is expected due to load within the stipulated time of shipment of this contract, shall be accepted in full execution of the same."

Methods of Payment

There is no uniform method of payment adopted in the various coal-producing districts. In the Bristol Channel ports, the f.o.b. terms are fixed by

the Chamber of Shipping's Welsh form of charter
of 1896; they provide for payment in 30 days
net, the former practice of giving $2\frac{1}{2}$ per cent.
discount for payment in 30 days having recently
been abandoned. The Scottish conditions are net
cash in 30 days, less 1d. per ton discount when payment is made in 7 days. In the coal ports of the
north-east coast payment is required to be made
net in 14 days after shipment or on the second Friday
after the day of shipment; and a discount of 1 per
cent. is allowed for cash with order. As regards
c.i.f. and f.o.t. contracts, payment may be made
within 30 days of shipment, 10 days after delivery,
three months' acceptances, or 3 months' open credit,
or by cash against documents. The terms vary
with the business status of the foreign customer,
and with the economic or political condition of the
country to which the coal is being shipped. New
or uncertain customers, or customers in countries
where there is great political unrest, are usually dealt
with only on a cash basis. During war periods cash
transactions are the rule.

Other systems of payment varying from those
mentioned above are also sometimes used. Cash
against invoice is one method; cash with order
is another; both being adopted either in order to
get additional discount, or because the seller is
doubtful of the buyer's credit. The particular business methods of the firms engaged in the transactions sometimes require the use of other systems of
payment. One firm, for example, pays for its con-

tract supplies by a two months' draft dated on the first of the month following the deliveries. In the Midlands the usual system in the domestic coal trade is to pay net cash on the tenth of the month following delivery.

Various Brands of Coal

The coals produced in the United Kingdom vary much in quality, and each kind has its own special designation, which in some instances is registered as a trade mark. In South Wales, and generally in other coalfields, the names are usually those of the localities from which each particular brand is now, or was first, obtained. The following list gives the names of the chief kinds, and in brackets the chief ports at which they are supplied :—

South Wales Coalfield

STEAM.—Cardiff Smokeless (Admiralty List) dry and other steams ; Monmouthshire Black Vein and Western Valleys, semi-bituminous. (Cardiff, Newport, Swansea and Port Talbot.)

BITUMINOUS.—No. 2 Rhonddas ; No. 3 Rhonddas ; Eastern Valleys (Monmouthshire) ; and "through" bunker coals. (Cardiff, Newport, Swansea and Port Talbot.)

HOUSE COALS.—Ffaldau ; International ; North's Navigation ; Monmouthshire Red Ash.

ANTHRACITE.—Stanllyd hand-picked malting, and other malting qualities ; Red Vein, French and German nuts ; Beans ; Peas ; Duff (slack coal) ; (Swansea.)

Great Northern Coalfield

STEAM.—Davisons, Cowpen, Bothal (Blyth) ; Bentinck (Blyth), Howards, Bowers, Buddles, Ravensworth, East Hartley, Hastings, West Hartley, Main, and Carr's (Tyne) ; South Hetton (Sunderland) ; Lambton, Hetton (Wear).

GAS COALS.—New Pelton, Holmside (Tyne) ; Waldridge, Washington Pelaw Main, Deans Primrose (Tyne) ; Thornley (East Hartlepool) ; Wearmouth (Wear) ; Londonderry (Seaham).

COKING COALS.—Weardale, Pease's West, Priestman's, Dunston, Garesfield, Towneley, Tanfield Moor, Burnhope, West Stanley, Redhough, Walbottle, and Consett (Dunston).

Yorkshire Coalfield

HOUSE COALS.—Haigh Moore, Wallsend, Silkstone, best Barnsley softs.

STEAM COALS.—South Yorkshire, best hards; West Riding Hartleys.

GAS COALS.—Screened Silkstone, rough through and through.

Midlands and Forest of Dean

HOUSE COALS.—Cannock Chase (Midlands) ; Block (Forest of Dean).

STEAM COALS.—Nottingham shale, and Dealy steams.

Scotland

Hartley, Hamilton Ell, Splint, Main, and Steam (Glasgow) ; Navigation steam and other coals

(Methil) ; Jewel, Hartley, steam and other qualities (Leith).

Admiralty Coal

The steam coals are those of the greatest value, the lead being held by the famous Welsh smokeless class. The following is a list of the coals approved and regularly purchased by the Admiralty for use by the British Fleet :—

Albion Merthyr.
Burnyeat's Navigation.
Cambrian Navigation.
Cory's Merthyr.
Cyfarthfa.
Dowlais Cardiff.
Dowlais Merthyr.
Ferndale.
Graham's Navigation.
Great Western Naviga-
tion.
Harris' Deep Navigation.
Hill's Plymouth Merthyr.
Hood's Merthyr.
Imperial Navigation.
Insole's Cymmer.

Insole's Merthyr.
Lewis' Merthyr.
Locket's Merthyr.
McLaren Merthyr.
National Merthyr.
Naval Merthyr.
New Tredegar.
Nixon's Navigation.
Ocean Merthyr.
Oriental Merthyr.
Penrikyber.
Powell Duffryn.
Rhymney Merthyr.
Standard Merthyr.
Tynybedw.
Ynisfaio Merthyr.

The British Admiralty also buys much coal both in South Wales, and in the North for use at its depôts and dockyards. In addition to the above standard qualities, much use is now made of smalls which, after grading and washing, find a ready sale. The proportion of small to large is yearly increasing. Through the introduction of elaborate machinery for grading and washing, the small coal is now some-

times made more valuable than the large coal produced at the same colliery.[1]

Commercial Organisation of Sale of Coal

Let us now consider in detail the commercial
organisation necessary for the proper disposal of the
coal ; and the various classes of persons engaged in
arranging for its sale and transport. As a rule
in the home trade, the large consumers purchase
their supplies direct from the colliery at pit-mouth
prices ; small consumers derive their parcels from
the coal merchants and retailers. Amongst the
large purchasers might be mentioned railway companies, steamship lines, and large factories, works
and hotels.

A much more complicated machinery prevails in
the case of the export trade. The collieries sell their
supplies either directly to consumers, or indirectly
through the agency of middlemen. Most of the
large colliery companies now do their own selling
through salaried salesmen stationed at the different
markets, who sell either to the shippers, or direct to
the foreign consumers. Others have acquired an
interest in shipping firms ; and the production and
export of the coal is centralised under one control.
The smaller owners, however, transact their business
through mercantile firms who do the work on a commission basis.

Amongst sales which are arranged directly between
coalowner and consumer might be mentioned those
to the Admiralty and the big steamship lines, rail-

[1] See Chapter IX, " Preparation of Coal for the Market."

way companies, etc. As a rule the colliery companies tender their prices directly—sometimes for small parcels, and sometimes for the delivery of large lots spread over long periods. Usually contracts are placed with the lowest tenderers. In the case of the British Admiralty and the Egyptian State Railways it is considered good policy, however, to distribute the contracts over a number of selected collieries producing a good quality of coal. For this purpose the British Admiralty has prepared, and revises occasionally, the list of approved coals quoted above ; and the inclusion of any name on the list is regarded as evidence of the superior character of the coal produced at that colliery. The distribution of contracts is necessary, also, because no single colliery is able to fulfil an order of one or two million tons. Contracts for large or small quantities, entered into as a result of tendering, generally stipulate for f.o.b. prices ; in some cases, however, the price covers delivery into ships' bunkers. The superintendence of the shipping of coal sold directly under contract is usually undertaken by the colliery agent ; but sometimes by local firms at the port of shipment who specialise in the chartering of tonnage and in loading.

Coal Shipping Agents

Although large, the volume of trade carried on directly with consumers is small as compared with the trade done through the medium of merchants or middlemen, mostly on a c.i.f. basis. The risks

connected with freights and foreign liabilities have
deterred colliery companies from undertaking the
shipping as well as the production of coal ; and for
this reason most of our foreign coal trade passes
through the hands of an enterprising class of merchants who have no pecuniary interest in collieries
unless as an investment, but are prepared to
undertake all risk connected with the sale and transport of the coal. This class of business is, of course,
more or less speculative in character. Limited
liability companies, private partnership firms and
individuals, undertake the business. Some of the
firms of higher status have a regular *clientèle* and
act mainly for the same people. On account of
their old-established connection with large bunkering
depôts at home and abroad, they have no difficulty
in disposing of the coal which they have contracted
to purchase. The big business done by such
merchant firms, and the reputation for commercial
soundness they have acquired, gives them a great
advantage over competitors in obtaining favourable
tonnage. The term " legitimate middlemen " is
sometimes applied to them ; and the name indicates
to some extent the settled and non-speculative
character of their operations.

Speculative Middlemen

Many other firms, however, carry on a more
precarious business. The old-standing merchant
firms can usually command certain sales for their
purchases, and they are content to accept a moderate

U

profit on their transactions.   The speculative middlemen, however, aim at buying low and selling high,
and in order to obtain a wide margin of profit they
willingly undertake risks which are avoided by the
merchants.   These speculators generally have a
world-wide knowledge of the state of the markets,
and base their operations on reports received from
correspondents in various parts of the world.   If a
big demand is anticipated later, they seek, by some
artificial means, such as withholding orders, to
depress the prices ; and when low water-mark is
reached they buy in order to sell again at a considerable profit to foreign or, sometimes, local
consumers.   The conditions of the British coal
trade do not allow of markets being actually cor-
nered ; but the tactics which lead to the formation
of corners in wheat, cotton, and other commodities
abroad are the tactics which guide the coal speculators in their operations at the various coal
exchanges.   As a rule these men taboo the markets
when prices of odd parcels are high, as is the case
when an unusually large volume of the output has
been covered by long contracts, and during such
periods content themselves with buying and selling
small quantities only at current prices on day to
day risks.   Their big operations are reserved for
periods of slump, when they sell forward and short,
calculating on buying later at a lower price.   The
practices of these speculators are disliked by coalowners in South Wales, and efforts have been made
to restrict their activities by the insertion in contract

papers of clauses forbidding re-sales.* An example
of such a clause is contained in the form of f.o.b.
Sale Note given above (p. 297). Some clauses
prohibit re-sales in certain foreign ports as well as
at home ; but such clauses have not been successful
as there are many ways of evading them, and only
very rarely are the specified penalties enforced.

In addition to legitimate middlemen and speculators there are also other operators who confine
themselves to buying and re-selling small lots, only
rarely, however, for shipment.

The Mixing of Coals

Many of the best grades of Cardiff coal are obtained by mixing different kinds and qualities in
definite proportions, as it is found by experience
that the disadvantages of certain coals are corrected
by the opposite qualities of others, and that it is not
only possible to improve the best coals by mixing
but also practicable to use cheap coals so as to
produce a result as good as any obtained from
higher priced coals. The practice of mixing coals
destined for export is rapidly growing, and it is
causing some embarrassment in shipping the coal—
at any rate in South Wales ports. The method
generally adopted is for truck-loads of the different
coals to be successively tipped into the ship's hold,
one or two trucks of each sort at a time, two, three,
or even four sorts being mixed. Trouble is occasioned at the docks in marshalling the various

* See Appendix 4, p. 809.

mineral trains, and in sorting out the empty trucks to be returned to their respective collieries.

Mixing is, of course, legitimate business when mixed coal is ordered, but there is also fraudulent mixing. To counteract advances in prices or freight rates some of the speculators have adopted the practice of mixing larger proportions of low-priced coals with the better qualities so that the standard has been lowered. Foreign customers, in order to protect themselves, now often require the middlemen to produce written guarantees of quality from the colliery companies whose coals they purport to be selling, or they appoint different agents to look after their interests at the shipping ports.

Coal Shipping Charters

After arrangements for the purchase and sale of coal have been concluded its transport to its destination must be provided for. The chartering clerk now comes on the scene, and acting under instructions from his clients, he arranges terms between the shipowner or shipbroker and the merchant. Colliery companies doing a large business employ their own chartering clerks; but most of the chartering clerks are employed by shipowners and shipbrokers. They wait on the shipowners each morning, then ascertain on the Exchange what tonnage is required and report to their principals. They have very many considerations to take into account. The freight-rate is, of course, the decisive factor if other conditions are suitable, such as the size of the steamer relatively to the cargo and the ship's " position," by which is meant its present location on a voyage and whether it will be in port at just the date at which the merchants are able to supply the coal. The shipowner must also consider the proposed destination in relation to the closing of the Baltic and other northern ports in winter, and in regard to the profitableness of the probable return cargo. The shipowner selects the most favourable of suitable offers, and then gives his chartering clerk orders to " fix " his boat. Freights vary from week to week, and even from day to day, and are influenced by variations in supply and demand. The charters or contract forms under which coal is shipped are very complicated ; but there are standard forms adopted by the Chamber of Shipping and generally used. They vary according to the " market " in which the port of destination lies.

Markets of the World

The following may be regarded as the general classification of the principal foreign ports to which British coal is shipped :—

Baltic, Arctic, and North Seas (Ports north of Hamburg).

Coasting (United Kingdom, Channel Islands and North French and North Sea ports, Brest to Hamburg inclusive).

Bay Ports (Brest to Bilbao).

Spanish and Portuguese (Atlantic ports).

Near Mediterranean (Malta and west of it, including Algerian ports).

Upper Mediterranean (All ports east of Malta).

North-west Africa, Islands, West Indies, and North American Atlantic ports.

South Africa.

Red Sea, Persian Gulf, East Africa, and Indian ports to Singapore.

Far East ports (East of Singapore).

South American (Atlantic) ports.

North and South American Pacific ports.

Many shipowners specialise in particular markets, and the merchant's clerk knows the most likely owner or broker to apply to for any tonnage he may be wanting. As business is usually transacted with great rapidity and a local owner will hold his offer open, perhaps, for only half-an-hour or so, non-resident owners and their local brokers are somewhat at a disadvantage, owing to the time taken in telegraphing or telephoning for fresh authority to " fix " a particular boat. Vessels are chartered both for " spot " or " prompt " positions (i.e. at the docks) or for forward shipment. The commercial business of selling coal and chartering boats is becoming more and more concentrated at a central point for each coalfield. The business of Cardiff is growing at the expense of Barry, Newport and Swansea. Although they ship at other ports, firms are removing their headquarters to Cardiff owing to the general convenience of the Exchange.

Trimming

Trimming is the operation of removing the coal from the base of the chute and distributing it evenly throughout the ship's hold. It is a task requiring considerable physical strength and endurance and some degree of skill, so that trimmers are always intelligent and highly paid workmen. The trimmers are employed in South Wales usually by the shippers, that is, the merchants or agents selling the coal; but in the north of England trimmers are sometimes employed by the Harbour Commissioners or other public authority, or in the case of large colliery companies which do their own shipping, by the colliery company. Trimmers are always paid by the piece on the group-contract system, upon a complicated tonnage-rate scale which varies considerably for different types of ships. Trimming charges are, indeed, rather more complicated even than colliery price-lists.

The trimmers always work in gangs controlled by a foreman who is one of themselves, and with whom the shipper settles the payment according to the agreed scale. The Coal Trimmers' Union is really a close corporation very highly organised and quite capable of holding up any port or series of ports, if necessary. In some ports it is said that men wait years in the hope of gaining admittance to the Union, and are only successful if they have the influence of friends. So close is the monopoly that trimmers certainly earn much more than competitive wages. In cases where the sum agreed for trimming the ship is favourable (for the scale does not work evenly with all ships), each trimmer of the gang may take home £5 as a result of 20 hours work. When a gang takes a ship in hand they continue until loading is completed, except that all work stops from 2 p.m. on Saturday till 6 a.m. on Monday, except in very urgent cases. At Cardiff, the trimmers in a gang sleep and eat at the dock as best they can, working in turns until they have finished a ship ; and the loading may take anything from 18 to 60 hours. The trimmers employ assistants called " hobblers," to whom, however, they only pay labourers' wages at 6d. to 7½d. an hour, whilst the work is irregular. As trade unionists, the trimmers often take a good deal of interest and part in the local government of their town, but they are not generally politicians or socialists.

Costs of Loading Coal

It may be of interest to show what it costs to get coal transported from the colliery and loaded on board ship. The charges for various services at Cardiff, which are a fair average example, are as follows :—

		Cardiff charges per ton.
1.	Railway rate, colliery to port	10d. to 1s. 6d.
2.	Wharfage	2d.
3.	Mixing (usual mixings) ..	1½d. to 2½d.
4.	Weighing	¼d.
5.	Tipping	2d.
6.	Trimming (according to type of vessel and quantity) ..	1½d. to 8d.

Of the above charges all those from the railway carriage down to tipping are included in the f.o.b. price and are paid by the shipper, but the trimming is usually paid for by the shipowner out of the freight rate.

CHAPTER XIII

Forms of Trade Combinations

A FEATURE of industrial organisation during recent years has been the tendency towards the combination of undertakings with a view to securing certain advantages to the owners. These combinations may be classified broadly into two divisions :—

(1) Those in which each firm retains the management of its own undertakings but in which an arrangement is made for regulating output, preventing competition, and pooling profits. The Unmarked Bar Association is a good example of this form in the United Kingdom, and the Wire Nail Association in the United States.

(2) Those in which the management of all sections of a business is centralised either in a "trust," or by amalgamation, or by means of a "holding corporation," which has a controlling interest in all the separate companies. Examples of this form are the Bradford Dyers' Association, in which no fewer than 35 firms were amalgamated, and in America the Standard Oil Company and the United States Steel Corporation.

314

" *The Limitation of the Vend* "

In this country the principle of combination has not been so widely applied in the coal trade as in other industries. An arrangement known as the " Limitation of the Vend " was made for regulating the price and output of collieries in the Newcastle district as far back as the seventeenth century, and the precedent thus created has been followed during later periods. The Newcastle policy was in active operation from about 1771 to 1844, and in 1830 the Town Clerk of Newcastle described the method of working to a committee of the House of Commons in the following terms :—

" The proprietors of the best coals are called upon to name the price at which they intend to sell their coals for the succeeding twelve months. According to the price, the remaining proprietors fix their prices. This being accomplished each colliery is requested to send in a statement of the different sorts of coal they raise, and of the powers of the colliery, that is, the quantity that each particular colliery could raise at full work ; and upon these statements the committee, assuming an imaginary basis, fix the relative proportions as to quantity between all the collieries, which proportions are observed, whatever quantity the market may demand. The committees then meet once a month, and according to the probable demand of the ensuing month they issue so much per 1000 to the different collieries.[1]"

[1] Report of the Select Committee on the State of the Coal Trade, 1830.

In case the actual demand was greater than that anticipated, the "issue" or portion of the total vend allotted to each colliery was increased in the same ratio all round. The effect of this arrangement was to maintain a high price level, and so save from extinction many mines which had little or no competitive power and would otherwise have to close down. The agreement worked well in so far as it limited the output. The success of its operation, however, in the direction of inflating profits led to a desire on the part of many of the interested companies to increase their own allotment. This they did by acquiring large royalties and by increasing the capacity of their pits or opening new ones. With an increased capacity basis came increased output and lower prices, and finally on account of the disagreement which arose between large and small mine-owners the arrangement was abandoned after seventy years' working, and the committee dissolved. An interesting fact about this combination was that the restrictions imposed applied only to coal shipped for British consumption, it being found convenient to dump surplus output abroad at low prices. The consequence of this was that often coal was sold to foreign markets at 40 per cent. under the prices in the London market; "to such an extent was this carried that English coal was sometimes to be purchased in St. Petersburg at half the price of the same coal in the River Thames." *

* Royal Commission on Coal, 1870 ; Report, Vol III., p. 12.

Later attempts have been made by mine-owners to " limit the vend." In 1893, for example, such a combination was undertaken by the Lancashire owners and lasted for several months. Outside competition, however, caused it to be abandoned. In 1896, Mr. D. A. Thomas suggested the adoption of a similar scheme in South Wales, but the proposal did not mature. In 1904 was established the Durham Coal Sales Association. This body, however, proved unable to command the loyalty of its members, and after a year's working the Association came to an end.

We may contrast with the failure of pooling and sales associations in this country the extraordinary success of such an organisation in Germany, the Rhenish Westphalian Coal Syndicate, which is described further on.

It should be noted that in the home market for house coal there is a sales association or " ring " for nearly every district. That for London regulates the retail price of coal most effectively, and colliery proprietors are not permitted to sell to merchants outside the ring who might cut prices. This is a combination of merchants, however, and not of colliery owners.

Amalgamation of Colliery Undertakings

When we come to the various schemes which have been put into force for securing economies in production by the amalgamation of many small concerns into one large one, we have greater success to record. During recent years a large number of huge syndicates have been formed by combining together smaller units, and there are indications that the greater part of mining enterprise in the future will come to be carried on by corporations with practically unlimited capital at their disposal.

A comparatively early example of combination in the British coal trade was that brought about in 1896 by the acquisition by the firm of Sir James Joicey & Co., Ltd., of Lord Durham's collieries, steamers and plant in the county of Durham at a cost of one million pounds. This concern now owns 27 collieries with an annual output of between four and five million tons of coal, and a fleet of 50 steamships. Several amalgamations have also been effected in the Yorkshire area. The firm of Henry Briggs, Son & Co., Ltd., formed in 1865 to acquire the Whitwood Collieries near Normanton, has been considerably enlarged by the absorption of other undertakings and the opening of new mines. In 1896 were purchased the adjoining Spydale Collieries and the Newmarket Haigh Moor Colliery near Leeds. Horden Collieries, Ltd., again, is a concern that has made considerable growth. Formed in 1900 to acquire at a cost of £95,000, 16,000 acres of leasehold coal royalties in Durham, it purchased two further freehold areas of 1,280 and 138 acres, and has opened new collieries. Pease & Partners, Ltd., is another example of a big Yorkshire merge. Several large collieries were bought by this firm and the aggregate output of coal now exceeds $2\frac{1}{4}$ million tons.

Scottish Combinations

One of the most successful mining ventures of recent years is the Fife Coal Company. This company was first registered in 1878 to work the Kelty-Beach Colliery. Later were added the Hill of Beath, Dalkeath, Lever, Pirnie, and Wellsgreen collieries. Afterwards, in 1895, the company was reconstructed. In 1896, the Cowdenheath Coal Company's interest was purchased. Then followed in succession the following purchases : in 1900, Lockore and Capledrae Cannell Coal Co. ; in 1901, Fife and Kinross Coal Co., and Blacradam Coal Co. ; in 1906, Rosewell Gas Coal Co., Ltd. ; in 1908, Doni-bristle Colliery Company ; in 1909, Bowhill Coal Co. In addition the Fife Coal Co. has recently opened other collieries on its own account. It now owns or leases 21 mines and employs over 14,000 workers. Kelty Pit, the first mine opened, is now the largest mine in Scotland, and has an annual output of seventy thousand tons. More than 40 per cent. of the working population of Fife are employees of this huge concern. The total output of coal in 1908 and 1909 was 3,185,129 tons and 3,754,864 tons respectively. The authorised capital of the company is £1,234,075, and the average dividend paid on the ordinary shares during the past fifteen years was 24½ per cent. Another large Scottish concern is United Collieries, Limited. This was formed in 1898 to acquire the business of the following companies : Belhaven Estate Collieries ; Bredisholm

Collieries, Ltd. ; Dunn, etc., Glen, Ltd. ; Larkhall
& Fairholm Collieries ; Newhouse Colliery, Ltd. ;
Litehill Collieries, Ltd. ; Thornton, Peter, Fauld-
house, Whitelaw Gavinent, Wishaw. In 1902, the
company further acquired the collieries of 23
other firms, including James Wood, Ltd., Colen
Dunlop & Co., J. Nimms & Sons, Ltd., Clydeside
and Calderbank Collieries, W. Black & Sons, Ltd.,
Loganlea Coal Co., Ltd., Larkhall Collieries Ltd.,
Maruelrigg Coal Co. Ltd., etc. It was considerably
over-capitalized, through purchasing collieries at
inflated prices during a trade boom, and has been in
financial difficulties. During 1908, eight of the
mines were sold and five closed and dismantled.
The authorised capital of the concern is two million
pounds, and the annual output exceeds $2\frac{1}{2}$ million
tons. Several other large mining ventures exist
in Scotland. The Edinburgh Collieries Co., registered in 1900, bought out in 1907 the Forth
Collieries, Ltd., R. & J. Durie, Ltd., and also acquired
Messrs. Jas. Waldie and Sons' interest in the
Tranent Collieries. These properties comprise six
leases of coal-bearing land with an aggregate of
2,662 acres. Its mines are situated in East Lothian.
The authorised capital of the company is £650,000.

The Cambrian Combine

Numbers of large undertakings exist in South
Wales also, of which the most important is the
Cambrian Combine, formed on the initiative of Mr.
D. A. Thomas, a director of the Cambrian Colliery

Company. This company acquired a controlling interest in the adjoining properties of the Glamorgan Coal Co., Ltd., the Naval Colliery Co., Ltd., and Britannic Merthyr Coal Co., Ltd., but the control was maintained by forming a holding company— The Cambrian Trust, Ltd. Mr. D. A. Thomas was made chairman of each of the companies, and other directorships were held by Cambrian Trust nominees. Each of the firms retained its own identity and disposed of its output by its former agents, but the control of the policy was exercised by the Cambrian Board. In March, 1913, a movement was initiated to consolidate the interests of the shareholders in the four associated concerns, and a new company was formed known as the Consolidated Cambrian, Limited, with a share capital of two millions, of which £1,900,000 was to be issued in exchange for existing shares in the separate undertakings. The advantages anticipated from a still closer combination of interest were thus described in a circular issued to the shareholders of the four companies :—

" (1) The larger capital of the new company, and more general distribution of the shares, would give to any holder who desired to sell a wider and freer market than is now enjoyed by shareholders in the existing companies. This in itself would naturally add to the market value of the shares.

" (2) Greater facilities would be afforded for providing fresh capital for further extension of

X

the operations of the new company if this were at any time considered desirable.

" (3) The risks inherent to mining speculation would be distributed over a wider area; and, consequently, the return to capital would be more regular and less liable to fluctuation than in the case of any one of the individual companies standing alone."

The output capacity of the four companies is about $3\frac{1}{2}$ million tons. This is obtained from nineteen pits and three levels; and it is anticipated that in a very few years over four million tons will be produced annually. In addition to the above collieries the Combine is also interested in the coal-exporting and patent-fuel making firms of Gueret, Ltd.; Thomas & Davey; and Amaral, Sutherland & Co.; the French ship-owning firm of La Société Générale de Houilles et Agglomérés; the pit-wood importing concern of Lysberg, Ltd.; and the ship-owning firm of Pillson & Co., Ltd. The Cambrian Combine divides honours with Sir James Joicey, Ltd., for being the most complete self-contained organisation in the coal trade of the world.

Other South Wales Amalgamations.

The amalgamation movement in the Welsh coal-field still continues. Mr. D. A. Thomas has considerably extended his interests and has become chairman of the Fernhill Colliery at the top of the Rhondda Valley, and of the Cynon and Duffryn Rhondda Collieries in the Afan Valley. He is also

on the board of the Ebbw Vale Iron and Steel
Company, which is now mainly a colliery enterprise with a capital of nearly a million.   Another big
firm, Messrs. D. Davis & Son, Ltd., formed in 1890,
to purchase the Ferndale and Bodringallt Collieries,
acquired in 1894 the adjoining colliery undertaking
of A. Tylor & Co., Ltd., covering 1,830 acres, for
£298,000.   These properties comprise nine winding
pits in the Rhondda Valley.   In 1911 the company
acquired the whole of the share capital of the
Welsh Navigation Steam Coal Co. (owning 2,400
acres in the Ely Valley, with an output capacity of
500,000 tons per annum) at a cost of £314,850.   The
firm has also acquired full ownership of the Griffin
Nantyglo Colliery at Blaina.   The subscribed capital
of the amalgamation is £864,063.   The output of
the amalgamated collieries is about 3¾ million tons.
The dividends payable on ordinary shares for
several years have been at the rate of 10 per cent.
As a result of thirteen years' trading a profit of
£1,694,253 has been recorded.

Another large undertaking in South Wales is the
Powell Duffryn Steam Coal Company, Ltd., which
has a paid-up capital of £1,135,483 and controls an
output of 3½ million tons.   The properties of this
concern are situated in the Aberdare and Rhymney
valleys.   The total mineral area is 16,000 acres, of
which 1,070 acres are freehold as regards surface and
two-thirds of the underlying minerals.   The dividend paid during recent years has been at the rate
of 20 per cent. per annum, and liberal bonus distri-

butions of shares representing capitalised profits
have also been made.

Among other large concerns might be mentioned
the Ocean Coal Co., the Rhymney Iron and Coal Co.,
and the Tredegar Iron and Coal Co. The latter
concern has large mining interests, and a paid-up
capital of over one million and an annual output of
about 1,800,000 tons of coal. In addition to its
own six pits, it has a large interest in other important
companies. In the Oakdale Navigation Collieries,
Ltd., for example, it holds 199,992 paid up founders'
shares, and an additional 160,000 shares partly paid
in addition to loan capital. Much of its mineral
estate is as yet undeveloped.[1]

Attempted Formation of Anthracite Trust

In the anthracite coal trade of South Wales the
advantages of combination are well appreciated, but,
up to the present, attempts to amalgamate a number
of small concerns have not proved successful. An
important effort was made in 1903–4. The promoters hoped to arrange for the purchase by a proposed Anthracite Trust of twenty-eight collieries
and actually acquired options over a considerable
number. This proving abortive, a further attempt

[1] Since the above was written another big merge has been brought
about in South Wales. Two large concerns, the United National
Collieries Ltd., with a capital of £663,500 and Burnyeat, Brown &
Co., Ltd., with a capital of £300,000 have been amalgamated, the
former name being retained. The authorised share capital has been
increased to a million pounds. The combined output from the
various mines owned by the amalgamation is in normal times about
2,750,000 tons per annum.

was made in the following year, and an amalgamation on the following basis was suggested :—

(1) An independent valuation was to be made of all the properties proposed to be acquired.

(2) The interests of the various owners were to be acquired on the basis of this valuation.

(3) The promoters were to find all the capital necessary for the amalgamated concern.

The capital of the Trust was to be £1,600,000, and the aggregate area of the takings to be acquired was 15,662 acres with estimated resources amounting to $227\frac{1}{2}$ million tons. The total output at the time the proposal was made was 1,110,500 tons per annum. As, however, the parties interested could not agree, the project was ultimately abandoned. There is a special difficulty in valuing anthracite mines owing to the disturbed character of the seams,[1] and so the owner's estimate of his property is apt to differ widely from the purchaser's valuation. The industry was also, perhaps, hardly sufficiently developed, there being so many small mines whose owners could have found no employment with the trust. The South Wales anthracite field is, however, the only source of high quality anthracite in Europe, and the field not being extensive, there is a splendid opportunity for a combination to secure a monopoly and artificially maintain the price by restricting output. As a monopoly is certain to be established sooner or later, I strongly urge that an

[1] See Chapter XXIII.

experiment in the nationalisation of mines should be made by securing the anthracite coal-field for the nation and working it through a Board or Trust.

The Rhenish Westphalian Syndicate

Perhaps the most powerful mining combination in existence is the Rhenish Westphalian Coal Syndicate, the famous Kartell of Ruhr or Dortmund coalfield in Germany. This coalfield has a larger output than any single British field, and raises more than half of the output of the German Empire. Over 90 per cent. of the production of Westphalia is the result of the activities of the seventy or so companies comprised in the syndicate. The greater distance of the other coalfields of Germany from the seaboard contributes greatly to the success of the Westphalian arrangement. The Kartell was established after many abortive attempts in 1893. Ninety-nine owners of separate concerns agreed to sell the whole of their commercially disposable output to the company, which thus at its outset controlled 87 per cent. of the output of the field, much of the remainder being that retained by the producers for consumption in their own steelworks. The company has a nominal capital of £45,000, entirely subscribed by the contracting mine-owners ; its business is conducted in large offices at Essen by an executive committee of four, under the control of a Board of Directors with the assistance of an administrative and clerical staff numbering about 240 persons.

Organisation of the Kartell

The object of the Kartell is to regulate output and maintain a remunerative average price level. The specially elected Participation Committee referred to above assigns to each firm in the syndicate a "participation figure," which represents the maximum allowable output from the mines of that firm during a specified time, as, for example, 100,000 tons monthly. It is, in fact, following just the same practice as the Committee of the Vend at Newcastle. The constituent firms meet periodically in a General Assembly, and decide what percentage of the "participation figure" may be produced from each mine during a given period. Their decision as to how far short of the maximum the production during the next interval must be is dependent upon their anticipations of trade development and prosperity. The reductions of actual output below the "participation" figure have ranged from nil in 1900 to 23 per cent. in 1908. As a rule, however, the estimates are conservative ones, and in practice it usually happens that the actual output of each colliery is greater than that stipulated at the last previous meeting of the General Assembly. These relaxations are allowed by the Participation Committee when it is found that the demand for the coal at the Kartell price is greater than that allowed for by the Assembly. Firms exceeding their "participation figures" without consent are penalised, and those producing less are compensated, about one shilling per ton in each case.

Firms meeting in General Assembly have one vote for each 10,000 tons production.   The basis of voting power for representation on the Advisory Council which determines the prices to be paid for the coal from time to time is one million tons annual output. This gives the right to elect one representative. To secure that their interests are not ignored, the smaller firms combine to secure joint representatives on the same basis as the larger concerns.   The Advisory Council prepares a schedule of normal prices for the various qualities of coal, and this forms the basis on which the executive committee fixes the " accounting prices," that is, the prices credited by the syndicate to the mine-owners for their sales. The committee also determines the actual sale prices at which the coal is sold in a large part of Germany adjacent to the coalfield, the so-called non-competitive areas ; and in these areas the sale prices vary only to a slight extent from the accounting prices. One of the aims of the syndicate is to overcome the competition of English coal both in the German Empire and adjoining countries.   Where the English market is strong the Syndicate cuts its prices very low.   The syndicate adopts a cutting policy also when confronted with the competition of the German coalfields or of Westphalian firms who have declined to join.   In some areas, for example, it comes into conflict with a smaller combine in the Silesian district or with the Prussian Government mines in the Saar district.   At times also when the home demand is slack the syndicate " dumps " its surplus coal into other countries at a considerably lower price than that which is maintained at home. The syndicate recoups itself for the loss thus incurred by a levy upon the constituent firms in proportion to their " participation figures."

Production of the Westphalian Syndicate

The power of the syndicate has been threatened during recent years owing to the secession of some important members, the growth of outside competition and other causes, and in 1912 it was thought that the combination would come to an end. The following table giving the latest obtainable figures is useful, as showing the growth of non-syndicate production in the Ruhr district :—

Year.	Syndicate Production (1) Tons.	Output of Non-Syndicate mines (2) Tons.	Percentage of (2) to (1).
1904 .	67,255,901	1,204,845	1.79
1905 .	65,382,522	1,324,157	2·02
1906 .	76,631,431	1,649,214	2·15
1907 .	80,155,994	2,107,915	2·62
1908 .	81,920,537	3,046,908	3·71
1909 .	80,828,393	4,167,015	5·15
1910 .	78,159,834		
1911 .	86,904,550		
1912 .	93,798,000		
1913 .	101,652,000		

Success of Kartell Policy

The Kartell policy has met with considerable success ; and although unpopular with consumers, it is doubtful whether its maintenance of a monopoly price has been seriously detrimental to German industry.   During periods of boom no attempt has been made to exact the utmost price which could be obtained, while in times of depression the prices have not been allowed to fall to the lowest competitive level.   Its policy has been to keep prices steadier, and to avoid the great fluctuations of price which characterised the period prior to the inauguration of the Kartell.   Such stability is a great advantage, but it must be admitted that it has probably kept the average level of coal prices somewhat above the competitive level, for it would not otherwise be able to sell at such low prices in the competitive areas.   The syndicate was temporarily strengthened in 1911, when the Prussian Government, failing to influence prices to any considerable extent, arranged for the sale of the output of the Westphalian State mines to be effected by the syndicate.   The decision of the syndicate to raise prices in 1912, however, led to the termination of this arrangement.

CHAPTER XIV

METHODS OF PAYING WAGES

Conditions of Work

THE conditions of employment in coal mining are unlike those of any other industry, and the details of how wages are determined are extremely complicated and varied. As there are well over a million persons employed in coal mining in this country, however, it is desirable that the public should be better acquainted with the methods of remuneration of so important a section of the community, particularly in view of the agitation for the nationalisation of mines, which the miners are likely to bring into prominence before long.

The difficulty of the problem of how to find a satisfactory basis for remunerating miners for their labour, arises from the variability of human nature on the one hand and the variability of coal seams, and the floor, roof, and other conditions, on the other hand. We all know that some men are more skilled than others ; also that some will work harder than others, either because they are physically stronger or have greater courage, or because they are more in want of money. Again, most men are honest in their intentions, and will do their best when they have agreed to do so ; but some others, unfortunately, will not.

The coal seams are just as perplexing as man himself.   Not only are they liable to vary from place
to place in thickness, but also in the jointing, so
that in some places the coal comes down easily and
in others it is hard to work.   Sometimes the roof is
firm, making work easy ; in other places it is so
loose that the collier must be always on the watch
for a fall, and he often has much " dirt " to clear
away.   Again, a band of stone or " clod " may
occur just over the coal, or actually in it, and the
stone must be sorted with care from the coal.   Some
places are wet, and the man sometimes has to work
more or less in a pool of water, and with the water
trickling or dripping on him.   It is the innate differences of men, and such vagaries as these on the part of
the coal, which have caused friction between the miners
and coalowners, and have caused numerous strikes,
culminating in the great National Strike of 1912.

Workers in all trades are paid either for the time
they are employed, or by piece-work, that is to say
in accordance with the quantity of work done.
Thus, a bricklayer may be paid simply at the
rate of 10d. per hour irrespective of the amount of
work done ; or he may be paid 1s. 6d. per 100 bricks
laid irrespective of how long it takes to lay them.
In the latter case the bricklayer would want to be
paid more per hundred bricks if there were many
corners and windows in the wall than if he were
simply building a straight wall ; but if paid by
time wage, it makes no difference to him upon what
work he is engaged—he gets his wage until he is

dismissed. Although there is a good deal of elasticity about a time-wage, there is always a tacit assumption that the workman will perform a certain minimum of work for his wage ; and, provided the supervision is good, he will lose employment if he does not do this minimum.

In factories, or in building, supervision is not difficult, and the workmen are often paid by time ; but piece-work is generally preferred by the employer for simple operations which have become standardised. The workmen also often prefer it, as it gives each man remuneration in proportion to his efforts when the task for all is the same.

In coal mining, however, there are special difficulties for the workman in the conditions of his work, and for the foreman in maintaining supervision. All who work away from the bottom of the shaft and the main roadways are confined to the use of the very dim light given by the safety lamp. To inspect the face of coal, or the roof, it is necessary to raise the lamp. In many mines the men work constantly in the wet ; and if the seam is a thin one they hew the coal lying down, and for hours together cannot stand upright. There is also the constant watch to be maintained for accumulations of gas, and for falls of the roof or face. These are the ordinary difficulties of the miner ; but in abnormal places, and also with bad ventilation, or bad management of other kinds, these difficulties become very much accentuated. A foreman in charge of only fifty working places would have to walk over half a mile in inspecting them ; and in practice he could not visit them more than four or five times a day. There is, therefore, plenty of opportunity for the man who likes to sit down and look at his work to have his own way.

The result is that colliery proprietors have never considered the ordinary system of direct employment upon the time wage satisfactory ; and it is now rarely used, except in special cases. In former times the difficulty was got over by a system of sub-contracting, or what the miners call the " butty-gang " system, which it has always been one of the first objects of the miners' trade unions to abolish. The colliery owners used to make contracts with men of the class of foremen to pay them so much per ton for coal worked and sent to the surface from a given number of places. Such a gang boss would then hire 15 to 20 men on day wages ; and if he could keep them up to a rapid rate of work, he could make a very profitable business of working the coal by contract. His main object being to keep the men going continuously, working the coal as fast as possible, he would also work in hewing coal himself, and often with prodigious energy whilst at it, so as to set the pace. His success depended very largely on his capacity for " driving " his men. In every coalfield the system was objected to by the miners ; and by a series of strikes it was abolished in one part of the country after another about the middle of the last century, and now exists only in Staffordshire.

The system introduced to take its place has every-where been that of direct employment of the hewers by the colliery company on a piece-work basis. It is usual for each hewer to employ one assistant or a boy, either upon the basis of sharing the earnings, or more commonly, by paying him a day wage if two are employed together on the same shift. The wage of such an assistant is usually paid by the colliery office and deducted from the hewer's remuneration when on piece-work ; but in the case of a boy the hewer usually pays him himself. The other employees, such as roof-rippers, repairers, and men engaged in driving headings or cross-cuts, are paid upon the piece-work basis whenever possible ; and when no acceptable rate or basis of payment can be arranged they are paid by the day. All the employees connected with the haulage of the coal from the face to the shaft, the men and boys attending to the ventilation, as well as the firemen and overmen, banksmen and enginemen, and most of the surface workers, are in general paid a time wage ; the result is that about half the employees work on piece-work and the other half on time wage, the latter half being mainly the unskilled and low paid workers, with the exception of the enginemen, firemen and overmen.

Standard Day Wages

The payment by time is usually spoken of as working on day wage ; and the workman is paid so much per shift of $7\frac{1}{2}$, 8 or $8\frac{1}{2}$ hours, or of 10 hours for surface workers. If he is paid 4s. 6d. per shift and misses a day, he, of course, misses his 4s. 6d., and occasional absence is permitted, although the

usual period of notice is from a week to a month in
different districts.[1]

The day wage rates at which the men are engaged
are, like the piece rates, subject to increase or
decrease as settled by the Conciliation Board of the
district under an agreement between the Colliery
Owners' Association and the Miners' Federation for
the district. In this manner the wages are subject
to an increase during periods of prosperity, while
the price of coal is high, and are decreased again
when the price of coal falls or the volume of trade
diminishes. Thus, if the wage agreed to be paid to
a haulier is 3s. per day plus percentage, he will get
4s. per day when the percentage is 33⅓ per cent.,
4s. 6d. per day when it is 50 per cent., and 4s. 9½d.
per day when the percentage is at 60, which is now
the maximum under the agreement for South Wales
which expires, together with all the agreements
throughout the country not already terminated, at
the end of March, 1915.

The standard wage rates are not determined upon
the same basis all over the country, there being
three principal standards. The lowest and the
most widely adopted is that of 1879, and the next
is that of 1877, which is higher by such an amount,
that the equivalent percentages above the 1877
standard are 15 per cent. less than percentages
above the standard of 1879. Thus, if the percentage
is 55 per cent. above the standard of 1879, it will be
40 per cent. above the standard of 1877, and 60 per

---

[1] See Chapter **XX.** " Coal Mines—Minimum Wage Act."

cent. above the standard of 1879 is equal to 45 above that of 1877. There is also extensively used the standard of 1888, which prevails throughout the Federated districts of the Midlands. Changes in the percentage above standard are usually made every few months by the meeting of Conciliation Boards under the presidency of an independent chairman, and the following table shows how the percentages above standard have varied from year to year, since the system was adopted :—

PERCENTAGE CHANGES IN COAL HEWERS' WAGES IN THE FEDERATED DISTRICTS SINCE 1888.

Year.			Plus or Minus.	Per cent.	Total Percentage above Standard of 1888.
1888 ..	..	..	+	10	10
1889 ..	..	..	+	10	20
1890 ..	..	..	+	20	40
1890-1894	..	..	..	..	40
1894 ..	..	..	—	10	30
1894-1898	..	..	..	..	30
1898 ..	..	..	+	$2\frac{1}{2}$	$32\frac{1}{2}$
1899 ..	..	..	+	$7\frac{1}{2}$	40
1900 ..	..	..	+	10	50
1901 ..	..	..	+	10	60
1902 ..	..	..	—	10	50
1903 ..	..	..	—	5	45
1904 ..	..	..	—	5	40
1904-1907	..	..	..	..	40
1907 ..	..	..	+	15	55
1907-1909	..	..	..	..	55
1909 ..	..	..	—	5	50
1909-1912	..	..	..	..	50
1912 ..	..	..	+	5	55

The fluctuations are striking and are caused by alternate booms and depressions of trade which follow one another in accordance with the well-known cycle of trade.

Detailed information as to the workings of sliding scales and Conciliation Boards is given in the next chapter. It is sufficient for the purpose of this chapter to explain that all wages in collieries are standard wages, subject to increase by the current percentage ; but it may be well to utter a warning against a common misapprehension, which is fostered even by some of the Board of Trade Statistical publications, that the changes of percentages as indicated in the above table are a measure of the earnings of the miners. It will appear from what follows that the percentage above standard is only one of the many factors influencing the actual amount which a miner paid by the piece has the opportunity of earning. As a matter of fact, the average earnings of hewers have not risen by anything like so much as the percentage increase above the standard ; and, particularly in Lancashire and South Wales, since the Eight Hours Act came into operation, there has been an actual decrease of the money earned by many hewers as compared with some years ago, which, with the general rise of the cost of living, has created a good deal of hardship.

Price Lists

The payment of wages in proportion to the amount of coal won or of work done is accomplished

by means of a colliery price-list and by weighing the
coal which each man hews, and "measuring-up"
any other kind of work.

The workmen of a colliery who form themselves
usually into a "lodge," or branch of the Miners'
Federation of the coalfield, negotiate with the owner
of the colliery a series of piece rates or "prices,"
to be paid for work done. The principal item is the
"cutting price," or standard rate per ton paid for
"getting" coal from the face in a properly opened
stall or place. The "cutting price" may be anything from 1s. 1d. up to 3s. per ton, and is subject
to the addition of the percentage. If the latter is
60 per cent. and the "cutting price" of a particular
seam is 1s. 8d., and if the man sends up 2½ tons of
coal in one day, he earns 6s. 8d. Thus :—

$$
\begin{array}{lll}
 & \text{s.} & \text{d.} \\
\text{1s. 8d.} \times 2\tfrac{1}{2} & = 4 & 2 \\
\text{Add } \tfrac{60}{100} \times \text{4s. 2d.} & = 2 & 6 \\
\hline
 & 6 & 8
\end{array}
$$

The amount of coal worked by each man in charge
of a stall is ascertained by weighing his trams at
the pit head. He chalks his number on each tram
when filled before it is taken away by the "putters"
or "hauliers."

As it was unsatisfactory to the hewers to have to
depend upon the accuracy of the weighing and the
book records made by a clerk appointed by their
employers and carried out entirely in their absence,
they sought the right of appointing a representative

to watch the weighing and record the weights of coal sent up by each man. One of these workmen's representatives is now stationed at every pit where coal is wound, and they are called " checkweighers." Their rights, duties, and appointment are governed by an interesting series of legislative enactments dating from 1860. It was the Act of 1887 which finally defined the position of checkweighers as it exists to-day. The statutes provide for the workmen of a colliery electing a committee, and the checkweigher is also usually elected by a poll of the workmen, though he may be appointed by a committee. This statutory workmen's committee is in practice and nearly always consists of the same, or almost wholly of the same men as the committee of the lodge of Miners' Federation for the same colliery.[1]

A specimen of a price-list of a large South Wales Steam colliery is reproduced, and it is one under which the " colliers," as the hewers are called in South Wales, can earn good wages averaging about 8s. 5d. per shift per man whilst the percentage is at its maximum.

LIST OF PRICES

For working the Nine Feet Seam of Coal at a South Wales Colliery.

Coal Cutting, including the cleaning of the
    Brass from the coal   ..   ..   ..   1s. 8d. per ton of
                                                              large coal
Brush Coal ..   ..   9½d. per ton

---

[1] See Chapter **XVII**, " Miners' Trade Unions," p. **472**.

LIST OF PRICES—*Continued.*

Clod—Under 6 inches in thickness .. Nil
      When 6 inches in thickness .. 1d. per ton extra
      ,, 9   ,,   ,, ... 1½d. ,,   ,,
      ,, 12 ,,   ,, ... 2d.   ,,   ,,
      ,, 15 ,,   ,, ... 2½d. ,,   ,,
      ,, 18 ,,   ,, ... 3d.   ,,   ,,
      Above 18 inches in thickness to be
      paid by consideration.
Headings—Narrow ..   .. 4s. 6d. per yard
      Wide ..   .. 3s. per yard
      Level ..   .. 2s. per yard
      Through Faces   ..   .. 2s. 10d. per yard
      Through Gob 9 feet wide by 5ft.
        6in. high under Collars .. 9s. per yard
      For every additional 6 inches
        in thickness ..   ..   .. 9d. per yard extra
Working Double Shift in headings .. 1s. per yard extra
Extra men working in any heading per extra
    man above two..   ..   ..   .. 6d. per yard extra
Working Double Shift on the coal in Stalls
    when two shifts are regular for a week
    or more ..   .. 2d. per ton extra
Straight Cut ..   .. 1s. per yard
Two Straight Cuts ..   ..   .. each 1s. per yard
Opening Faces at Right Angles   .. each side 1s. 6d.
                         per yard
Airways   ..   .. 2s. 3d. per yard
Ripping Top Coal on Roads   ..   .. 1s. 6d. per yard
Ripping Clod up to 12 inches in thickness 1d. per inch per yard
Cutting Bottom   ..   .. 1d. per inch per yard
Walling Waste   ..   ..   .. 5d. per yard
Double Timber, 6½ feet   ..   ..   .. 1s. 6d. per pair
Double Timber, 9 feet   ..   ..   .. 1s. 9d. per pair
Repairing Timber, 9 feet ..   ..   .. 2s. 1d. per pair
Standing Posts   ..   .. 5d. each
Drawing out Posts ..   .. 1d. each
Drawing out Double Timber   ..   .. 3d. per pair

LIST OF PRICES—*Continued.*

Standing Flats	1s. per pair
Standing Cogs	1s. 9d. each
When stood in the full thickness of the Seam, 3 cogs to be paid for 2 ..	
Drawing out Cogs	1s. each
Colliers' day wage when required to work on occasional day or days for the Company	4s. 5d. per day
Unloading and Walling Stones	5d. per tram
Unloading Rubbish	4d. per tram
Filling Water	3d. per cask

When two turnings are made off any Stall, Road, Level, Heading, prices to be paid on such Road from the second turning.

All measurements on Roads to commence from the crossing.

All the above prices are now subject to 33¾ per cent. advance, and are subject to further reductions or advance as may be agreed upon hereafter by the Sliding Scale Committee, or its Successor. No notice in respect of same to be demanded by either side.

Signed on behalf of the   Colliery Co., Ltd.
                                                    Agent and Manager.

Signed on behalf of the Workmen employed at the above Collieries.

Witness to the above Signatures.

Dated April 25th, 1900.

It will be observed that there are numerous piece-rates agreed besides the cutting price of coal. These are all for doing " dead-work," as it is called, which is required to keep the working places and roadways in safe and efficient condition or to get the coal in clean condition. For instance, in a thin seam, the ripping of the roof to provide sufficient headroom over the roadways is a constant item of expense and is paid at so much per yard. The collier constantly has to " stand " new props as the face advances ; and in many pits, if the roof is good, the props are taken out and used again, when the collier is paid 1½d. or so for " drawing " each prop, which is rather a dangerous operation in unskilled hands. Again, in pits where they are working under towns or villages and there is risk of subsidence which would lead to expensive claims for compensation, the " gob " or " goaf " is packed with walls, either with stone found below or brought from the surface ; and in every pit a certain amount of walling may be necessary for the ventilation and other matters. Here again it is done at so much per yard. In working the coal it is often found that there is a band of stone in it or a loose layer of stone just over the coal which always comes down with it. It is necessary that the hewer should have remuneration for the trouble of separating the fragments of stone from the coal and so an allowance is made at the rate of so many pence per ton of coal cut, the rate per ton depending on the thickness of the " clod " or band of stone.

The payments to be made for dead-work, extras and allowances on piece-work are determined generally by the under-manager, who goes round each place once each week on the " measuring up " day, together with his clerk, the overman of the district, and the fireman. At each place the dead-work done since the last measuring up is carefully noted, and the rates of allowances are agreed with the hewer in charge of the stall, or at least they are argued, for a bargain is not always struck. If on reflection the collier thinks the price which he finds allowed when he gets his pay-ticket is inadequate, and he fails to get it rectified on the next measuring-up day, he will probably report the matter to the lodge secretary, who will bring it before his committee. They sift the case, and, if they think there is a real grievance, take it up with the manager. In a large pit there may be 30 or 40 cases a week dealt with, especially if the manager is trying to cut down expenses.

The wages of each individual hewer are often composed of many items, and subject to so many deductions, that clerical work in calculating the wages for payment must occupy several days. It is, therefore, customary for miners to be paid wages for a complete week on the Friday or the Saturday of the following week, so that, practically speaking, they must be always a week's wages out of pocket. Before the Coal Mines Act, 1911, the period was a fortnight, and it was sometimes a great hardship for men commencing work at a new pit to receive no money for three weeks.

When the wages are actually paid to him the hewer receives a pay-ticket, setting forth how the net amount paid him is arrived at. A specimen of such a ticket is reproduced on the opposite page; it shows first of all, on the right-hand side, the

PAY ENDING..........191

Dr.	£	s.	d.	T.	C.	Q.		£	s.	d.	Rate	£	s.	d.
Doctor							Large Coal							
Coal							Brush							
Rent							Allowance Bast							
Stores							,, Clod							
Fund							,, Stowing Clod							
Checkweigher							,, Cleaning Large							
Nurses							Headings							
Cash advanced							Ripping Top							
Nat. Health Insurance							Cutt. Bottom							
							Trams Rubbish							
							Face							
							Airway							
							Walling							
							Boxing							
							Dble. Timber							
							Cogs							
							Flats							
							Props							
							Allowance							
							Days							
							Rib							
							Per Cent.							
Total Deductions														

Deductions

Balance Due

number of tons of coal sent up during the week, and this is priced at the standard cutting price. Then come the payments for dead work or allowances according to the price list. If the amount is less than the legal minimum wage for the number of shifts worked, and the collier has given notice of working in an abnormal place, there will follow here the allowance to bring the total up to the minimum wage ; but in this case there is none. At the bottom on the right-hand side is the percentage addition to the standard. On the left are various items of deductions for tools, lamp cleaning, colliery doctor, national insurance, and perhaps a contribution to a colliery institute or library. The total of such deductions may amount to as much as 2s. 6d. per week ; and they have tended to grow heavier during recent years. When pit-head baths are adopted there will probably be another $1\frac{1}{2}$d. per week deducted. Where a man is working with a boy or helper, his wages are sometimes deducted by the colliery company, and paid to the boy directly, and so they appear upon the pay-ticket ; in such a case, however, the deduction is a reproductive one, as the boy's service has enabled the collier to send up more coal.

Variety of Price Lists

One of the greatest difficulties of the wages question in the coal trade arises from the total want of system which has always prevailed in the negotiation of price lists except in the North-Eastern coalfield. In each colliery the workmen's committee has negotiated with the manager and agent ; and in a lengthy negotiation the miners' agent of the district has been called in on the one side, and the Board of Directors on the other. There is often a splendid exhibition of bargaining, more or less drawn out ; but the ultimate result depends upon the relative skill in bargaining of the two parties, and also on their potential staying power in case of a rupture leading to a protracted strike or a lockout. It will be understood, therefore, that very different results have been attained in different collieries ; so that there may be as much as 20 per cent. difference in the cutting price of coal for the same seam, under practically identical conditions between one colliery and the next. There are most extraordinary variations, too, in the payments of dead-work. In some pits in South Wales they have been fixed so high that it pays the collier better to do dead-work, such as ripping the roof, and standing props, than to cut coal. Not unnaturally, advantage is taken of this peculiar situation ; and the colliery company is involved in extra expense in supervision to see that only the necessary dead-work is done. On the other hand, sometimes the dead-work is paid at such a low rate that men kept continually at such work would make less than half their normal wage. In some pits where such price lists prevail, a custom has arisen, which in some cases has extended to many neighbouring collieries in a district, of paying for dead-work in excess of that actually done in order to make up for the low price. Thus there will be an understanding that ten props set or drawn count as 12. It was such local customs as these which caused infinite trouble to the joint district boards in fixing the first minimum wage rates and determining the working rules ; indeed, many of these difficulties are not yet wholly over-come.

It will be wondered why such differences of price lists between one pit and another should be allowed to continue. It would seem so easy to meet and agree that the rates be raised or lowered. In fact, however, there are few things more difficult to alter than a price list once agreed and established. It is sure to become precious almost beyond imagina-tion to one side or the other, according to whether it is found ultimately to favour the men or the owners. The practical result is, that no alteration can be made without a long and bitter struggle. Each side has been for many years so well organised that, if the miners strike to increase a price list, the owners will be supported in their resistance by the Coal Owners' Association. If the owners take the step of demanding a reduction of the price list, the men affected will probably be supported inde-finitely by the Miners' Federation. The consequence is that once a price list is fixed for a given seam in a given colliery, it remains almost as permanent as the very nature of the coal seam itself. So expensive an operation is it in South Wales for the owners to alter a price list if it is too high, that there have been cases of the owners selling the pit rather than face the loss of a prolonged strike. The buyers have either had to make profit possible by expensive improvements of machinery, or to face a long strike in reducing the price list. Presumably they will have taken into account the possibility of much expense in reducing the price list when valuing the colliery for purchase.

It is indeed becoming more and more difficult to effect any change at all in price lists once established ; and it is also becoming more difficult to fix a price list in the first instance. An interesting case is that of the Gelli pit belonging to Messrs. Cory Brothers, in South Wales. The pit was a comparatively new one which Messrs. Cory Brothers had taken over and wished to open. The price list they offered proved unacceptable to the men, who went out on strike in October, 1910. After some months the executive of the South Wales Miners' Federation took control of the case, and they furnished the strike pay. In spite of various attempts and negotiations, neither side has seen its way to offer sufficient for the parties to meet and make a bargain, with the result that the pit has remained closed for more than three years, and there is still no prospect of its being opened. The men on strike gradually drifted away to find employment in other pits, and in September, 1913, the Miners' Federation ceased payment of strike pay. The result is that the pit is practically of no commercial value, because no member of the South Wales Miners' Federation can be got to work there ; and as this includes practically all the experienced men, the company has no prospect of opening the pit at the price list it offers.[1]

The negotiation of a price list is indeed a most dramatic event in the life of a mining town, the prosperity and character of the town hanging in the balance on the " give and take " of a 1d. or $\frac{1}{2}$d. per ton. The usual practice is for a number of working places to be tried by men paid by the day in order to test the productiveness of the seam ; and this system is continued for many months. During the negotiations there are usually at various stages accusations and counter-accusations, the men asserting that the management have opened up only the better places so as to get a high average yield of coal per day per man, whilst the management assert that the men are purposely slacking to keep down the output. If the manager is a sporting man, he will do a day's work in a stall, and show that he can cut twice as much coal from the seam as the average man is sending out ; to which the men retort that he has probably chosen an easy place, and in any case, if he liked to exhaust himself any skilled man could double his output for one day. Sometimes, as in the great strike at the Rhondda in 1910 and 1911, over the Ely Colliery price list, the dispute far transcends in local public interest all other questions of local or national politics. Each side alternately gets out leaflets stating its case, and the

---

[1] Since the commencement of the Great War the dispute has been settled.

public is bewildered by the contrary statements of fact as well as the irreconcilable claims. Every legitimate art of industrial warfare is brought to bear by one side or the other in the colossal bargain which means thousands of pounds per annum going to either one side or to the other for a long period of years, perhaps for the whole life of the colliery. It is asserted, indeed, that illegitimate means, such as bribery of the miners' officials, have occasionally been instrumental in settling a price list. The miners' agent responsible for such negotiations may be put to a very heavy temptation indeed from an unscrupulous owner, to whom a satisfactory settlement is worth many thousands of po' ..ds ; and it is a matter of congratulation that no suggestion of malpractices occurring at the present day is ever made.

Social Effects of varying Price Lists

Very interesting are the results of variations of price lists between one colliery and another. The fact that different prices persist in the same market for the same labour naturally leads to practically all offers of service going first to the collieries offering high prices. They have their pick of the most skilful and steady hewers—men who know how to work safely and fast, and will go down regularly. On the other hand, where the price lists are low there congregate the unskilful, lazy, and irregular men. This accounts for the peculiar fact that in South Wales in one colliery the average earnings of hewers may be as high as 8s. 9d. per shift, whilst in a neighbouring pit, men apparently doing the same work earn on the average only 7s. 2d. per day.

Every colliery creates its own small town of 6,000 or 7,000 inhabitants ; and where the workers are well paid the whole character of the town is different. In South Wales and Monmouthshire, in those towns where wages are high, there is always a large Workmen's Institute with a fine hall, occupying a commanding position and costing from £6,000 to £20,000. It is equipped with reading rooms and library, billiard room, committee rooms, and class rooms, besides the meeting hall and a temperance refreshment room, and sometimes a gymnasium and baths. The cost is usually defrayed almost entirely out of the miners' wages, the contributions being sometimes deducted for the purpose at the colliery office. There may be a loan from the colliery company to finance the building, and officials and directors usually contribute subscriptions.

Other features of the high wage places are :— (1) a large percentage of men owning their own houses, sometimes two or three houses ; (2) cooperative stores flourishing ;   (3) A scarcity of public-houses ;   (4) numerous   chapels   solidly built ; (5) in the streets one sees many serious and intelligent workmen of the best type.

On the other hand, in the low wage towns there is usually no institute, or only a poor affair, established in one or two houses.   There are more public-houses in evidence, and comparatively few men own their houses. A large proportion of men in such towns are fairly recent immigrants from various parts of England, having left the most various callings, or they are labourers from the agricultural districts of South Wales, and of Devon and Somerset. Often they have left a wife and family behind them, some remitting half to three-fourths of the wages, others not sending anything, having deserted their families. They are usually not fixed residents like the workmen of the high-wage towns, and stay in each place but a few months. They leave sometimes, perhaps from mere restlessness, or a tiff with the overman ; but often in search of work in a more remunerative place. In the low-wage towns the men are of a rougher appearance, and coarse language may be heard. There are fewer clean and neat cottages, and the children are dirty and untidy. The purely Welsh miner is on the whole a " temperance " man, and revivals which have taken place from time to time have practically killed the liquor trade in certain places ; insomuch that the numerous valleys in the South Wales coalfield are classed by some into " dry valleys " and " wet valleys," the latter being those in which the ordinary amount of alcoholic liquor is consumed.

Scotland

These differences of average earnings of the whole colliery in neighbouring collieries working the same seam are to be found in other parts of the country besides South Wales ; and they are well known in

Scotland. In part they are explained by variations of a seam from place to place, so that it is more difficult to work and requires more skilled and experienced men in one mine than another. In Scotland, as in Durham and Northumberland, there is far more division of labour than in South Wales ; and the hewer devotes himself almost entirely to getting down the coal, other men being provided for breaking, cleaning and filling the coal into the trams and for setting props, effecting repairs and laying tram rails. The South Wales hewer rather prides himself on being able to do all these various tasks, and they have the advantage of providing some variety in the work. As regards repair work, it is, however, mainly different physical conditions which necessitate a different practice, as it is necessary in South Wales for the hewer to keep a constant watch on the roof and to be ready at any moment to take measures for his own safety. The long price lists of South Wales are not, therefore, paralleled in other coalfields.

In Scotland the cutting price of coal is not usually quoted as a standard rate, always at a given level for each seam or district of a colliery, and to which the percentage in force for the time being is added. The Scotch practice is to add the current percentage to the cutting price itself, so that instead of men's minds being fixed on the percentage as 50 or 60, etc., above standard they have to remember that the cutting rate now stands at 1s. 7d. or 1s. 9d., etc. All the calculations of wages are thus made directly, and no percentage is added on the wage tickets. It is possible, and often necessary, to get at the standard in a Scotch pit by a little calculation. Thus, if the cutting price is 1s. 9d., and wages are 80 per cent. above standard, the standard itself for that seam is

$$\frac{100}{180} \times 1s. \ 9d. = \frac{5}{9} \times 21d. = 11\tfrac{2}{3}d. \text{ per ton.}$$

The cutting rates in Scotland are generally adjusted as nearly as possible to be equivalent to a standard or basis of 4s. per day for hewers, whilst day-wage men are on a basis of 3s. 6d. or 3s. 9d.

Variety of earnings as between different collieries exists ; but it is somewhat checked in Scotland by the system of a " common " or " field " price, for a district or coalfield embracing a number of collieries, which the miners insist upon as providing the minimum earnings for the district, and which defines the lowest figure acceptable for the cutting price of any seam in the district. Thus, if the minimum of average daily earnings in a district is fixed by the miners at 4s. per day as standard, it is an accepted policy that no colliery can be allowed to fix a price for cutting coal which shows a lower average of earnings over a period of some months than this district minimum. By numerous strikes the employers have learnt that they must accept this principle ; and it has the support of the general body of colliery proprietors as they realise that it prevents newly opened mines or seams undercutting them in the market by getting cheaper labour.

The Scotch system of the "common price" appears to be a half-way stage to the elaborate "county average" system of Durham and Northumberland shortly to be described. It has the great advantage that in Scotland it is by no means so hard as in South Wales to change a cutting price once it has proved to work out unfairly to owners or men, because there is a recognised standard of minimum earnings in a district ; and although there is a good deal of variety at different collieries in earnings above the minimum, the workmen will not be unreasonable about lowering the cutting price if the earnings are very much above the minimum. This "common price" is, of course, only a minimum of average earnings, and not an *individual* minimum such as was established afterwards at a lower figure by the Minimum Wage Act.

Movement in South Wales for Unification of Price Lists

A new policy on the lines of the Scotch "common price" system is being tried by the South Wales Miners' Federation in regard to new price lists, whether in new collieries, or in new seams that are being opened in old pits. Instead of leaving the negotiation of a new price list to the local lodge, supported by the miners' agent of the district, it was arranged that from October 1, 1913 onwards, no new price lists should be fixed without the sanction of the Executive Council of the Federation, which in effect means centralising the negotiations.

There was also adopted a uniform basis for new price lists intended to be everywhere adopted as the irreducible minimum which the Federation would allow miners to work for. The very different conditions of different seams render it almost impossible, however, to ensure exact equality of earning power to the miners under different new price lists ; so that the Federation are thrown back upon requiring that the day wage for colliers shall be the same in each price list. The cutting price of coal and the various rates for dead work would then be fixed so as to bring the earnings by piece-work to a figure which, for a man of average skill, would be at least equal to, and generally slightly more than, the day wage rate. It is, therefore, not possible to get absolute uniformity of new price lists, but merely an approximation thereto by getting uniform rates of time wages for any work which is paid by time.

There is much doubt, however, whether the South Wales Miners' Federation will be able to achieve its object, at any rate, by the present movement. It has, indeed, set itself a hard task by adopting a day wage for colliers higher than that paid under most of the existing price lists. This means that every price list now negotiated, and there are several dozen being negotiated, or likely to want negotiating very shortly, will be hotly disputed by the colliery owners. Each particular owner will have the support of the whole of the Coal Owners' Association, so that a strike in any particular locality could be continued indefinitely so far as the owners are concerned. If only two or three new price lists were being disputed the Miners' Federation could also keep the men out in those localities indefinitely, and might, therefore, enforce these claims; but it is obviously impossible for the Miners' Federation to keep men on strike for a long period in forty or fifty different mines. Its only weapon would be a general strike throughout the whole South Wales coalfield; and it is unlikely that it would be prepared to go to this length upon this particular issue. If the principle is to be recognised it would seem that the only practicable means is to demand a moderate day wage, so that the task resolves itself for the next few years into merely levelling up a few price lists in which the owners would have obtained low rates, had it not been for the support of the Central Executive.

County Average System

The constant difficulties attending the fixing, or attempted alteration of price lists in South Wales throws into strong contrast the advantages of the *county average system* of Durham and Northumberland. Here they have gone much further than the common-price rule prevailing in Scotland, and have established averages of earnings at the standard rate for each county, from which the average in each district of a colliery is not to diverge by more than 5 per cent. above or below. The system in the county of Durham is administered by a Joint Committee of the Durham Coal Owners' and Miners' Associations,

the object of which is to " take into consideration
and determine local disputes arising at any particular colliery." The " county average " is a fixed
figure ascertained many years ago for each class of
workers, being in Durham 4s. 2d. for all grades of
workers and all kinds of collieries, and in Northumberland 4s. 9½d. for steam coal collieries, and
4s. 7½d. for soft coal collieries. The standard
hewing prices and other prices in each colliery, and
district of a colliery, are then adjusted from time to
time so as to bring the average earnings of that colliery, or of one of its districts, within 5 per cent. of
the county average. If the average is more than
5 per cent. below, the workmen will apply to the
Joint Committee ; if it is more than 5 per cent.
above, the owners will apply, unless they think that a
reduction of earnings would mean curtailing the
necessary supply of labour. The procedure worked
out in Durham appears to me so important that I
make no apology for quoting several clauses of the
agreement here *in extenso* :—

" 11. *Five per cent. Margin.*—Before any application for an advance or reduction in the wages of
hewers (including kirvers and tub loaders) shall be
entertained it must, except as provided in Rule 12,
be clearly shown that the average wage earned by
the same class of persons in the seam (or portion
of a seam if cavilled separately) is at least 5 per
cent. above or below the recognised county standard
rate, but there shall be excluded from the averages
the earnings of any hewers who are paid an extra

price in addition to the ordinary rates in consideration of their working at night under any special arrangement.

" Before any application for an advance or reduction in the wages of any other classes of workmen paid by the piece shall be entertained it must be clearly shown, if the application relates to workmen employed underground and paid by the piece, that the average wage earned by the same class or classes of persons in the seam is at least 5 per cent. above or below the recognised county standard rate, or if the application relates to workmen at bank paid by the piece that the average wage earned by the same class or classes of persons employed at the pit is at least 5 per cent. above or below the recognised county standard rate.

" 12. *Objections to Cases.*—No demand from the workmen at any colliery shall be entertained where, within the three pays supplied as per Rule 18, or, where averages are not required, within the three pays prior to the day appointed for hearing, losing before time or restriction of work by the class of workmen seeking the change has been a general practice. No demand of either party shall be considered by the Committee if the party submitting the demand is not carrying out or has within the previous three months failed to carry out any county or local agreement or any previous decision of the Joint Committee, or any award.

" 17. *Counter-Claims.*—All applications by one party for advances or reductions of piece-work prices shall entitle the other side to raise the question of the prices paid to the same class of workmen throughout the whole of the pit, and all applications for advances or reductions of the datal wages of any workmen shall entitle the other side to raise the question of the datal wages paid to workmen of the same class throughout the whole of the pit, provided that, in either case, not less than seven clear days before the day appointed for hearing by the Joint Committee, a statement of any counter-claim intended to be made, together with a statement of average earnings or datal wages as provided for in Rule 18, shall be handed to the manager, or, as the case may be, to the workmen of the colliery, and, in the case of hewing prices, specifying the names of districts where the advances or reductions are sought. The provisions of Rule 11 shall apply to counter-claims in the same way as to original applications.

" 18. *Statement of Average Earnings to be furnished*.—No request for advance, reduction, or revision of piece-work prices, or datal wages, shall be entertained unless a statement showing the average earnings per shift, or of the datal wages, has been supplied by the Secretary of the Owners' or Workmen's side of the Committee, as the case may be, to the other Secretary at least nine clear days before the date fixed for hearing by Joint Committee. In the case of piece-work the averages thus supplied shall be those of at least three recent consecutive pays (each given separately) actually received by the workmen, excluding the first and last pays of each quarter, and the two pays immediately following the date of a decision, award or agreement becoming operative. The averages of seams, and of the districts where the advance, reduction or revision is asked for, shall be supplied in each case, and also the numbers and dates of the pays. The Joint Committee, in considering and determining claims for advance or reduction of piece-work prices, shall be guided by the difference between the average earnings of the workmen during the pays for which averages have been supplied and the county standard rate for the class concerned, and in considering and determining claims respecting datal wages shall be guided by the difference between the datal wage of the workmen concerned, and the county standard wage for that class."

The numerous local strikes which occur in South Wales and other fields on questions of price lists are thus avoided; and it would seem very desirable that the system should be widely extended. One difficulty, however, is that in the newer coalfields most of the colliery owners have their time so fully occupied with sinking new pits, and the multitude of questions which attend the development of a col iery in its early stages, that they have not time to give to establishing complicated arbitration machinery for the benefit of the coal trade as a whole. It would appear to be most desirable that a director of each of the larger colliery companies should specialise on employment and labour questions, and leave them less in the hands of an overworked pit manager.

There is another movement on foot which has not yet, however, attained prominence, though it probably will do so in a few years' time. I refer to the demand which is being made by the miners in several districts that a new standard basis should be adopted. It must be admitted that the standard rate of 1879, involving the addition of 60 per cent. above standard, is somewhat archaic ; and the further fact that there are also standards of 1877 to 1888 in use which, of course, each require a different percentage, gives reasonableness to the proposal that a new standard should be adopted. It is proposed that this should be uniform throughout a large area if not throughout the whole country ; and that it should be upon a higher level than any of the existing standards, so that the standard itself should form the minimum. At present under the above conciliation agreements, the minimum stands at anything from 15 to 50 per cent. above standard.

Theoretically, there is everything to be said for a new and uniform standard. It would simplify conciliation agreements, and for a short time, at any rate, would somewhat simplify the calculation of wages. It is unlikely, however, that the direct merits of the change will be the real motives for advocating or opposing it. At present the employers absolutely and unanimously object to any change. They evidently fear that it will be used as a kind of " stalking horse " for obtaining an actual increase of money wages. The men at large, having become accustomed to high percentages above standard, may think themselves ill used if they have only 5 or 10 per cent. above standard, although it really means the same or more money than 60 per cent. above standard now does.   Again, it is customary to make advances in multiples or sub-multiples of 5 per cent. of the standard ; but, if for example, the standard day wage is raised from 5s. to 7s., an increase of 5 per cent. would really mean getting a larger increase than 5 per cent. on the old standard. The psychological advantage of the change appears, therefore, to be all in the direction of stiffening the bargaining power of the workmen ; consequently we may expect that the proposal will not be countenanced at all by the employers, unless possibly under a long-term agreement.

In the course of many years such a change will no doubt be made, and with the further spread of education, the grounds of the employers' objections will be decreased.   The experienced hewer is already unusually competent at arithmetic, and understands percentages a good deal better than most college students, as I know by practical experience.   At the same time a large proportion of the miners, particularly those newly arrived from other trades, have learnt too little arithmetic to fully grasp the meaning and the facts of the change of standard.   Recent improvements of teaching, and the continued discussion of the subject, will gradually make it evident that the change, when made, should be merely for the sake of arithmetical convenience and apart from any bargain in the actual rate of money wages.

CHAPTER XV

Accidents in Mines

THE dangers which attend the coal-miners' work are of several different kinds, and unless the most strenuous and constant precautions to prevent accidents are taken both by the management of the mines and by individual miners, the hazard to life and limb is greater than in any other occupation. The attention of the public at large is arrested by a great disaster, like the recent explosion at Senghenydd; but it is not generally realised how dangerous is coal mining from a variety of causes other than explosions. The tragedy of the mines is familiar enough, however, to anyone living in a colliery district. Small accidents are frequently happening to one's friends and neighbours who are miners, and now and again someone in the village is seriously injured—perhaps disabled for life. In the local newspapers of the coalfield one sees almost every day grim little notices like the following :—

> Thomas Morris, single, collier, of Winifred Road, Skewen, was killed at No. 7 Pit of the Main Collieries at Skewen early this morning. Deceased was buried under a fall of about two tons of coal.

If the reader will think of the bereavements that these accidents entail, the Government statistics will have some meaning. They will tell what suffering is caused by the winning of coal—much of it avoidable, as I shall point out.

A complete return of all accidents which injure any person so as to disable him for more than seven days must be made each year to the inspector appointed by the Home Office, and these figures are collated and published in the General Report on Mines and Quarries. From Part I I can quote some interesting figures which illustrate the number and kinds of accidents which occur, and they are set out in the table on p. 367. Since the statistics for 1913 include the most disastrous explosion ever known in this country, that of Senghenydd, in which 439 lives were lost, I will quote also the figures for 1912. In passing I would like to say that these General Reports on Mines and Quarries, especially Parts I and II, should be far more widely read than they are at present. Part I contains mainly tables of statistics. Part II contains the interesting and informative matter describing certain accidents and their causes, giving comparisons of death-rates and explanation of new regulations, and so forth. They can be had through any bookseller for a few pence each.

PERSONS KILLED AND INJURED IN AND ABOUT COAL MINES

	1912. Killed.	1912. Injured.	1913. Killed.	1913. Injured.
Explosions . . .	124	145	462	131
Falls of Ground . .	567	53,185	620	62,094
At Working Face . .	358	42,358	401	49,311
On Roads while repairing or enlarging. . .	112	5,554	136	6,535
On Roads while otherwise working or passing .	90	5,226	79	6,172
In shafts . . .	7	47	4	76
Shaft Accidents . . .	71	895	98	825
Overwinding . . .	4	71	16	48
Ropes or Chains breaking	3	2	2	6
Whilst ascending or descending by machinery .	6	60	14	95
Falling down shaft . .	19	28	29	24
Things falling down shaft	8	204	14	196
Miscellaneous, etc. . .	31	530	23	456
Miscellaneous,—Underground .	339	84,983	400	100,434
By Explosives . .	23	236	30	263
Suffocation by Natural Gases	8	8	5	11
By Underground Fires .	1	8	25	10
Irruptions of Water. .	1	2	9	5
Haulage (a) Ropes or Chains breaking	23	190	8	191
(b) Run over or crushed by Trams or Tubs :—	(190)	(21,223)	(214)	(25,862)
Mechanical Haulage .	86	3,486	83	4,272
Horse Haulage . .	55	11,034	61	14,026
Hand Haulage . .	14	6,179	16	6,896
Runaway Trams or Tubs .	35	524	54	668
(c) Other Haulage Accidents	26	14,183	29	17,940

TABLE.—*Continued.*

	1912.		1913.	
	Killed.	Injured.	Killed.	Injured.
Electricity . . .	7	18	13	24
By Machinery . . .	10	471	17	500
Sundries	50	48,644	50	55,628
TOTAL UNDERGROUND .	**1,101**	**139,208**	**1,580**	**163,484**
TOTAL ON SURFACE .	**175**	**11,444**	**173**	**13,705**
By Machinery . .	39	688	28	822
Boiler Explosions . .	—	14	—	20
On Railways, Sidings, or Tramways :	(78)	(3,885)	(81)	(4,102)
While engaged in moving Wagons . . .	20	1,319	19	1,500
While engaged in coupling or uncoupling wagons .	3	195	6	271
Run over while passing along or across Railways or Tramways . .	14	139	24	208
Crushed between Wagons and Structures . .	8	600	14	440
In other ways . .	33	1,632	18	1,683
Electricity . . .	4	15	3	19
Miscellaneous . . .	54	6,842	61	8,742
Gross Totals . . .	1,276	150,652	1,753	177,189
Gross Totals, 1911 . .	1,265	166,616		

The totals of the principal classes of accidents are shown in heavy type and these totals are subdivided in the figures beneath them in small type. Amongst the injured are included only cases of accidents which disable the man for more than seven days. Accidents are included in the " surface " group if they occur anywhere in and about the surface works of a colliery, including washing and screening plant, coking ovens, tramways, and private branch railways. The classification and persons included are the same as in the statistics of the numbers of persons employed above and below ground.

It will be observed that of the four principal classes, " falls of ground " cause the largest number of deaths and a very large number of injuries. Fortunately the proportion of deaths to the total of killed and injured is not so large in this class as it is in the cases of explosions and of shaft accidents. If we take the average of a number of explosions, about half the persons involved are killed.

By reading the reports of the chief inspectors of the eight inspection districts into which the United Kingdom is divided, which are issued separately each year, it will be seen that the accidents are due to the most various causes, some of them apparently quite unforeseeable. The impression is left, however, that quite a large proportion of the accidents are preventable. In some cases they are clearly traceable to the colliery management having failed to comply with certain requirements of the Mines Acts, or Home Office regulations. In many cases it appears that the workman has failed to comply with the statutory regulations or with the rules of management of the colliery. In a few cases pure rashness or negligence on the part of the workman is responsible. It is marvellous how some men will risk their lives to save themselves a little trouble.

2 A

Such men when detected taking risks should be most severely dealt with, as they are often not only a danger to themselves but to a large number of their fellow-workmen. The important fact is, however, that a large proportion of the accidents are of a preventable character, if the regulations and well understood precautions are properly carried out. I shall consider later in this chapter how the more thorough inspection of mines may be expected to decrease the number of serious accidents.

Decrease of Mining Mortality

Serious as is the loss of life and limb in our mines at the present day the conditions show a great improvement on those prevailing fifty or sixty years ago, and a vast improvement as compared with the shocking mortality of a hundred years ago when miners' lives were of little account. I am chiefly concerned to show the progress which the application of legislation to safety in mines has already effected, because it shows the efficacy of such legislation and proves conclusively that with increasing stringency of the law and better inspection, improved conditions have been brought about. There is thus every hope that legislation, still better considered and enforced by a larger staff of inspectors, will prevent a great many of the accidents which still unhappily occur. As the number of employees in mines has greatly increased during the last fifty years, it will be well to study a death-rate table representing the number of persons who have died from accidents in or about mines each year per thousand persons employed. A complete table of this kind is given in the Home Office Report.[1] and I quote from it here the average death-rates for every period of five years :—

DEATH-RATES IN MINES FROM DIFFERENT CAUSES

Year.	Underground Workers taken separately.					Surface Workers taken separately.
	Explosions of Firedamp or Coaldust.	Falls of Ground.	Shaft Accidents.	Miscellaneous.	All Causes Underground.	
1851–55	1·28	2·02	1·30	·56	5·15	1·01
1856–1860	1·23	1·85	·90	·65	4·63	·99
1861–65	·62	1·71	·67	·79	3·79	1·11
1866–70	1·16	1·58	·53	·73	4·00	1·26
1871–75	·52	1·21	·44	·57	2·74	·90
1876–80	·81	1·32	·32	·50	2·71	·85
1881–85	·41	1·11	·26	·53	2·31	·85
1886–90	·31	1·02	·20	·52	2·04	·91
1891–95	·24	·80	·19	47	1·70	·83
1896–1900	·12	·79	·13	·46	1·47	·80
1901–05	·12	·75	·12	·42	1·41	·80
1906–10	·24	·75	·11	·48	1·59	·74
1911	·04	·72	·12	·42	1·31	·73
1912	·14	·66	·09	·39	1·27	·81
1913	·51	·68	·11	·44	1.74	·79

Looking first of all at the death-rate from all causes underground, we see that this has fallen from

[1] Part II (1913), p. 148.

an average in 1851–55 of 5·15 to as low as 1·41 in
1901–05. There was a slight increase during the
next five years to 1·59, and in the last three years
the figure was approximately the same, although
1912 had the lowest death-rate ever recorded. It
is very interesting to see how this total death-rate
is made up in the four principal classes. Falls of
ground are responsible for more deaths than explosions. In both classes there has been a great
diminution of death-rate progressively in the falls
of ground group, but not so regularly in the explosives group because of the effect upon the death-rate
of individual big explosions. Comparing the average
of 1906–10 with the average of 1851–55, we find that
the death-rate has been reduced in the different
classes as follows :—

Explosions	by 81	per cent.
Falls of ground	by 63	,,
Shaft accidents	by 92	,,
Miscellaneous	by 14	,,
All causes underground ..	by 69	,,
Surface workers	by 27	,,

The reduction has been enormous in regard to shaft
accidents and explosions, and the mortality from
falls of ground also shows a very striking diminution.

It has been argued that the reduced death-rate
is not entirely or even mainly due to safety legislation and better inspection, but mainly to the in-

creased education and intelligence of the workman.
Whilst there has undoubtedly been a very great
change in this respect, and one which would
undoubtedly reduce the number of accidents, this
certainly cannot be the main cause of the diminution. The reduction of accidents to surface workers
and in the miscellaneous group may well be accounted
for very largely by the increased education of the
workman, but this percentage is relatively small
compared with the three classes of accidents against
which legislation has been principally directed.

Mining Mortality in other Countries

It will be highly instructive to compare the
mortality in mines experienced in recent years in
this country with that recorded by the Government
statistics of the other principal coal-producing
countries. For that purpose I have prepared the
following table, which gives the mortality per
thousand employed in the United States and the
principal European coal-producing countries compared with the United Kingdom. In the first
column is given the average mortality of the fifteen
years ending 1911 and the next three columns give
quinquennial averages for the same period. The
first column gives us a fair comparison of the
safety of mines in the different countries, averaging
out the effects of great explosions. The three
columns of the five-year averages give us a pretty
good indication as to whether the mortality in the
mines is on the increase or decrease.

FOREIGN DEATH-RATES IN MINES

	1897–1911.	1897–1901.	1902–1906.	1907–1911.
United Kingdom ..	1·32	1·30	1·28	1·39
United States ..	3·31	2·89	3·26	3·79
Germany	2·21	2·35	1·95	2·34
France	1·52	1·21	2·28	1·08
Belgium	1·03	1·09	1·00	1·00
Austria	1·28	1·43	1·24	1·18
Natal	4·84	5·29	3·94	5·28
India	0·90	0·65	0·86	1·19

Several very interesting facts emerge from an examination of this table. The highest mortality is in the mines in Natal, where the very large majority of underground employees are Kaffirs and other natives. The next highest is the United States. Here the five-year averages show a progressive increase of a most significant character, which is probably to be attributed partly to the increase of deep mining, but mainly to the growing use of electricity for haulage work underground and machine coal-cutting. The figure is very high, but human life is valued lightly in the States. Germany comes next on the list, and the United Kingdom next, followed by Austria, whilst Belgium has the lowest rate of mortality of any European country. The figure for India is remarkably low, but the quinquennial figures show a progressive increase which may be due either to the extension of deep mining or possibly to an improvement in the completeness of the returns. The exceptionally low figure of 0·65 for a five-year average certainly suggests that not all the accidents were returned. The figure for Belgium is remarkably low for a country where the mines are so deep, and reflects credit upon the mine engineers and the legal regulations for the protection of the workers. In Austria we note a progressive improvement, but the same cannot be said of Germany or of our own country.

Growth of Safety Legislation

The improvement in the safety of mines recorded by the statistics from 1851 in this country and from later dates in other countries, has been achieved mainly by a continuously growing body of legislation requiring of both the mine management and the miners almost innumerable precautions. Safety legislation has become exceedingly technical, and only a slight idea can be given here of the nature and objects of the various rules. The various provisions of the present law can, however, be appreciated in their true perspective with a little knowledge of the history of safety legislation ; for it is generally true that the rules most necessary and efficacious in preventing the commonest serious accidents have been adopted first.

The Industrial Revolution, and the application of steam power to pumping and to winding machinery, led to very extensive developments in coal mining. In the early part of the nineteenth century many serious explosions took place ; and in 1813 the

Sunderland Association, with Sir Ralph Milbanke as president, was formed to inquire into the causes of colliery accidents, and to devise means for their prevention. Previously to 1814 it was not customary to hold inquests on miners killed underground. It was at the invitation of this philanthropic body that Sir Humphrey Davy devoted attention to explosions and invented his famous safety lamp. This, however, was not the first safety lamp, for Humboldt in 1750, and Dr. Clanny in 1815, had invented similar appliances which had met with more or less success. The Davy lamp, while it undoubtedly reduced the risk of colliery explosions, did not prevent their occurrence, and in 1835 a Select Committee of the House of Commons was appointed " to inquire into the nature, cause, and extent of those lamentable catastrophes which have occurred in the mines of Great Britain, with the view of ascertaining and suggesting the means of preventing the recurrence of similar fatal accidents." The Committee made a number of suggestions, of which the following are the principal :—

(a) Adequate ventilation.
(b) The use of safety lamps.
(c) The preparation of complete and correct plans of all collieries.
(d) The better education of miners.
(e) The appointment of competent and trustworthy officials.
(f) The enforcing of discipline.

Little of practical effect was done, however, till 1842.

Enquiries made in connection with the earlier Factory Acts were extended to the hardships of women and children employed underground. Many women and girls were found employed dragging loaded trams along narrow roadways, often going on all fours in a kind of harness ; or they had to carry heavy baskets of coal up the shafts on slippery ladders in small pits where there was no winding gear, and many fell. At the instance of Lord Shaftesbury, a Royal Commission was appointed in 1840 to inquire into the employment of children and young persons in mines and quarries, and it proved that children of five years old and upwards were employed underground and constantly being injured, and that the presence of young women in the mines led to gross immorality. The revolting revelations of the report were a great shock to the public, and led to the immediate passing of the Act of 1842, which wholly prohibited the employment of women and girls underground, and of boys under ten years old. Other safety provisions of this first of the more important Mines Acts were the limitation of the apprenticeship period to eight years, the prohibition of persons under fifteen years old being left in charge of winding engines, and the appointment of inspectors. That it was necessary in 1842 to prohibit mine owners leaving the lives of scores of men in the care of a boy under fifteen years old in charge of an engine speaks volumes as to the horrible conditions still prevailing in some of the mines. The appointment of inspectors was furiously opposed, and their

powers of inspecting underground were unfortunately
not made sufficiently definite in the Act.

Only one inspector, Mr. Tremenheere, was
appointed, and his name deserves all honour for
the zealous way in which he discharged his difficult
duties. His first report appeared in 1844, and in
succeeding annual reports he dealt not only with the
causes of accidents in the mines, and the enforcement
of the Act, but also in a faithful and interesting manner
with the social conditions of the people engaged in
the mining industry. His work followed a useful
report by the Midland Mining Commission, which
was appointed after extensive riots in the Midland
coalfields. This Commission discovered a whole
series of grievances connected with the " truck "
and " butty " systems, fines and forfeitures, and the
frequency of accidents. So difficult was it for one
inspector to enforce the simple provisions of the Act
of 1842, that Mr. Tremenheere found that women
were actually still being employed underground in
South Wales in 1851.

A number of serious colliery accidents happened
about this time, and a Commission consisting of three
well-known scientists, Messrs. Michael Faraday,
Charles Lyell, and Lyon Playfair, was appointed to
report on their causes. In 1845 Messrs. Lyon, Playfair and De la Bêche were asked to enquire into the
nature and action of mine gases, and to suggest
preventives. Public opinion was now thoroughly
roused, and a general desire was expressed for the
safer working of coal mines. Between 1846 and

1849 the workmen employed in the mines loudly
demanded the appointment of a stronger staff of
mines inspectors, a demand which they have continued to advance ever since. The results of the
numerous enquiries into individual disasters which
had been made were apparently not satisfactory ;
and in 1849 a Select Committee of the House of
Lords investigated firedamp explosions and their
causes ; and in the following year appeared two
valuable special reports on the ventilation of mines.
In 1850 was passed, after much opposition, " an
Act to provide for the Inspection of Coal Mines in
Great Britain." The inspectors were given rights
of free admittance to all parts of a colliery, of calling
for correct plans of workings, to question and notify
the management regarding defects, and to report
to the Home Office. Owners were required to give
notice of all fatal accidents to the Home Secretary,
and coroners to give two days' notice of inquests ;
the object being to enable inspectors to fully investigate accidents and attend inquests. Accurate
plans of mines were required to be kept, and to be
submitted when required to the inspectors for examination. This Act is a landmark in mining law, for
it provided the basis of the system of inspection
which has since been so much extended and strengthened ; and secured the means of thoroughly investigating the causes of fatal accidents.

By 1853 there were six regular inspectors at
work, in addition to Mr. Tremenheere ; but the
mere appointment of inspectors did not suffice

to prevent colliery accidents. In 1852 and again in 1853 new Select Committees were appointed to deal with the questions of mine ventilation, management, and inspection. The Second Committee, which reported in 1854, recommended the drafting of rules for the safe management of mines and the adoption of various appliances and arrangements for securing the safety of the miners. In 1855 was passed an Act to further strengthen the law in regard to inspection and notification of accidents, which embodied some of the recommendations of the Committee and contained the first code of General Rules. The latter required the provision of adequate ventilation (but did not define it), also the fencing and lining of shafts, means of signalling up and down a shaft, and the provision of an indicator for the winding engine-man of the position of the cage in the shaft, chiefly to prevent the terribly frequent accident of overwinding. Without this device the cage, when wound up against the pulley wheel, is broken off and falls to the bottom of the shaft, instantly killing the occupants. The opposition to this Act had been organised in a skilful and determined manner by the colliery proprietors, and workmen and reformers supporting them were bitterly disappointed with what they regarded as a " Masters' Act."

The agitation therefore continued, and with Alexander Macdonald as one of the leaders, it became more successful, so that in 1860 another Act for the Regulation and Inspection of Mines was passed. This measure repealed the Act of 1842, except the clauses prohibiting the underground employment of females, and enacted a number of new provisions. The limit of age of boy labour was raised to 12 years, except in the case of boys between 10 and 12 who had obtained a certificate of competency in reading and writing, or of school attendance. The qualifying age for persons controlling steam was raised from 15 to 18 years. The number of General Rules was increased from eight to fifteen, and important new provisions were included requiring the fencing off of dangerous and gassy parts of the mine, the examination and locking of safety lamps, and the putting of boreholes in advance of a heading approaching old work ngs, so as to avoid sudden inundation by accumulated water. This Act also sanctioned the appointment of checkweighers by the workmen ; and altogether it marked a great step forward. In 1862 an important Act was passed prohibiting the working of coal mines without two outlets from each seam, so that if an explosion wrecked one shaft the uninjured men would not be hopelessly entombed.

Another epidemic of colliery accidents now occurred, in one of which, that of the Hartley Colliery, 204 lives were lost. These calamities had a profound effect on the mining community ; and in 1863 the Miners' National Association was founded[1] for the purpose of securing " the interests of operative miners as regards legislation, the inspection of mines, and compensation for accidents." In 1864

[1] See p. 481.

the Association petitioned Parliament, and in the following year another Select Committee was appointed to enquire into the working of the Mines Acts. The Committee laboured during three sessions and issued voluminous reports, the final one appearing in 1867.

Meanwhile 1866 was a particularly disastrous year for colliery explosions, a total of 1,484 lives being lost in the mines. Mr. Kinnaird, speaking in the House of Commons, claimed that it should be made an offence against the Mines Act if 5 per cent. of gas were found in a mine ; but this was generally derided as a counsel of perfection !

Safety Legislation

A number of Bills embodying the Committee's recommendations were introduced in 1869, 1870, and 1871 ; but owing to severe opposition and the pressure of other Parliamentary business, the Government withdrew or blocked them. In 1872, however, a Bill was passed after a conference of representatives of employers and men with the Government. This Act is of great importance, as it may be said to be the foundation of modern legislation to secure safety in mines. It embodied most of the fundamental principles which subsequent Acts have elaborated and enforced more adequately. The outstanding reform of 1872 is the certification of managers. Henceforward no incompetent man was to be allowed to have charge of the working of a mine, and competency was to be certified by a board of examiners appointed by the Home Secretary. Other important new provisions were : the daily inspection of the whole mine by officials recognised for the purpose (known as firemen or deputies) ; permission to the workmen to appoint two of their number to inspect all parts of the mine and report on its condition ; strict regulations as to the use of explosives for blasting underground ; the provision of refuge-holes in the walls of roadways where coal is drawn by mechanical traction ; more rigorous conditions as to the use of safety lamps ; and further safeguards for checkweighers. With the passing of this Act the provisions of the law had become so much more detailed and varied that it was necessary to increase the staff of inspectors.

The reforms did not seem to have much effect, however. Between 1872 and 1879 a number of serious explosions occurred, and in the latter year a Royal Commission was appointed to enquire into Mines Accidents and their prevention. This Commission presented an Interim Report in 1881, and its Final Report in 1886. It established the dangers of coal dust, in causing, or at least assisting in the propagation of explosions ; and it recommended the use of high power explosives in place of gunpowder for blasting, and electrical firing appliances. The Commission also advised the experimental testing of lamps and oils, and the systematic inspection of each mine by the workmen's representatives under the Act of 1872.

In 1886 another Mines Act was passed. It

enacted that checkweighers need not necessarily be employees at the mines to which they are appointed, adjusted the law relating to inquests on the victims of mines accidents, and provided for special inquiries into mine accidents. This was only a temporary measure and, with the Act of 1872, was repealed in the following year, 1887, by the Coal Mines Regulation Act, which was designed to give effect to the principal recommendations of the Royal Commission. Its main provisions were the following :—

No boys under 12 to be employed underground.

Employment of boys above 12 regulated.

Extension of the privileges and powers of check-weighers.

Classification of managers holding certificates of competency as first and second class.

Raising of age of engineers to 22.

Persons in charge of minor machinery to be above 18.

Eight new General Rules prescribed.

Further enquiries were made in 1891, when another Royal Commission was appointed " to inquire into the effect of coal dust in originating and extending explosions in mines, whether by itself or in conjunction with firedamp ; and also to inquire whether there are any practicable means of preventing or mitigating any dangers that may arise from the presence of coal dust in mines." The Commission reported in 1894 ; and in 1896 a Bill based on the Coal Dust Commission's Report was introduced

and became law. By this measure the Secretary of
State was empowered to revise and modify the
special rules in use with respect to the use and storage
of explosives, the use and management of safety
lamps, the prevalence of gas or coal dust in mines,
etc. Workmen were secured the right to representation at arbitrations concerning safety provisions
in mines. The steady levelling up of the age limit
in mines was carried one step further in 1900 by an
Act restricting employment underground to males
over thirteen.

In 1902 the Home Office issued a very comprehensive code of general regulations prescribing
the details of safety precautions under the Acts in
force. These made several important provisions
for safety in working and management. A characteristic of recent mines legislation has been the
growing insistence on qualification of officials by
education and experience. In 1903 an Act amend-
ing that of 1897 was passed. This allowed managers
to substitute in part for academic training a certain
period of practical experience. In 1906 the " Notice
of Accidents Act " was passed in order to require
the immediate notification to the district inspectors
of mine accidents, which, although not fatal, cause
serious injury to any one.

The most recent, and perhaps most useful and
exhaustive enquiry into the health and safety of
mines, and the administration of the Mines Act,
was made by the Royal Commission on Mines
appointed in 1906, and extended in 1907. This com-

2 B

mission considered a variety of questions. The first
Report, published in 1907 dealt with Rescue Stations
and Breathing Appliances for Rescue Purposes, and
made recommendations as to how rescues could
most satisfactorily be effected. A second Report,
issued in 1909, dealt with Mines Ventilation, the
examination of mines, and the causes and means of
prevention of accidents from falls of ground, underground haulage, and in shafts. The main conclusions of the Commission may be summarised as
follows :

1. The staff of mines inspectors should be considerably increased.
2. Inspection districts should be rearranged, and
   the system of inspection revised. Practical
   miners should be appointed on the inspector-
   ate.
3. The respective responsibilities of owners and
   their agents should be definitely fixed.
4. Firemen or deputies should be qualified by
   examination or experience.
5. Inspections of mines should be more regular and
   more frequent.
6. A higher standard of ventilation should be
   adopted.
7. New ventilation arrangements were suggested.
8. Methods of minimising the quantity of coal dust
   in mines were investigated and described.
9. Precautions were devised for adoption during
   shot-firing.

10. Rules for the proper testing and use of safety lamps, and the effective timbering of mines were recommended.

11. Winding-enginemen should be medically inspected regularly, and should not be allowed to work two shifts following.

12. The organisation of rescue stations and the provision of rescue appliances, etc., in colliery areas should be attempted.

13. Pit-head baths and dressing-rooms should be provided.

14. Colliery plans should be kept accurate and up-to-date.

The third Report of the Commission, issued in 1911, dealt with various matters, including ventilation and the treatment of pit ponies.

As a result of the investigations of this Commission further important legislation has been enacted. In 1910 an Act was passed which empowered the Secretary of State to order for any district or group of mines the provision of rescue and ambulance appliances, and the organisation and training of ambulance and rescue brigades. This measure was repealed in 1911, and its provisions incorporated in the very important general Act of that year, which, subject to the short Act of 1914, at present governs the whole working of coal-mines.

Causes and Prevention of Explosions

The foregoing outline of the efforts continually exerted during the past hundred years to secure safety in the working of coal mines will be enough to prove the enormous difficulty and complexity of the subject. Including the numerous and voluminous reports of Parliamentary Committees and Commissions, the literature on safety in mines would by itself form a library of no mean proportions, so that I may be pardoned for being unable to give here more than the slightest discussion of this most important subject. Many of the provisions of the Coal Mines Acts (1911 and 1914) which I shall proceed to consider, are more or less obvious precautions to secure safety in the use of machinery, which experience has shown it to be necessary to enforce by law; but the provisions to secure safety from explosions and fire are of a special character, and are the result of a most extensive series of investigations and inquiries as to how explosions and fires are originated and propagated. I shall, therefore, explain as best I can in a popular manner the generally received opinions as to the causes of explosions and the best means of preventing them, and of mitigating their effects when unhappily they occur. This will render clear the purpose of many sections of the Act of 1911, and will enable us to understand why even the present law is defective in certain respects.

What is called " fire-damp " is a gas consisting of a mixture of hydrogen and various hydrocarbons, chiefly methane, or " marsh-gas "; and it is not unlike the coal gas used for domestic purposes, except that it has no smell. This natural gas is pent up in the coal itself and in the strata immediately overlying it, often at great pressure. It is given off all over a freshly exposed surface of rock or coal; and accumulations of gas may occur in fissures and may be released by slips and movements of the strata caused by the removal of the coal or rock in working the mine. A considerable release of gas in this way at one spot, constitutes what is termed a "blower." When the inflammable gas mixes with air only in small proportion it does not form an explosive mixture; but if the gas reaches more than a certain percentage it will explode immediately it comes in contact with any flame or spark. If there is a considerable further concentration of gas the proportion of oxygen becomes too small to combine with the gas, and it cannot explode. At this strength, however, a man would be immediately asphyxiated, and there would always be an explosive mixture where the concentrated gas meets a current of fresh air.

The methods of taking precautions against explosions are two-fold : first to prevent the accumulation of gas anywhere in proportion strong enough to form an explosive mixture, and secondly, in mines known to be "gassy," to wholly prevent the presence of any naked flame or spark in every part of the mine where it is possible for gas to occur. The first is accomplished by efficient ventilation, the second by the use of the locked safety lamp for illumination underground, by prohibiting the taking of matches below ground, and by very particular precautions in shot-firing, and in the use of electricity for coal-cutting or haulage underground. A further set of regulations is designed to mitigate the effects of explosions should they occur.

To secure ventilation there must always be two shafts with which every seam is connected, a downcast and an upcast. In the old-fashioned method the upcast shaft was simply used as a chimney, a furnace being constructed at the base so as to pass hot air into it, which by ascending drew a current of air down the downcast shaft and all through the passages of the mine. It is quite impossible, however, safely to obtain by this means a current of air powerful enough to ventilate a large mine with many miles of roadways, so that mechanical means are now universally used. There are many kinds of ventilating fans in use, and whilst in a few cases air is forced down the downcast shaft, the usual procedure is to extract the air from the top of the upcast shaft. For this purpose enormous fans are made, up to 25 feet diameter, capable of extracting 200,000 cubic feet of air per minute.

The current of air would tend to go almost entirely from the downcast shaft straight to the upcast shaft, which is usually only thirty or forty yards away, if arrangements were not made to distribute it throughout the mine. To secure an adequate supply of air to all working places in the remotest parts of the mine is no easy task. Some of the roadways must be planned to take the currents outwards in various directions from the downcast shaft, whilst other roadways, or if these are not available, specially constructed airways must be used for the return current. The intake airways contain nearly pure air for the first three or four hundred yards ; and naked lights can be safely used in them, especially if they are built up with a brick arch ; but the return airways in a gassy mine generally contain a fair amount of gas and explosive mixture. The greatest care must be taken not to allow any of the return current to get mixed with the intake air, and to secure that the current goes with adequate force to every working place.

Sketch plans of the roadways of two mines in which explosions have occurred are reproduced here. On studying the arrows, which show the directions of the air currents, it will be realised that it is no simple matter to get the air distributed. Currents must be split time after time, and sometimes the return must cross the intake, whilst the haulage roads must pass from intake to return airways, or vice-versa. All this is managed by wooden partitions, or by wooden frames, on which the impervious " brattice-cloth " is stretched. A weakness of such partitions is that they are all instantly destroyed when an explosion occurs, so that it is very difficult to re-establish the ventilation and keep alive men cut off in remote parts of the mine.

When the proportion of firedamp in the air exceeds 6 per cent. the mixture is explosive ; and it is impossible to avoid the accumulation of gas in higher proportion than this in particular spots in the

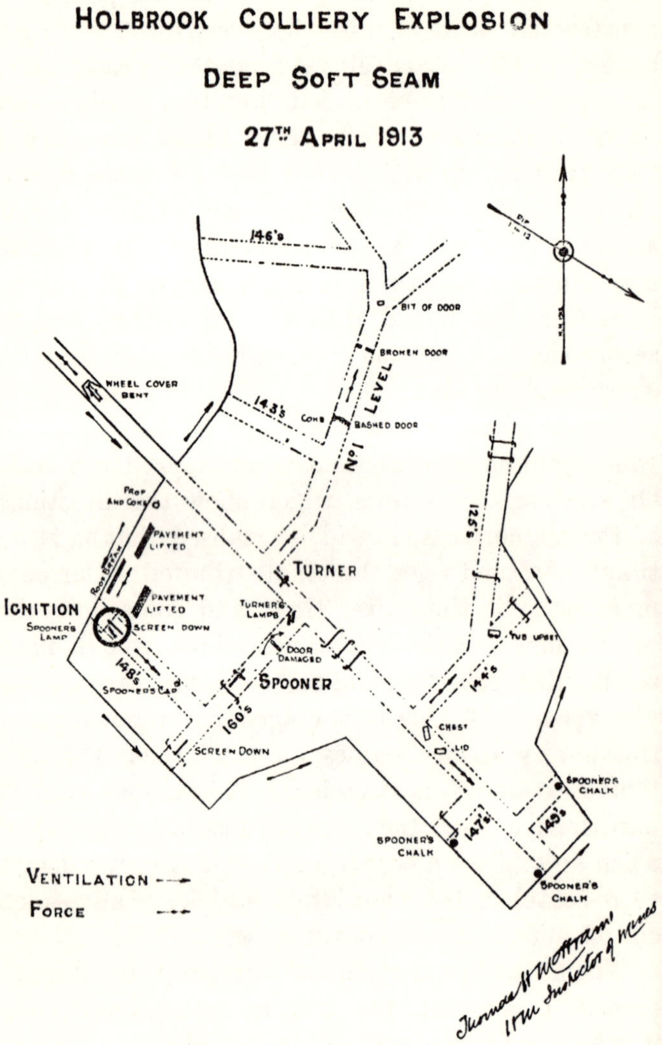
HOLBROOK COLLIERY EXPLOSION
DEEP SOFT SEAM
27TH APRIL 1913
146's
BIT OF DOOR
BROKEN DOOR
No 1 LEVEL
143's
COKE
BASHED DOOR
WHEEL COVER BENT
PROP AND COKE
PAVEMENT LIFTED
ROOF BREAK
TURNER
PAVEMENT LIFTED
IGNITION
SPOONER'S LAMP
SCREEN DOWN
TURNER'S LAMPS
DOOR DAMAGED
SPOONER
148's
SPOONER'S CAP
160's
SCREEN DOWN
125's
TUB UPSET
144's
CHEST
LID
SPOONER'S CHALK
SPOONER'S CHALK
147's
149's
SPOONER'S CHALK
VENTILATION
FORCE

Fig. 13.

LODGE MILL COLLIERY.

WORKINGS WHERE TWO MEN WERE SUFFOCATED ON THE 28ᵀᴴ JANUARY, 1913.

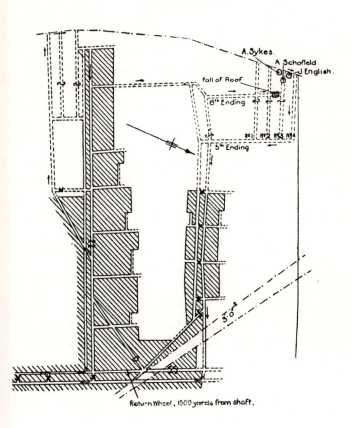

Fig. 14.

mine. The gas is lighter than air, and consequently is always found in greater concentration towards the roof. Where falls of the roof have occurred, a dome-shaped cavity is left, and in this gas is very likely to accumulate. If there are any passages, blind headings, or stalls, into which the air current is not directed freely, gas is certain to accumulate to a dangerous proportion. It is quite unsafe, therefore, to take a naked light into a gassy mine, and the miners must use the safety lamp. The principle upon which Sir Humphrey Davy relied in constructing his famous lamp, and which is still employed, is the rapidity with which wire gauze conducts heat and thus cools the surface of a flame in contact with it, so that the flame cannot pass through. It is a pretty experiment to bring down a piece of fine wire gauze over a gas flame, such as the bluish flame of a gas-ring. It will be found that the flame will not pass *through* the gauze. If the gauze is brought low down towards the burner, unburnt gas actually passes through the gauze, but does not burn on the upper side unless a match be applied to it, or the gauze be allowed time to become very hot.

In all the varieties of safety lamps now in use the oil flame is entirely enclosed by metal, glass and gauze, the air needed for combustion passing in and out through the gauze. The lamp is so arranged that it can be locked with a special key, which is in the hands only of the lamp-man at the pit-head, and of the man in charge of the lamp station under-

ground, the latter being usually situated about 300
or 400 yards from the shaft bottom along the main
roadway.   At this lamp station every miner, as he
passes in to work, must hand in his lamp for inspection to see that it is undamaged and properly locked.
This station is also the only point underground
where the miner can have his lamp relighted, should
it be accidentally extinguished.   The lamps are, of
course, made to stand all manner of ill-usage without
breaking, the glass being specially toughened.[1]

The original miner's lamp had no glass, and the
available light was only such as filtered through the
gauze.   The successful insertion of a glass has greatly improved the light, and thus the rapidity and
safety with which the miner can work.   A good
lamp gives only about half the light of a candle,
however, so there is probably a future for the
portable electric lamp which is already being used to
some extent because it gives a more powerful light.
According to the Home Office return, out of a total
of 749,177 miners' lamps in use at December 31,
1912, 10,727 were electric, the remainder being
various patterns of safety oil lamps.   One great
drawback of the electric lamp is that it does not
serve to indicate the presence of gas, as does the
gauze-lamp.   The flame of the latter is affected at
once by the presence of gas in the air.   A specially

---

[1] The Departmental Committee appointed by the Home Secretary to inquire into the testing of safety lamps found that good
glasses will stand plunging into cold water without breaking,
although they are heated up to 250° F., whereas inferior glasses crack
on immersion when heated only as high as 176° F.

constructed lamp will indicate the presence of even one-half per cent. of gas by a slight lengthening of the flame ; and an ordinary lamp clearly shows the presence of any undesirable concentration of gas by a lengthening and swelling of the flame. When the lamp is in an explosive mixture a bluish flame fills the entire interior of the lamp. It is an imperative danger signal, and the lamp must be carefully lowered to the ground. The electric lamp can give no such indication, so that there is the risk of the miner unconsciously breathing foul air. Various forms of firedamp indicators have been invented for use where the miner works with an electric lamp, but so far they have not fulfilled all the practical requirements of certainty and durability.[1]

In some mines accidents have occurred from the sudden outrush of gas which was pent up in the rock at high pressure. It has actually blown out some tons of coal from the face, and liberated such an enormous volume of gas that the ordinary precautions have been insufficient and an explosion has occurred. Such a sudden inrush may completely upset the ventilation and even drive the air back along the intake. Another occasional difficulty is a persistent " blower " of gas which is sometimes met with. In some cases the volume of gas emitted has been so large that, as it showed no signs of

[1] For further information regarding safety lamps and the detection of fire-damp the reader is recommended to consult a useful little work entitled *Safety Lamps and the Detection of Fire-Damp in Mines* (Broadway Text-Books of Technology, Messrs. Geo. Routledge & Sons, Ltd.).

diminution, the colliery owners have found it advisable to wall it in and lead from it a special pipe to the surface, rather than attempt to deal with it by dilution by an increased ventilating current. In some cases these supplies of natural gas have been used to commercial advantage for heating purposes about the pit-head. On the other hand, there are a number of mines where gas is almost unknown, or quite insignificant in quantity, and in which it is the custom to use naked lights, because of the far better light thus obtained, and the greater ease with which the roof can be inspected. Mines which are free from gas are mainly those of shallow depth, or those which, like the house-coal collieries of South Wales, are merely levels driven into the hillside, the coal seams lying nearly horizontal in the hills and cropping out on the slopes. In the Forest of Dean and Somerset even fairly deep mines are free from gas and are worked with naked lights. The explanation appears to be that in such cases the natural gas has had an opportunity of escaping before the mining operations commenced. In the coal measures there are numerous impervious beds of clay, shale, or shaly sandstone, so that the gas formed during long ages has been unable to escape, and thus exists at high pressure in the coal seams and certain beds of porous sandstone. Where the seams are mainly above the valley bottoms, or as in the Forest of Dean, are covered solely by fissured sandstone, the gas has almost entirely gone.

Atmospheric conditions undoubtedly exercise an

influence on the amount of gas found in a mine at any time. It has been proved by long continued observations that when the barometer is falling gas is detected in larger quantities, and for this reason it is now required by law that a barometer should be kept near the head of every mine, and its readings be recorded. It is easy to understand why the atmospheric pressure draws out the firedamp; for every gas expands in proportion to the reduction of pressure. When there are slow currents of gas making their way through numerous fissures to the working places of the mine, the gas in the fissures expands when the barometer is falling and the currents flow out faster; on the other hand, when the barometer is rising there is compression of the gas and the currents of gas flow out more slowly.

Firedamp is not the only source of explosions. In comparatively recent years it has been proved beyond the possibility of doubt that coal dust is responsible for many explosions, either by itself or in addition to firedamp. Almost all mines are dusty, except the shallowest and a few exceptionally wet mines. The reason mines are so dry is because the air is considerably increased in temperature in the deeper mines, and the amount of moisture which saturates the air at, say, 45 degrees, would be far less than the air would absorb at 75 degrees.

Much dust is inevitably formed both in cutting the coal and in getting it down, also in loading it into the trams, and by the jolting in conveying the trains of trams to the shaft bottom. When the

tram lines are roughly laid, and the haulage is
rapid, there is doubtless a great deal of coal dust
formed in transit, and it is well distributed and
agitated by the brisk currents of air needed for ventilation.   When the dust suspended in the air is
sufficient in quantity, any large hot flame will set it
alight and cause an explosion.   This has been proved
abundantly by many experiments carried on upon
a large scale representing as nearly as possible the
conditions of the galleries underground.   The larger
the amount of volatile matter in the coal from
which the dust originates, the more easily does it
ignite ;   and there is practically no danger of an
explosion of anthracite dust.   In addition to the
evidence of experiments, it has been shown that
disastrous explosions such as those at Altoft's in
Yorkshire, and Camerton in Somersetshire, as well as
the great disaster at Courrières in Northern France
occurred in mines which had been, and have been
since, entirely free from firedamp, so that naked
lights were in use.   It is, therefore, now well recognised that coal dust is a source of great danger in
mines, for even if it is not present in such quantity
as to cause an explosion itself, it tends to increase and
propagate any explosion of firedamp which may
occur and which might otherwise be localised.   It is
believed that explosions of coal dust have frequently
been caused by shot-firing.[1]

[1] It should be explained that in a coal mine " shot-firing " means
a blasting operation on a small scale.   A hole is drilled over the coa
which is to be brought down, and a charge of some permitted explosive

It is unfortunate that the brisk ventilation needed to reduce danger from firedamp tends, if anything, to increase the danger from coal dust. It is, therefore, desirable to take steps both to reduce the amount of coal dust formed, and to prevent such as cannot be avoided from escaping to the air in the haulage roadways. This can be accomplished by using completely closed wagons, and spraying with a jet of water the coal put in them before it starts on its journey. At present the trams have low sides on which the coal is built high, and have holes in the bottoms or corners so that there is ample opportunity for the dust formed to escape into the air. An efficient preventive of danger from coal dust would be to keep the whole of the mine well sprayed with water ; but this is hardly practicable. The dust settles upon all the little ledges on the sides and the roof, both on the stone and the timbers quite as much as on the ground. It would be necessary, therefore, to water the sides and roof as well as the ground ; but this would be a very costly and even dangerous operation, as the stone and timbers would tend to decay and constant watching and repairs would be necessary. A policy which is being tried in certain mines is to keep certain sections or zones of the roadways watered. In these sections the roadway is arched with a smooth arch in brick or

to the amount of from one-quarter to one-half of what would be used for blasting rock, is inserted and fired by an electric spark. It is said that wedges are not powerful enough to bring down the coal without a good deal of work in many seams.

cement, and by means of a fine spray, the roof and sides are kept constantly wet. Throughout such a section there is little dust in the air, and should an explosion occur it would not be propagated through such watered sections, provided they are made long enough. Another idea for rendering coal dust inno-cuous, which has been adopted in many mines within the last two or three years, is to dilute the coal dust with very fine stone dust, which is specially brought into the mine for that purpose. It is found that in a dusty atmosphere in which the larger proportion of the dust consists of any inert matter like stone, no explosion can be propagated. Stone dust is, there-fore, discharged into the air in the main roadways, and it settles over all the ledges and forms a coat-ing, so that when a gust of air blows up any dust it must blow up mainly stone dust, and only a small proportion of coal dust. The main objection to this system has come from the miners, who fear, probably not without reason, that breathing an atmosphere impregnated with stone dust will be in the long run prejudicial to health, predisposing to miners' phthisis, which is a common and serious disease in metalliferous mines where there is much stone dust from rocks composed largely of silica (quartz or sand).[1]

The dangers of shotfiring in gassy mines are so

[1] Since the above was written the fifth Report of the Depart-mental Committee on Explosions in Mines has been issued. It deals very fully with the use of stone dust and recommends the dust of argillaceous (clayey) stones as being free from danger to the miners' health.

2 C

obvious and well known that it might be thought
that the practice would be prohibited. It is stated,
however, that many collieries would have to be
abandoned if shotfiring were discontinued, so the
Home Office has chosen the next best course, by
regulating the mode of firing so as to minimise
accidents. If its regulations were always strictly
carried out both in the spirit and in the letter, the
numerous explosions due to shotfiring would be prevented. Only certain high explosives are permitted
to be used, which are mostly mixtures of nitro-
glycerine, nitrates of ammonium and other bases.
Each brand of explosive is exhaustively tested
at Woolwich Arsenal, and passed only if twenty
shots fired successively in an explosive mixture of gas
have failed to ignite it. Some remarkably interesting experiments with various explosives have been
carried out in the well equipped testing station of
the Westphalian Mining Association. It appears,
however, still to be a matter of controversy as to
why " high " explosives do not ignite an explosive
mixture of gas. The most plausible theory would
seem to be that the air wave caused by the shock of
the explosion is so violent that it extinguishes any
incipient combustion of the firedamp.

The use of electricity in mines is attended with
great potential dangers because of the liability to
sparking and to fusing of wires. The Home Office
has issued elaborate regulations as to cables, wiring,
etc. The main roadways of mines are now generally
lighted by electric lamps fixed to the roof ; and in

many of the larger and newer mines electricity is
used not only for coal-cutting machines but also
for main-road haulage, driving conveyers at the face,
and other purposes. The principal precautions
taken are to enclose all parts liable to sparking in an
air-tight cover ; to use a current of low electrical
potential which renders any sparks which may occur
extremely less likely to fire an explosive mixture ;
and to make all conductors amply sufficient to carry
more than the intended load, besides making the
insulation thorough, so as to avoid fusion.

CHAPTER XVI

MINING LAW AND INSPECTION

THE great Act of 1911 was a codification of the law regulating coal mines in regard to safety, so that it embodied the safety provisions of all previous Acts and repealed them, and also added a number of new and much more stringent provisions. All the Coal Mines Acts prior to that of 1887 are now entirely repealed. The Act of 1911 is too long and technical to examine in detail ; but as it is probably the most extensive and detailed of any form of government regulation of industry, it will be interesting to examine its main provisions and the objects and principles underlying them.[1]

The first part of the Act deals with the management of the mine with a view to securing safety. It provides very definitely both for the responsibility and training of the managers and subordinate officials. Every mine has to be under the control of a manager " who shall be responsible for the control, management, and direction of the mine." The owner, or his agent, must appoint himself or some one else ; but the owner or agent is prohibited from taking part in the technical management of the mine unless

---

[1] Extracts from the Coal Mines Act, 1911, are given in Appendix V.

STEMMING THE SHOT.

ASHINGTON COLLIERY.

PONY HAULAGE.

he is duly qualified to be a manager, which means being at least twenty-five years of age, and holding a first-class certificate. The manager is required to give personal supervision daily, but in his absence, or if his post becomes vacant, the under-manager or other person with a first or second class certificate may be put temporarily in charge, such period not to exceed four months in the case of the post of manager being vacated. I should think that this is an unnecessarily long period during which a man with only a second class certificate can be in charge of a mine. A very useful and necessary provision is that which prevents a manager taking charge of a number of different mines, unless he has the approval of the inspector of the division. No person may be manager of more than one mine if the aggregate number of persons employed underground exceeds one thousand, or if all the shafts of the mines do not lie within a circle having a diameter of four miles ; and if a man is manager of more than one mine there must be an under-manager for each mine who may, of course, have only a second class certificate. Notice has to be sent to the inspector of every appointment of a manager or temporary delegation of his duties.

The Act prescribes fully how certificates of competency are to be granted, and regulates the examination of candidates for certificates. The Home Secretary appoints a " Board of Mining Examinations " on which there are equal numbers of representatives of the owners, or their agents or managers,

and of workmen employed in mines, besides inspectors of mines and outside persons eminent in
mining or scientific knowledge. The examination
and other qualifications of applicants for second
class certificates are to be suitable for practical
working miners. Every applicant for a certificate
must be at least 24 years of age, and must have had
five years' experience of working in a mine, unless
he has received an approved degree or diploma in
mining, when the period of practical experience is
reduced to three years ; he must also give satisfactory evidence of his sobriety, experience, and general
good conduct. There is a *viva voce* examination
which has to be conducted by a person possessing
practical acquaintance with the local conditions of
mining. There are, of course, a number of universities and colleges in the country which grant degrees
or diplomas in mining, and practically all of these
are " approved " by the Home Office for the purpose of granting certificates. The Act also gives the
Secretary of State power after holding public
enquiry to revoke or suspend the certificate of any
person on the ground of incompetency or negligence
in performing the duties of manager or under-
manager, or on account of any conviction of an
offence against the Coal Mines Regulation Acts.

The manager is made responsible for appointing
competent persons as firemen (also called examiners
or deputies) to make inspections and tests as to the
presence of gas, as to the sufficiency of the ventilation, the state of the roof and sides, and the general

safety, besides checking and recording the number
of persons under his charge.   These are his statutory
duties, and he is prohibited from being given other
work except in measuring up the work done by
persons in his district, and firing shots in his district,
if these duties do not interfere with carrying out his
statutory duties in a thorough manner.   This
restriction of the duties of the firemen does not
extend to Durham or Northumberland, owing to the
special customs of the district, and the inspector of a
division can give exemption elsewhere on the
ground of the special circumstances of a mine.   In
view of the constant assertion by the workmen that
the firemen are not able to carry out their daily
inspections with the thoroughness desirable, the
provision that the district assigned to a fireman
shall not be so large as to prevent him carrying out
his statutory duties in a thorough manner is of interest.   It would appear to be a point on which
the law should be more precise, for it would
probably be extremely difficult to get a conviction
of any owner or manager on the basis of this vague
prohibition.   Candidates for the position of fireman
are obliged to fulfil certain conditions, which do not,
however, require any special study ; for if a man is
25 years of age and has had at least five years'
practical experience underground, of which two
years have been in working at the face, he has only
to obtain a certificate from a mining school or other
approved institution of his ability to make accurate
tests with the safety lamp for inflammable gas, and

to measure the quantity of air in an air current, and that his hearing and sight are good. It would not be thought by most people that good hearing is a necessary qualification for a miner ; but, as a matter of fact, deaf men are never employed underground, mainly because the only warning which a miner gets of an impending fall of roof is through the preliminary cracking and creaking of the timbers.

A very important provision of the Act is that which allows the workmen to " appoint two of their number, or any two persons not being mining engineers " who have had five years' experience underground as practical miners, to inspect the mine on their behalf. They have the right to inspect once at least in every month all parts of the mine, roadways, airways, ventilating apparatus, machinery and even old workings ; and they must be allowed to go to the place where any accident has occurred, accompanied by the legal adviser or any mining engineer selected by the workmen to make such inspection as may be necessary for ascertaining the cause of the accident. These workmen-examiners have the right of demanding to see the certificates of all firemen employed in the mine, and they are required to make a full report of the result of their inspections in a special book kept at the mine for the purpose. The owner, or manager, of the mine must at once send a copy of any such report to the inspector of the division. This is a most important section of the Act ; and although the right of appointing workmen-examiners had existed since the Act of 1872, it was only in 1911 that such examiners were put in a strong and independent position. Since the Senghenydd explosion, there has been a movement amongst the South Wales miners for the appointment of permanent examiners on their behalf, and several districts have appointed them. In order to give the workmen the opportunity of knowing the condition of the mine, it is provided that reports made by such examiners, and the reports of the daily inspections before commencing work and during the shifts, shall be posted up at the pit-head not later than 10 o'clock in the morning of the day following, and remain there till 10 o'clock the next morning. As the men are all down the pit long before 10 o'clock, they could not see the reports until they come up, when they will be more than 24 hours' old. It is difficult to see why a report should not be required to be made and posted up within 8 hours of every inspection, which would give the workmen the opportunity of knowing before they went down what was the condition of the mine on the previous day.

I need not stop over the elaborate provisions as to the plans required to be kept of the working and ventilation of the mine, and as to the deposit at the Home Office of plans of abandoned mines. It is obvious that up-to-date plans of the mine should be always available, not only for the use of the management, but also for the inspectors when they call. The preservation of accurate plans of abandoned mines is most important, so that when a newly-opened mine or seam is being worked in a direction

towards old workings, which are generally full of
water, the manager may know when he is getting
sufficiently near for precautions to be necessary.   In
the past many fatal accidents have come from the
flooding of mines due to a new working being carried
quite unsuspectingly close to an old working of an
abandoned mine.   The coal has suddenly been burst
through, and the mine flooded by a huge volume of
water.   The preservation of plans of old workings is
not the only necessary precaution against such an
incursion.   The Act provides in a later section that
when any working has approached within 40 yards
of a place likely to contain an accumulation of water
(which means any disused workings which have not
been examined and found to be free of water) there
should be constantly kept at least 15 feet in advance
of the heading at least one bore-hole in the centre of
the working, and bore-holes on either side at intervals
of not more than five yards.   Any bore-hole 15 feet
long would tap the water and reveal the danger
before the wall of coal was thin enough to give way.

It is still the custom here and there in South
Staffordshire, and to a small extent elsewhere, to
let out the working of the coal by contract to small
contractors or " butties," who employ the miners.
When these contractors have had to provide the
pitprops and employ firemen, the system has been a
fruitful source of danger, for the contractors are over
anxious to effect economies and have less sense of
responsibility than the owner or manager.   The
Act of 1911 made it obligatory on the owner himself

to provide all necessary materials and all firemen
and officials required for the safety of the mine ;
and made it impossible for the owner to appoint in
such capacities the contractor or any person employed by him.

Coming now to Part II of the Act, which is
entitled " Provisions as to Safety," we find it starts
with elaborate requirements as to ventilation.   In
order to explain what is meant by these I reproduce
the plan of a mine, which shows clearly the intake and
return airways.[1]   The main provision is that all
roads, levels, stables and workings of the mine on the
intake airway shall be kept normally free from
inflammable gas up to one hundred yards from the
first working place at the face ; and "normally
free " is defined to mean that the average percentage
of gas found by an inspector in six samples taken at
intervals of not less than a fortnight must not
exceed a quarter of one per cent.   This is a delicate
test, and involves the use of laboratory apparatus,
besides six visits by the inspector.   It has not yet
been applied even once to every mine in the country,
and the Senghenydd mine had unfortunately not
been tested in this way when the explosion took
place.   The Home Office is required to make
General Regulations under the Act so as to define the
details necessary to carry out the general provisions
of the Act, and may also make special regulations applicable to particular mines, as the inspectors may think desirable, and subject to appeal
by the owner to arbitration as to the reasonableness

[1] See page 204.

of the special rules. As regards the standard of purity of air in return airways this has to be secured by increasing the current of air entering and flowing through the mine, or any separate ventilation district of the mine, to such an extent that the return air after passing through the working-places contains only a safe proportion of inflammable gas. Obviously the more gas a mine gives off in its working places the greater must be the volume of air passing through to secure the requisite dilution. Further, it is evident that as the working of a seam expands in course of time from the shaft bottom, not only is there generally a larger range of working places, but the length of roadways along which the air currents must pass continually increases. This means more friction of the air to be overcome by the ventilating fan, so that a fan of given power will not draw so much air through long passage-ways as through short passages of the same width.

This consideration has an enormously important bearing upon the safety of mines, and indicates a growing danger which should receive adequate recognition. It means that the fan which is installed and serves perfectly for the first fifteen or twenty years of the life of a colliery, gradually becomes less and less equal to the work required of it as the working places recede from the shafts. To replace it by a larger fan, and perhaps a new engine, or to add an additional fan and motor, costs much money ; and some colliery proprietors not unnaturally postpone making the change, especially when trade is depressed, until their attention is strongly drawn to the need of more ventilating power. The colliery engineer has, perhaps, worked the fan faster from time to time until it is normally used at its highest capacity. Then there may come one day a sudden and long continued fall of the barometer which almost floods the mine with gas, with which the ventilation cannot cope. A prudent manager would bring the men out of the mine ; but this would mean serious loss to the colliery. The general regulations of the Home Office are perfectly satisfactory in theory : the danger is, that with the best intentions, or at worst a certain degree of negligence, they may not be, and often are not, observed in practice. It would be a wise precaution if the law were to stipulate for a wide margin of reserve ventilating power always to be ready to be put into operation ; say, fifty per cent. beyond that normally required for ventilating the mine. If the machinery were always ready, it would be used whenever the mine became exceptionally gassy. Further, there is the great advantage that the fulfilment of such a requirement is easily tested by an inspector, whereas if the obligation is simply to maintain a certain low percentage of gas, the inspectors must make much more frequent visits for taking observations than is possible with anything like the present staff.

Two requirements of the Act of 1911 are of great importance in case of an explosion. It is provided that the ventilating apparatus must be situated at the surface so that it may not be destroyed by an explosion—or at least sufficient apparatus must be at the surface to secure the safety of all persons underground—and that there must be provided means of immediately reversing the direction of the currents of air throughout the mine. Had the last provision been enforced at Senghenydd there would have been means of getting fresh air down to the men cut off by the fire from the intake air current.

The next section of the Act deals with safety lamps, and it begins by defining exactly the conditions when safety lamps, and not naked lights, must be used in the working of any seam. These are :— (1) When the air current in the return airway normally contains more than one-half per cent. of inflammable gas ; (2) When an explosion causing any personal injury whatever has occurred during the past twelve months ; (3) in, or near to, any working or place where there is likely to be any quantity or accumulation of gas. Exception is made in favour of electric lamps which, if enclosed in airtight fittings, may be used in main haulage ways, and elsewhere in accordance with the regulations of the mine approved by an inspector. These lamps are fixed to the roof in roadways ; the portable electric lamps are a form of safety lamp, and several patterns have been approved by the Home Office. The Act requires that all safety lamps used in a mine shall be provided by the owner, and be of an approved type. About thirty different patterns are at present approved by the Home Office. Extremely strict regulations have to be made as to the use of the safety lamps provided,

for there are always some workmen prone to careless-
ness, and a single thoughtless act may nullify the
whole advantage of using safety lamps.   A man will
intend no harm ; but, if his lamp should be blown
out, and by chance he finds it unlocked and discovers he has some matches, there is a great temptation to open it and relight it, and so save walking
perhaps a mile with a borrowed lamp to the relighting station.   He would, of course, be careful and
would probably go out of his working place two or
three hundred yards down the roadway where he
believed it to be usually quite free from gas.   But
that day it might not be free, and the catastrophe
might happen.

The requirements of the Act as to systematic
locking and examination of safety lamps are, therefore, most rigid, and of the greatest importance.
Every lamp on being returned to the lamp-room at
the surface is to be examined, and if found damaged
a record of the damage is to be made in a book kept
for the purpose, " and the damage shall be deemed to
have been due to the neglect or default of the person
to whom the lamp was given out, unless he proves
that the damage was due to no fault of his own, and
that he immediately gave notice of the damage to the
fireman," or some other official appointed for the
purpose.   Until a returned lamp has been examined,
and found to be locked and undamaged, it may not
be given out again ; and a record must be kept of
the man to whom each lamp is given out.   As a rule
each man has his own lamp kept on his own peg.

No person other than a lampman appointed by the
manager in writing for the purpose may unlock a
lamp or possess any key or contrivance for unlocking
a lamp.   I have already referred to the relighting
station underground, but it is interesting to note that
the law requires that this shall not be situated in a
return airway—an obvious precaution which unfortunately has not always been observed.   Machines
have been invented lately which relight a safety
lamp without its being opened, by means of an
electric spark produced only within the lamp.   As
the whole machine is also enclosed in an air-tight
cover, it could be safely used even in an explosive
mixture ; and if the use of such machines were made
obligatory the lamp stations could be placed
conveniently near the working places.

The next section (35) prohibits all persons, except
such as are authorised to relight lamps or fire shots,
from taking matches, or any apparatus for making a
light or spark underground in any mine or part of a
mine where safety lamps are required to be used,
and likewise any cigar, pipe, or cigarette.   To make
this important prohibition effective the manager is
required to cause all persons to be searched immediately before or after they enter the mine, or at least,
such persons as may be selected according to a plan
approved by the inspector.   The general practice is
to search ten per cent. of the men, but choosing them
a different way each day.   Men are not infrequently
caught with matches or cigarettes, and a prosecution,
conviction and fine of £2 or more invariably follows.

It is extraordinary that men should not be careful to go through their pockets, even if they think they have no matches. As regards the cigarettes or pipes often found on men, it is to be noted that it is rather a grievance that they have no place to leave anything at the pit heads. It should be made compulsory for a colliery owner to provide lockers or a kind of cloak-room. In this business of searching, the law is no respecter of persons, and it is provided that before searching any workmen the official must allow himself to be searched by two workmen employed in the mine. If found with matches or smoking apparatus he cannot himself search any workmen, and presumably he can be summoned by the workmen for an offence against the Act.

We pass now to the all-important stipulation that there must be at least two shafts or outlets from every seam which is being worked, and that they are not to be nearer to one another than fifteen yards, and are to be connected by a passage of adequate size. The principle is even applied now to every district and subdivision of a mine, for the section proceeds :—
" Every part of the mine in which ten or more persons are employed at the same time shall be provided with at least two ways affording means of egress to the surface, and so arranged that, in the event of either becoming impassable at any point, the other will afford means of egress to the surface." The only exemptions are in the case of new mines when not more than 20 men are employed underground, or, by order of the Secretary of State,

2 D

whilst a new outlet is being made, or any shaft or outlet which has been rendered unavailable by an accident is being repaired, the conditions of such order of exemption being duly observed.

We come next to stipulations intended to prevent shaft accidents. Every shaft or entrance to a mine abandoned or not in actual use is to be securely fenced ; and every shaft which is being sunk is to be securely cased or lined. The requirements as to winding apparatus are very comprehensive, and are too technical for us to examine in detail. The important principles are that the provision of ladders only in any shaft used for ingress and egress is prohibited, and a proper winding apparatus with cage must be provided ; that persons must be lowered or raised only in the cage (except for special purposes), and that the winding apparatus for the two shafts which the Act requires for every mine must be entirely independent. The Act even requires that every winding engine used for raising or lowering persons must be separated by a substantial partition from every other such engine, and any other kind of machinery, the object being to guard against an accident to one machine wrecking or damaging a winding engine whilst in use. If the shaft is more than 300 feet deep an automatic contrivance to prevent overwinding must be fixed ; and every shaft over 150 feet deep must be provided with guides to keep the cage from rocking or swinging.

A prolific source of accidents in past time has been

the imperfect closing in of the cage. Men have been killed by things falling down the shaft and through the roof of the cage, which is now required to be completely covered, and many others have been killed by falling out of the cage, which is usually entirely open on two sides except for a light movable bar, which is lifted up when trams are pushed in on one side or out on the other. One of the inspectors describes a fatal accident which happened through the cage having a slight swing, so that the movable bar was knocked up and the man standing next it instantly fell out. The Act now requires the provision of proper gates or fences at each of the open sides of the cage. It is a wise precaution that no persons are allowed to travel in the cage with coal, materials, timber, or any tools other than scientific instruments. There are other provisions as to brakes, signalling, and safety catches which I need not detail.

There follow now provisions as to the roadways. With certain exceptions, every seam in every mine opened after July 1, 1912, is to have two main intake airways, one not used for haulage, and both always maintained in such condition as to afford a ready means of ingress and egress. Similarly in new mines opened after the same date, all stoppings between main intake and return airways and all air-crossings are to be so constructed as not to be liable to destruction by an explosion. Both of these provisions, particularly the first, are expensive, and will add considerably to the capital expenditure in opening new mines and, in the first case, to the cost of maintenance ; but both will tend greatly to reduce the probable loss of life in the event of an explosion. It is highly improbable that both intake airways would become stopped, and with a ventilation system which would withstand the explosion, there should be little fear of a number of men being cut off. A further stringent condition applying to new mines is that if the air current of the main return airway normally contains more than one-half per cent of inflammable gas, it may not be used for haulage of coal. Of course it will be very many years before there are any great number of mines laid out and constructed on these highly satisfactory principles. In the great number of older mines, which altogether employ over 800,000 men underground, other precautions have to be relied on.

More than three pages of the Act are devoted to safeguarding men on the haulage roads. They are prohibited, without written permission, from riding on trains of trams drawn by mechanical power. Refuge holes must be provided every ten yards ; and there is to be a clear space two feet in width between the trams (or tubs) and the side of the road, or else the haulage must be stopped whilst the men are going in and out. Various devices to prevent and stop runaway trains of trams must be provided.

There follow detailed instructions as to supporting the roof and sides with props, which even go so far as to specify the minimum distance apart at which the hewer must put his props or sprags to hold up

CLEARING OUT THE CUT.

Ashington Colliery.

AFTER THE SHOT.

the coal while holing under it. For the dangerous
operation of withdrawing props from under the
roof where the coal has been worked out, a safety
contrivance is to be used.

As to the detailed provisions regarding signalling
and the use of machinery, I need only remark that
the lowest age for a person in control of winding
machinery is now 22 years, which is a great change
from the 15 years of the Act of 1842. No winding
engineman is allowed to work more than eight
hours in any day, nor may a winding engine be left
even for a moment without a qualified engineman
in attendance whilst there are men below ground.

After the passing of the Act no more internal
combustion engines were allowed to be installed
underground. No such prohibition could be applied
to the use of electricity, which is extensively employed underground in many large collieries,
although it is viewed with suspicion by many
persons and undoubtedly stringent precautions are
necessary. The Act prohibits the use of electricity
wherever the inspector considers it unsafe, subject
to appeal, and insists that the current shall be at
once cut off from any part of the mine where more
than $1\frac{1}{4}$ per cent. of gas is found in the general body
of air. There are detailed regulations made by the
Home Office in regard to wiring and casing motors,
switches, etc. There are similarly detailed regulations as to the use of explosives. None may be
taken into the mine except kinds on the " permitted
list " ; and to make the enforcement of this easier

all explosives taken into the mine must be provided by the owner of the mine, who must sell them to his workmen at cost price. In spite of precautions all sorts of accidents occur with charges of explosives, chiefly in " thawing " them, as many kinds cannot be used quite cold. The charges also sometimes get astray. If you happen to see a little paper-covered tube in your coal, looking like a firework, do not put it on your fire. I have heard of accidents occurring in this manner, through the hewer dropping one of his shot charges and its being shovelled up with the coal.

To prevent the distribution of coal dust in a mine it is required that screens for tipping or sorting coal must be 80 yards at least from the mouth of the intake shaft, that within five years all trams (or tubs) shall be so constructed as to prevent coal dust escaping, and that the floor, roof and sides of roads shall be systematically cleared of accumulations of dust, and roads watered where practicable.

The law as to daily inspections is interesting and most important. A station is to be appointed at the entrance to each district of a mine beyond which no workman may pass unless the fireman has made a complete inspection of the district and certified it free from danger of gas, or from the roof or other cause. This inspection must be made within two hours of a shift commencing work, and a further two inspections must be made during the course of each shift. The fireman has to make a written report of his first and last inspections for each shift.   All machinery used in raising or lowering persons must be thoroughly and completely inspected and reported on in writing once every twenty-four hours, and all other machinery in or about the mine and all airways at least once a week.   All written reports by firemen and other officials required by the Act must be made accessible to the workmen.

There are various miscellaneous provisions which I must pass over rapidly.   If the manager, or his deputy for the time being, finds that the mine or any part of it is dangerous, whether from gas or otherwise, he must immediately withdraw all the workmen from the whole mine or the part affected, except only such as are required to ascertain the cause of the danger and to rectify it.   The safeguard against an inrush of water by boring in front of workings, I have already referred to.[1]   No inflammable goods, such as oil, grease, or canvas, may be stored underground, and no combustible material, such as wood, may be used in the construction of any engine-house below ground or in the shaft or pit-head frame or roof over the pit.   This will prevent the disastrous shaft fires which have sometimes been caused by explosions and have made it impossible to rescue men.   The provision of fire-extinguishing apparatus and of meteorological instruments is also required, and readings of the latter must be taken and recorded.

Inexperienced workmen who do not understand

[1] *Ante,* p. 410.

the nature or meaning of the various precautions which must be adopted at the working face are a danger to many others besides themselves ; and consequently the Act insists upon every man being put through an apprenticeship before being employed in getting coal or ironstone in a place of his own. For the first two years at the face he must work as an assistant to a skilled workman, and the latter must at no time have more than one unqualified assistant.

This ends Part II of the Act. Part III deals with the health of miners. It requires the provision of pit-head baths when the workmen so desire,[1] the provision of sanitary conveniences according to regulations, and that notification is to be given to the inspector of the district of the occurrence of any industrial disease. The causes of diseases peculiar to miners, such as *ankylostomiasis* (a parasite) and *nystagmus* (an affection of the eyes), are becoming well known ; and although they are happily not prevalent in this country it is necessary to take every precaution against them. In order to prevent miners' phthisis it is provided that a spray of water must be used to prevent dust escaping into the air when siliceous rock is being drilled.

Part IV is concerned with accidents, their notification and investigation. Any accident must be notified which in the words of the Act either :—

" (1) Causes loss of life to any person employed in or about the mine ; or

---

[1] Coal Mines Act, 1911, Part III, Sections, 76 and 77

" (2) Causes any fracture of the head or of any limb, or any dislocation of a limb, or any other serious personal injury to any person employed in or about the mine ; or

" (3) Is caused by any explosion of gas or coal dust or any explosive, or by electricity, or by overwinding, or by any other such special cause as the Secretary of State specifies by order, and causes any personal injury whether to any person employed in or about the mine."

The notice must be sent not only to the inspector of the division but also to any person appointed by the workmen to receive such notices on their behalf. When there has been a serious accident, the place where it occurred must be left as it was immediately after the accident for a period of three days or until it has been visited by an inspector, excepting when so doing would create further danger or impede the working of the mine.   Full power is reserved for the inspectors to gather information, make reports to the Secretary of State and publish them, as is done in the annual Home Office reports, and in special reports on serious explosions.   Power is also taken for the appointment of a Court of Inquiry, and such a court made an investigation of the Senghenydd explosion in 1913.   An inquiry made by experts is likely to be much more useful than the coroner's inquest, and the Court of Inquiry has the fullest powers of compelling attendance of witnesses.

There are in addition to the foregoing twenty-five

more pages of the Act which I have not yet summarised. Amongst various formal matters, they contain the sections dealing with the maintenance of rescue appliances and the training of rescue brigades, with the employment of boys, girls and women, the appointment of inspectors, the payment of wages, and the protection of horses used in mines. This Act raised the age of employment of boys underground from 13 to 14 ; and besides prohibiting the employment of women and girls underground, it regulates their employment above-ground. No boy or girl may be so employed under the age of 13, and no boy, girl or woman may work more than 54 hours per week, or more than ten hours in one day, and not at all between 9 p.m. and 5 a.m. or after 2 p.m. on Saturday and all day Sunday. Other provisions are similar to those of the Factory Acts ; but "no boy, girl or woman shall be employed in moving railway wagons " or in lifting or carrying anything so heavy as to be likely to cause injury.

The Secretary of State is granted full liberty in regard to the appointment of any number of inspectors at salaries approved by the Treasury, but no inspector is to be otherwise employed in connection with mining or minerals, nor may he have any pecuniary interest whatever in any mine. The inspectors are given the fullest powers of visiting and inspecting and calling for plans, certificates, etc., and owners must provide facilities for the inspectors to do their work.

Criticism of the Act and Regulations

A good deal of adverse criticism has been passed upon the Coal Mines Act (1911) and upon the regulations made under it by the Home Office. Colliery managers complain that it is too " academic" or " scientific " ; they do not believe that all the measuring of percentages and many other provisions are of practical utility, and some say that they are no real test of the safety of a mine. The commonest complaint of managers and owners is that the requirements of the Act are so varied, so multifarious and detailed, that it is impossible for anybody to carry them all out. Many are of the opinion that if the officials carried out every requirement in the Act and regulations to the letter, they would have no time for any of the business management of the mine.

It is true that if anyone will compare the present Act and Regulations with the Act of 1887 and regulations made during the fifteen years following, which constitute the law with which every manager has become thoroughly acquainted, he will see what a great change has been effected in the increased number and stringency of the provisions. There is a multitude of matters, such as the appointments of all sorts of officials in writing and the making of written reports and observations, besides a number of the smaller safety provisions which have never been required before ; and managers who were generally overworked before the Act came into

force, having also the Minimum Wage Act and the
National Health Insurance Act to contend with, have
felt themselves overwhelmed. The opinion forces
itself upon me that it is just as necessary to force
colliery proprietors by law to provide an adequate
management staff, as it is to make them provide
new machinery and roadways. It would be wise, I
am sure, to provide that the manager responsible
for the technical working of the mine and its safety
shall not be concerned with the business management, which should be undertaken by a man specially
appointed for the purpose. Such a change would,
I am sure, in the long run prove itself of the greatest
advantage to colliery owners, not only by decreasing
the number of accidents, but also by giving the man
responsible for the business side time to concentrate
upon problems of cutting down costs. The man
made responsible by law for the safety of the mine
should have nothing to do with costs. It puts him
at once in a false position. It should be his business
solely to undertake the owner's obligations in respect
of the law and of the general safety of the mine.
The owner must employ someone else to look after
his business interests in the mine.

Another criticism of the Act and Regulations is
that although its provisions are excellent if carried
out in the spirit of the framer's intentions, many
requirements are open to a slipshod observance
which may seem to satisfy the letter of the law, but
in practice are nearly useless. As an instance, I
may mention the regulation which requires that

before a shot is fired the floor shall be sprinkled with water to a distance of 20 yards in every direction. Many firemen think they are fully carrying out the requirements of the Act if they take a bucket of water and ladle out splashes here and there, perhaps only with the hand. This not only actually raises dust where the water falls, but it leaves a large total area completely unwatered. It is obvious that what is wanted is a fine spray such as is obtained with a sprinkling nozzle on a garden hose or with one of the numerous forms of syringes for spraying fruit trees. Why should not the regulations stipulate precisely the kind of instrument to be used to make the spray ? In every case there should be no room for a slipshod interpretation of the provisions of the Act or of the regulations.

Another defect of the Act is the too frequent occurrence of the words " as far as possible " and " so far as practicable." Such qualifying words tend to provide an excuse for the unwilling. The framers of the Act should have made up their minds whether a given requirement was practicable or not, as for example, the provision of dust-tight trams. In some cases the saving phrase seems to be designed to cover cases where it is physically or commercially impossible to make the provision which is defined. If it is physically impossible, of course the Act cannot apply. If it is a monetary or management difficulty, the arbitration clause should be used and not merely a loose phrase which seems to leave it open for the management to claim impossibility or impracticability on all kinds of grounds.

From what I am told by miners' agents, I am afraid that there are many provisions of the Act in regard to inspections, measurement of air currents, reports, and so forth which are not yet being properly observed. Since the Senghenydd explosion there have been numerous strikes at individual collieries upon the question of the safety of the mine, the men refusing to go to work until their examiners were satisfied. In several of these strikes the men have insisted successfully upon the dismissal of particular firemen whom they accused of negligence in the carrying out of their inspections. As the result of one strike early in 1914, a deputation of the miners was received by the directors, and the latter were upon investigation convinced that the mine had not been kept in a safe and proper condition. The managing director and manager resigned, and nearly half the officials of the colliery were dismissed. Several of the districts of the South Wales Miners' Federation have appointed a full-time assistant to the miners' agent, whose sole duty is to act as workmen's examiner along with one local employee at each colliery, according to the powers under the Act. I have before me an interesting leaflet issued by the Blaina (Monmouthshire) District of Miners appealing for funds in aid of the men engaged in another of these " Safety Strikes." Whilst I cannot vouch that the statements of the workmen are correct, as no official report dealing with the case has been issued, it is worth while to print the following extract, if only as an illustration of the kind of dangers with which the workmen believe that they have to contend :—

" Our fight is one for a greater measure of safety ; for years past we have taken reasonable advantage of Section 16, New Mines Act, to take Workmen's Examinations, and our greatest complaint has been that we always find Coal Dust in abundance and report same, yet we cannot confess that it is ever satisfactorily dealt with. But during the recent 20 months there has been a distinct tendency to lag under great cavities in the roof instead of cogging up to the roof so that it can be ventilated and consequently kept clear of gas.

" The incident responsible for our present stoppage is briefly as follows :—That we charge the firemen of No. 3 district to have failed to correctly report the state of the said district on certain dates during last month (May). For instance, the two firemen who examined this district on the morning of May 25th (4 a.m. to 6 a.m.) preparatory to the day shift coming on, reported everything all right with the exception of a bit of gas found in one place which was cleared. Now a couple of our responsible workers at the close of the shift of Saturday, 23rd of May, being suspicious that these lagged off cavities were not examined and correctly reported, adopted a certain course which justified them in applying for lamps at 6.30 on the morning of May 25th (Monday), so as to endeavour to learn what reliance could be placed on the reports of the official examiners upon whom depends the safety of all underground workers ; they hurried on to the first of these suspicious cavities and found their worst fears realised, in finding an accumulation of gas to the approximate extent of 3,600 cubic feet, also three other lagged holes with large quantities of gas were found in the immediate neighbourhood with six smaller accumulations still in the same district.

" Let us again remind you, fellow-workers, that our mine is never free of large accumulations of dry and fine coal dust, so that you can fully appreciate, along with us, what awful possibilities existed for a repetition of another ' Senghenydd ' in the South Wales coalfield."

When, as in this case, 900 men will unanimously strike for three or four weeks, solely on the ground of safety, it is probable that there is some justification for their action. Further confirmation of the view that there is a good deal of laxity in the observance of the statutory requirements in some mines, is obtained by perusal of the report of the Court of Enquiry upon the Senghenydd explosion, which ought to be read by every one interested in this difficult question of safety in mines.[1]

This disastrous explosion, in which 439 men were killed, occurred on October 14th, 1913. The mine was known to be a " fiery " one. In its early history there had been an explosion with serious loss of life. So recently as October, 1910, there had been a

---

[1] Parliamentary Paper [Cd. 7346], 1914, 1s. 6d.

large outburst of gas, which was not got under
control for four days, during which time all the men
were drawn out.   In spite of this fact, the manager,
who was also the agent of the owners, had apparently
not taken any special precautions.   He had not even
taken any steps to carry into effect a great number
of the provisions of the Coal Mines Act, 1911.
The report of the Enquiry bristles with evidence of
infractions of the Act, and if many of them, like
failing to appoint officers in writing, were more or less
technical, others like the failure to provide apparatus
for reversing the current of air, and failing to deal
properly with coal dust, are of the greatest moment,
and are most probably responsible for a large proportion of the deaths.   I will not attempt to
summarise the report ; for I hope that readers will
form their own opinions from the well-weighed
words of the Chief Inspector.

Unfortunately it was impossible, from the evidence
obtainable after the fire caused by the explosion
was subdued, to be certain of the cause of the explosion, but the Chief Inspector inclines to the view
that heavy falls of the roof occurred liberating a
quantity of gas which may have been set on fire
either by a spark from the wires of the electrical
signalling apparatus, or by sparks caused by the
falling of stones.   In spite of a special circular sent
out a few months previously by the Home Office,
calling attention to the proved danger of unprotected
electric bell wires, the owners had taken no steps
whatever to avert danger from this cause.

2 E

We are not at the end of the story of the Senghenydd disaster yet, I hope, for there are many lessons which Parliament must still learn from it. The last event up to the time of writing is the judgment delivered in connection with the prosecution instituted by the Divisional Mines Inspector for Wales on behalf of the Home Office against the manager and owners of the Senghenydd Colliery for infractions of the Coal Mines Act. The case was tried by a bench of three of the local magistrates. There were 17 informations against the manager and four against the company. Seven of the former were dropped; and the magistrates dismissed several others, and all against the company. They also gave extraordinarily lenient sentences on the charges on which they convicted the manager. Here is a list of the offences and the sentences :—

CHARGES AGAINST THE MANAGER.

1. Failed to appoint lampman at the surface in writing. (Technical.)	Fined £2, or 14 days.
2. Ditto, underground lampman. (Technical.)	Ditto.
3. Failed to keep a book and enter therein readings of the barometer, thermometer and hygrometer. (Recorded readings can give valuable warnings of dangerous conditions.)	Fined £5, or 14 days.
4. Failed to provide means of reversing the air current. (Very serious infraction.)	Fined £10, or 1 month.
5. Failed to systematically clear coal dust. (Chief Inspector took serious view of this.)	Dismissed.

6. Failed to make daily report on condition as to coal dust. (Serious.)	Fined £5, or 14 days.
7. Safety lamps provided with glasses not of approved type. (Not serious.)	Dismissed.
8. Failed to measure air currents.	Dismissed. (See below.)
CHARGES AGAINST COMPANY. The above charges, Nos. 3. 4, 7, 8, repeated.	Held company not liable as they had properly appointed a manager, not interfered with him, and made all necessary financial and other provisions, and the company had no knowledge of the offences.

That is the sum total of the convictions obtained.[1] Failure to provide the means of reversing the air current meant the loss of perhaps 100 lives or thereabouts, which might have been saved had the apparatus been available. If the coal dust had been dealt with, the explosion might have been much less serious. No wonder that the local Labour paper headed its report, "Miners' Lives at 1s. 1¼d. each!"

It is interesting to note how advantage was taken of the consideration shown by the inspector, and of

[1] Since the above was written, it is stated that the Home Office have appealed against two of the decisions of the magistrates, namely, the dismissal of the case against the manager that he did not take steps to prevent the accumulation of coal dust, and the case against the owning company that they had not provided the means of immediately reversing the air current.

the letter of the law, when it seemed to be in favour of the management.   As regards means of reversing the air current, this should have been provided by January 1st, 1913, but four months later the manager first applied for an extension of time under the Act. After correspondence and a visit by the sub-inspector to the mine, the extension was granted until September 30th.   When the explosion occurred fourteen days later (October 14th), no work had been yet commenced for complying with this requirement. As regards measuring the air current, the magistrates decided that as the regulations made only came into force on September 16th, 1913, and only required the measurement to be made once a month, the manager had still two days to make the necessary measurements at the time the explosion occurred. It appears to have been overlooked that the Chief Inspector in his report says that though part of the mine resumed work in November, no measurements were, in fact, made until the following January. Why are these laws passed if Parliament does not mean them to be enforced ?

Let us assume that Parliament does mean them to be enforced ; and see how that may be done. It is evident that when certain people are so ready to evade the law, it is necessary to have a strong force of inspectors in a perfectly independent position who will be constantly in touch with the collieries, and be supported by magistrates who will convict on reasonable evidence and inflict adequate penalties.

A number of additional inspectors have been appointed since the Act of 1911 came into force, but they are not yet nearly sufficient in number. There has, however, been recently an entire re-organisation of the grades of inspectors and their districts. For instance, in 1907, besides the head inspector there were thirteen chief inspectors and twenty-six assistant inspectors. In 1913 the staff had been re-organised so as to consist of the chief inspector, Mr. R. A. S. Redmayne, C.B., eight inspectors-in-charge, eleven senior inspectors, thirty-two junior inspectors, twenty-two sub-inspectors of mines, and eight sub-inspectors of quarries. The number of the staff was thus increased from 40 to 82. The senior, junior, and sub-inspectors are allotted to particular districts, each under control of one of the inspectors-in-charge.

It is very questionable, however, whether the increased number of inspectors now at work is by any means sufficient for the work that is to be done. There are 3,289 collieries in the country which have to be inspected, and a thorough inspection of a colliery would take at least three working days spent at the pit. The inspectors, however, are not occupied only in visiting and going over the different collieries. They spend much time in investigating the accidents which are constantly occurring, attending inquests, and writing up their reports. It is, therefore, obviously impossible for the present staff to thoroughly inspect all parts of all the collieries in the country. They are forced to adopt a system of inspection

" by sample." They choose at random certain machinery or a particular district of the mine, and request to be taken to it. If what they select is found to be in good condition, it is assumed that the proper standard is being maintained in all parts of the colliery. But this assumption is fundamentally unsound as a protection to the miners, for any system of inspection by sample always assumes that a small percentage of the whole not coming up to required standard is not detrimental and may be neglected. This is not the case, especially with regard to the accumulation of gas in a mine, for the smallest accumulation not treated with care may produce a disaster in a dusty mine. The principle is clearly enough recognised by the Coal Mines Act itself in regard to safety lamps, for it insists in the most rigorous way that every lamp shall be carefully examined and locked, on the ground that a single lamp damaged or tampered with may involve the whole colliery in disaster. The inconsistency of securing perfection in regard to the lamps, but not in regard to the maintenance of the various parts of the mine, is so patent that I hope it will soon be taken to heart, and the necessary additional inspectors be appointed.

It is fortunate that a very useful article dealing with this subject has just been published by Mr. William Brace, M.P., President of the South Wales Miners' Federation.[1] Mr. Brace is very clear and forcible in his statement of the reforms needed, and

[1] *London Magazine*, October, 1914, p. 169.

I cannot do better than quote his remarks about
inspection :—

"The inspectors, I repeat, are not sufficient
in number. They have to cover too wide an area.
By the time an inspector leaves his residence and
gets to the collieries a long way up the valley, by the
time he has examined the plans and read the fireman's reports, a substantial part of the working day
has gone, and there remain only three or four hours
for his underground investigation. We want a real
inspection. There is not a colliery in the country
but ought to be inspected once a month, and by
inspection I do not mean that a man should turn up
for a few hours and run round the workings, but that
every section, whether working or not, shall be
examined, roads travelled over, timber and so forth
examined—that we shall have, in fact, a thorough
inspection so that men will be able to feel, after it is
all over, that their colliery is in a safe condition.

" We should survey and map out the whole of the
United Kingdom into easily-covered areas ; we
should have for a limited given number of areas a
first-class inspector in charge ; each area should
contain not more than 5,000 workers, in any case,
and while the chief district inspector's special
function would be to prepare reports for the Home
Office (reports which ought to be public documents
and posted up at the top of the colliery to which
they refer), he would in addition represent the
Home Office at the inquest, should, unhappily, men
lose their lives.

" He should also supervise the work of his staff of inspectors. He should have sufficient time at his disposal, when he is called on to inspect a colliery that has been reported as dangerous, to give it more than the superficial examination he gives it now. In present circumstances his inspection lasts two or three hours. It ought to last eight or ten, or even days.

" The inspectors acting under the direction of the chief district inspector should be selected by the workmen, but paid by the State. Instead of their time being taken up by railway travelling, they should be placed in an area easily ' get-at-able,' so that they can give the whole of their time to their particular work. These inspectors should be selected by a ballot of the workmen in the same manner as the checkweighers are selected."

The suggestion that there should be junior inspectors paid by the State but selected by the workmen, is one which would certainly lead to controversy ; but there is an insistent and growing demand by the workmen, which I think I may fairly summarise under the following three headings :—

(1) That there shall be at least one inspector paid by the State and permanently employed at each mine having more than 1000 workmen or any group of smaller mines making up 1000 workmen.

(2) That candidates for such positions must have had five years' experience as working colliers,

besides obtaining a certificate of competency by a theoretical and practical examination.

(3) That a sufficient salary should be paid to such inspectors to obtain candidates of a high standard and to prevent the likelihood of colliery owners being able to offer more remunerative employment as under-managers, etc., to inspectors who might retire.

There is a party amongst the miners who would go further than this. They desire that the whole of the firemen employed in inspecting every colliery should be State officials, and should be employed solely in their duties of inspecting the mine and instructing the workmen how to take precautions for safety. The cost of their salaries, it is proposed, should be recovered by the State from colliery owners, who now bear the cost of the inspection by being obliged to employ the firemen. At present the firemen are employed in measuring up piece-work as well as at their statutory duties ; and there is, further, the trouble that as they are employed by the owner it is part of their duty, implied or expressed, to keep down the cost of timbering and other deadwork, so that they are placed in an invidious position, where their duty to the workmen and to the State may appear to be in conflict with their duty to their employers. This question of obtaining really efficient inspection, like maintaining adequate ventilation, is bound to become a more and more serious one as the workings of all collieries are gradually extended in area.

Although improved inspection will be the most fruitful means of reducing the death-rate in mines, there are certain directions in which the law still needs to be strengthened. Now that we know the dangers of coal dust, it should be made compulsory to bring all small coal and dust out of the mine. In many mines it is the custom to throw dirty small coal into the " gob " or waste, where it is always a potential source of danger either from spontaneous combustion or in taking fire when an explosion occurs. As regards the removal of coal dust, I think experiments ought to be made with the use of a suction plant on the vacuum cleaner principle. Where stone dust is not used, the present practice is to either brush the dust off the ledges, or to blow it off with compressed air. But this simply raises the dust and allows it to settle again further on in the mine, making the air dusty in the meantime. If the dust was sucked into a receptacle it could be taken in closed bins to the surface, and there be put through a series of machines which would sift out the stone and shale dust by gravity in air currents, on the principle by which the different parts of the wheat grain are separated in a modern flour mill. The coal dust, cleaned in this manner, could be most economically used for firing the boilers at the colliery in one of the various forms of burners designed for the purpose. Once experiments have shown it to be practicable, Parliament should make the removal of coal dust by suction compulsory. Until this can be accomplished it should certainly

be made compulsory to water certain zones in the
manner described above.[1] The necessary air cross-
ings, doors, and so forth, for maintaining the minimum of ventilation in a mine should be required to
be made of permanent incombustible materials in
all mines which will not be worked out within ten
years, and not only in new mines. The system of
permanent emergency safety doors should also be
adopted to facilitate a change or modification of the
ventilation in case of an explosion, so that air might
immediately be brought to any part of the mine.

It ought to be compulsory to fill up or shut off
completely with some non-combustible material all
overhead cavities where falls have occurred. Fire
damp, being lighter than air, always accumulates
about the roof and particularly in these cavities.
For the same reason the use of overhead sheaves or
rollers on which the tail rope of the haulage system
passes should be prohibited. Such sheaves are often
attached to the timber overhead, and if they jamb
and fail to turn the rope sliding over them quickly
causes great friction and sparks of fire which occur
just where they are most dangerous.

The technique of mining is constantly changing
with the more extended adoption of machinery
underground. The use of electricity for motive
power, the installation of coal-cutting machines and
band conveyors and other forms of machinery, are
always giving rise to new sources of danger which
provide new problems for the inspectors and the

[1] Page 429.

legislature. But whilst new regulations and legislation will be constantly needed from time to time, the urgent need of the present moment is to have the many satisfactory provisions of the Act of 1911 properly enforced. Conscientious colliery proprietors have gone to great expense to meet the requirements of the law ; and it is most unfair to them that others should be allowed to continue working their collieries, perhaps under-cutting them in the market, because they have saved expense by failing to comply with the law. If we take this into consideration together with the paramount importance of increasing the safety of the workmen in mines, we shall understand the urgency of the demand for a more adequate force of inspectors.

CHAPTER XVII

MINERS' TRADE UNIONS

Early Conditions

THE trade union organisation of the miners of Great Britain is in many ways unique both in regard to the rapidity of its growth in recent decades and to its completeness and effectiveness, which is maintained without the aid of an auxiliary system of sickness and unemployment benefits. It is not my purpose to give a history of the growth of trade unionism amongst miners ; but it is eminently necessary to see by what stages it has grown to its present strength, what battles have been fought and won—what has been the power of the strike and how much has been gained by legislation—in order that we may comprehend what the present organisation has to maintain and what are its ideals for future progress.

In the former chapter [1] on Safety in Mines, the appalling conditions of labour in the coal mines in the eighteenth and early nineteenth centuries were described. Serfdom was finally abolished in Scotland in 1799 ; but the custom of yearly hiring, which replaced it and persisted for long in Durham and

[1] Chapter XV, p. 375.

445

Northumberland, together with the unjust law making a servant's breach of contract a criminal offence, was an agency by which a certain degree of bondage was maintained. It must be remembered, too, that until 1842 women and children were employed underground at heavy and degrading tasks and that accidents in mines from falls and machinery, as well as explosions, were horribly frequent before inspection was inaugurated under the Act of 1850, whilst the miners at many collieries were defrauded of their earnings by the truck system, unjust weighing machines, fines and confiscations. It is sad to think that evils and injustices were allowed to prevail in the darkness of the pits in remote mining districts which would never have been tolerated so late in the nineteenth century had they been known to the general public.

The miners, then, in their early efforts at organisation, had to struggle not merely to raise their wages but to protect themselves from a whole series of dangers and impositions. There were frequent strikes from the earliest times, but miners' unions were for decades of a weak and evanescent description. A wave of sudden resolve to resist oppression and seek better conditions spread over a coalfield or two at a time, and for a few months, sometimes for three or four years, a union existed embracing a varying, but usually small proportion of the men. For example, a union was formed in Lanarkshire in 1835, and broke up in 1837 after a four months' strike to resist a reduction. In North Staffordshire, after several small local unions had been formed and collapsed, an apparently strong union was formed in 1840, and it gained some important concessions. The depression of trade beginning in 1847 led, however, to increased hours and reductions of wages, and in 1849 the union broke up, the men being completely disheartened. The same happened in other coalfields.

Prior to 1860 there were no firmly established and permanently organised trade unions amongst the miners, and permanent organisation was only at length achieved after a long series of attempts which failed. A number of successive stages in the evolution of miners' organisations can be recognised, and an approximate name and date may be given to each period, though it must be remembered that the dates are more or less arbitrary in many cases, merely marking when a particular tendency or policy which had been growing for years began to mature in events or organisation.

The first period is that of organised strikes against grievances, but with no real attempt at forming a permanent union. Commencing at the end of the eighteenth century it continued till about 1835, at which time attempts to form local or district unions became numerous. Previously a strike had occurred in relation to particular grievances such as a reduction of wages or opposition to the truck system when the price of provisions was raised, or the quality of goods lowered ; but such a strike was organised by a committee formed for the occasion, which generally had no funds at its disposal but what it could raise from sympathisers. Many of these early strikes were hard fought, however, and lasted several months.

Ephemeral Unions

The second period is that of ephemeral unions, and it may be said to last approximately from 1835 to 1860. Its chief characteristic is the failure of organisations, but a decided growth in the spirit of combination, together with the emergence of leaders of great ability who devoted their whole time to organising the miners. County associations were formed about 1840 in Lancashire and Yorkshire, Durham and Northumberland, besides the Scotch and Staffordshire unions already mentioned. A most important step was taken in 1841 by the formation of the Miners' Association of Great Britain and Ireland at Wakefield. This was a federation of county and local unions ; and under the leadership of Martin Jude it became extraordinarily active in 1843–4, during a period of improving trade. The association carried on a great propagandist campaign, at one time employing as many as fifty-three paid organisers, who visited, it is said, every pit in the kingdom. The delegate meetings of 1844 represented practically the whole of the mining districts of Great Britain, and the membership of the federated unions is said to have risen to 100,000. A National Conference was held at Glasgow in this year, and support was given to a bitter conflict in Durham, where 30,000 miners struck against Lord

Londonderry and other employers. The strike was to secure fairer terms at the yearly hiring, but Lord Londonderry, who was Lord Lieutenant of the County, descended to every artifice of legal and industrial oppression to coerce the miners back to work on the old terms—wholesale evictions, importation of Irish strike-breakers, intimidation of tradesmen and others in any way assisting the strikers by giving credit.[1] The Miners' Association of Great Britain was obliged to spend its funds lavishly. The Durham strike failed after some months and so did other lesser conflicts. In 1846 came the turn of trade, and in 1847 a severe and general commercial panic, the coal trade soon after entering on a period of depression with general reductions of wages. All the high hopes of the last

[1] In their *History of Trade Unionism*, Sidney and Beatrice Webb quote the following manifesto issued by Lord Londonderry :—

" Lord Londonderry again warns all the shopkeepers and tradesmen in his town of Seaham that if they still give credit to pitmen who hold off work, and continue in the Union, such men will be marked by his agents and overmen, and will never be employed in his collieries again ; and the shopkeepers may be assured that they will never have any custom or dealings with them from Lord Londonderry's large concerns that he can in any manner prevent."

" Lord Londonderry further informs the traders and shopkeepers that having by his measures increased very largely the last year's trade to Seaham, and if credit is so improperly and so fatally given to his unreasonable pitmen, thereby prolonging the injurious strike, it is his firm determination to carry back all the outlay of his concerns even to Newcastle.

" Because it is neither fair, just, or equitable that the resident traders in his own town should combine and assist the infatuated workmen and pitmen in prolonging their own miseries by continuing an insane strike, and an unjust and senseless warfare against their proprietors and masters."

2 F

few years crumbled away, the membership of the union dwindled rapidly, and in 1848 the Miners' Association of Great Britain ceased to exist. The same fate overtook practically all the county and local unions in this and the following year.

It must not be supposed, however, that this association lived in vain. For a few years it fought a strenuous battle on behalf of individual miners who were persecuted by the conditions of their hiring and the magistrates' interpretation of the law of master and servant. At that time a workman who wilfully broke his contract of service, either by absenting himself or leaving work unfinished, could be proceeded against for a criminal offence, and committed for three months' imprisonment, whilst an employer who wilfully and without excuse broke a contract of service was only liable to be sued for damages, a proceeding which no unassisted workman could hope to undertake successfully, unless prepared to risk all his savings. In court it was the law that a master could give evidence on his own behalf when sued by his servant, but the servant could not give evidence for himself when sued by his master, and every industrial employee was a servant in the eye of the law and could be arrested on the warrant of a single justice and suffer sentence to imprisonment by a single justice from which there was no appeal.

Although all trades were persecuted under these laws, the miners suffered perhaps more than others, and the Miners' Association of Great Britain engaged the services of a zealous and active solicitor, W. P. Roberts, who had most successfully fought many cases for the Durham and Northumberland miners. Roberts, writing of his work in the latter counties, says : "We resisted every individual act of oppression, even in cases where we were sure of losing, and the result was that in a short time there was no oppression to resist." [1] There is no doubt that some permanent amelioration, besides an impression on public opinion, resulted from this work which Roberts continued in the wider sphere of the National Association. The association also carried on a good deal of agitation for safety legislation and for inspection of mines.

An attempt was made by Martin Jude in 1850 to resuscitate the association under the title Miners' National Association, and in October of that year a conference was held at Newcastle. The coal trade was, however, in a state of depression, from which it only began to recover two or three years later. At the close of 1855 unionism amongst miners had almost completely disappeared, and wages in many districts reached their lowest ebb in 1855–6.

In 1856 there entered upon the scenes a personality which exercised the greatest influence upon the trade union and legislative reform movements amongst miners, in the shape of Alexander Macdonald. He was a Scotch miner, who worked his way through Glasgow University in the winters, spending his summers in the pit, became a school

---

[1] Webb, *History of Trade and Unionism*, p. 166.

teacher, and quietly saved till he had a modest competence which enabled him to devote himself continuously to the miners' cause. " A florid style, and somewhat flashy personality, did him no harm with the rough and uneducated workmen whom he had to marshal. The main source of his effectiveness lay, however, neither in his oratory nor in his powers of organisation, but in his exact appreciation of the particular changes that would remedy the miners' grievances, and in the tactical skill with which he embodied these changes in legislative form."[1]

" ' It was in 1856,' said Macdonald, when addressing the Miners' National Conference at Leeds in 1873, ' that I crossed the Border first to advocate a better Mines Act, true weighing, the education of the young, the restriction of the age till twelve years, the reduction of the working hours to eight in every twenty-four, the training of managers, the payment of wages weekly in the current coin of the realm, no truck, and many other useful things too numerous to mention here. Shortly after that, bone began to come to bone, and by 1858 we were in full action for better laws.' The pit clubs and informal committees that pressed these demands upon the legislature became centres of local organisation, with which Macdonald kept up an incessant correspondence. An arbitrary lock-out of several thousand men by the South Yorkshire coal-owners in 1858 welded the miners of that coal-field into a compact district association, and enabled Macdonald, in the

[1] Webb, *History of Trade Unionism*, p. 286.

same year, to get together a national conference at Ashton-under-Lyne, at which, however, the delegates could claim to represent only 4,000 men in union."[1]

In 1850 there had begun in earnest the long series of legislative enactments to safeguard miners and provide organised inspection; and in 1855 the extent and efficiency of protection had been greatly strengthened. In the same year a clause in the Friendly Societies Act gave Trade Unions a legal status as such with registration and power to sue, thus providing the means of strengthening the cohesive power of every large trade organisation. In 1859 an Act was passed expressly permitting peaceful persuasion to obtain an alteration of hours or wages, which was rendered necessary by the harsh convictions of magistrates seeking to stamp out peaceful strikes; and in 1860 was passed the Mines Regulation Act, which by Macdonald's exertions gave miners the right to appoint at every pit their own checkweighmen. The necessity for this I have explained in the chapter on Mining Legislation; but in addition to protecting the hewers from fraud the Act is of the greatest importance in trade union organisation, and it ushers in the third stage of union growth, which was very largely concerned with securing complete independence for the checkweigher and perfecting the machinery for appointing him.

---

[2] Webb, *History of Trade Unionism*, p. 287.

Third Period—1860–1888

A noted and important conference of miners was
that held at Leeds in 1863, to re-establish the
Miners' National Association. There were 51 delegates in attendance, representing practically all the
coalfields; and the conference met in three sections,
sitting simultaneously and dealing with Law,
Grievances, and Social Organisation respectively.
A substantial volume of " Transactions and Results "
was afterwards published. The principal objects of
the conference were to improve the conditions of
employment and remuneration, to secure more
efficient administration of the Mines Acts, especially
by having additional and more practical inspectors,
and to limit the hours of juvenile labour. A
formidable list of grievances was drawn up, amongst
which it is interesting to note enumerated :—the
employment of girls on the pit bank, and of boys
under 12 ; the truck system, which still prevailed in
South Wales, Scotland and South Staffordshire ;
the contract or " butty-gang " system ; the for-
feiting of filled trams or " tubs " for such nominal
offences as not packing them full of coal, or sending
out dirt in the coal ; the ill-adjustment of weighing
machines which the workmen were not permitted
to test ; the payment of wages only monthly instead
of fortnightly ; ignorant men being employed as
managers of mines ; the yearly bond system ;
and so on.

A cleavage between the Durham and Northumberland miners and the rest of the country, which continued until the passing of the Eight Hours Act in 1909, first became evident at this conference. The issue was the reduction of the hours of boys under 14 to eight per day, which was advocated by all other districts, but strenuously opposed by Durham and Northumberland then and since. The explanation is that the hewers, fillers, repairers and other skilled workers in those two counties worked upon the two shift system, and had, therefore, already been able to reduce their own hours to eight, and even to seven and six and a half in some groups of pits. Whilst the hewers worked in two shifts, the hauliers and boys worked in one long shift of $10\frac{1}{2}$ or 11 hours to serve both hewing shifts. It was feared that the change would either throw a large number of the hewers out of work, or else so add to the cost of handling and hauling the coal as to affect the hewers' wages adversely.

The " butty-gang " system deserves more than passing mention, as it was one of the great evils which the miners' unions fought and remedied without legislation. It was particularly prevalent throughout Staffordshire, Shropshire, Wales, Derbyshire, Leicestershire, and Notts ; and at the time of the conference (1863) men from the last three counties were introducing the system into Yorkshire. The system is one of sub-contracting, and was applied wherever the long-wall system of working was adopted. A contractor would take a certain length of face from the owners and contract to send out a minimum of so many tons per week at so much per ton, himself employing ten or twelve, or even up to thirty, hewers with fillers, timbermen, boys, etc., paying them all day wages. The contractor was a small " working master," and spent some of his time hewing coal himself, when he would set the pace for his men and expect them to keep up to it. However much the men were driven by abusive language, and threats of dismissal, they got nothing more by their hard work, but the contractor pocketed a handsome profit. The unions in the building trade are at the present day still fighting a similar system, where speculative master builders, particularly in building small houses, let out the work by piece to small contractors, who are men of the status of foreman. What the miners desired, and have almost everywhere obtained by a succession of strikes, is direct employment on piece-work by the colliery company. Most miners' unions now restrict a hewer in charge of a stall to employing one man and a boy on time-wages.

The National Association as resuscitated at Leeds in 1863 was not so much a federation for trade purposes, *i.e.*, negotiations and strikes, as for education and propaganda, both amongst the miners and the public. Macdonald lent powerful aid to the movement, which resulted in the Master and Servant Act of 1867, and remedied the grossest injustices of the law. He also assisted in the continued agitation that led to the passing in 1871 of both the Criminal Law Amendment Act and the Employers and Workmen

Act, which together so greatly improved the position of trade unions and working men. In 1871, for the first time, the employer and his workmen became equal in their rights before the law. The association had established a " labour emergency fund " and heavy calls were made on this in support of numerous strikes and lockouts in South Yorkshire and elsewhere between 1864 and 1869, the issues being the rights of checkweighers, reductions of wages, opposition to the butty-gang system, and attempts of employers, who were now well combined in Yorkshire, to smash unionism.

These fights in its own special districts of Yorkshire and the Midlands left no funds available for other districts, and the miners of Lancashire and Cheshire started in 1869 a rival federation of miners' unions, under the presidency of Mr. T. Halliday, called the Amalgamated Association of Miners, which also gained great strength in South Wales in 1870–73. It did not draw members away from the National Association (or National Union, as it came to be called), but continued to add to the whole number of miners in union until the total for the two federations reached, in a few years, 200,000. In the same year, 1869, two important and permanent district unions were founded : The Durham Miners' Mutual Confident Association, and the North Staffordshire Association.

The milestones on the road of progress are now rapidly passed. In 1871 the first Parliamentary Committee of the Trade Union Congress was ap-

pointed, and Alexander Macdonald became its
Chairman. The Mines Regulation Act of 1872
contained important provisions for the certification
of managers and strengthening the position of checkweighers. In this year there was an extraordinary
boom in the coal trade, the price of coal going up to
a level never since attained, except perhaps locally
and temporarily during strikes. Wages were raised
very substantially, reaching probably a higher
general average than in any subsequent year until
1900. In 1872 the new Durham Association, under
the able leadership of W. Crawford, secured the
abolition of the yearly bond, substituting a fortnightly agreement; and they began the series of meetings
with representatives of the employers, which soon
turned itself into a conciliation and arbitration board,
when in 1874 arbitrators and an independent chairman were called in to make an award as to a proposed reduction of wages, a method which has since
been continuously and most successfully adopted in
Durham, excepting only during the sliding scale
period.

In 1873 came the collapse of trade, and of coal
prices; and consequently wholesale reductions of
wages by 15 or 20 per cent. at a time. The Northumberland miners, who were getting an average of
9s. 1½d. per day early in 1873, were reduced by stages
to 4s. 9d. per day in 1878, and to 4s. 4d. in 1880,
when at last the tide turned again. The piece-work
prices prevailing in the autumn of 1879, at the very
bottom of the depression, form the present basis

to which the percentage is added in many of the principal coalfields.[1]

An interesting and instructive episode in trade union history is the enthusiasm for the idea of co-operative production which seized the movement in the early seventies, as it had done before in 1852, and under Robert Owen's propaganda in 1833–4. It is frequently suggested at the present day that co-operative and co-partnership methods of production will solve the industrial problem, and that the miners' unions with their large funds could easily run two or three collieries of their own. The miners made their experiments once and for all in 1874, and with disastrous financial results. The Durham miners in the boom year 1872 became interested in the idea of retaining the profits of the trade for the workmen, and a propaganda was carried on up and down the country. Next year it bore fruit in the formation of a co-operative company, which decided to purchase a colliery at Monkswood, two or three miles north of Chesterfield. On April 4th, 1874, the Council of the Durham Miners Association resolved :—" That we take £5,000 out of the General Fund and invest it in the Co-operative Mining Company, as we believe productive co-operation to be the only solution to the many difficulties that exist between capital and labour." Altogether, the association subscribed £15,500 to the venture, which had a total paid up capital of £40,000, many private persons having taken shares, even to the extent of

[1] South Wales, Durham, Yorkshire

all their savings.[1]   By June 30th, 1875, there had
been a net loss of £10,863 ; the output had never
reached anticipations, and the cost of production
was very high.   The colliery was in a very unsatisfactory state when taken over, and a suit was filed
in Chancery for the return of the purchase money,
but unsuccessfully.

The South Yorkshire Miners' Association went
even more heavily into a similar venture, as it purchased a colliery with two pits at Shirland, ten miles
south of Chesterfield, in which it sunk and lost
£31,500.   The miners' leaders and others who formed
the committee of management had in each case
advice from a mining engineer before purchasing ;
and the main cause of the failure, in addition to the
falling price of coal, seems to have been inability to
properly direct and control the management.   The
co-operative idea was, in fact, never realised, as the
workmen were only to get one-third share after a
10 per cent. dividend to capital, which was never
paid.   None of the workmen became shareholders, in
spite of strenuous efforts which were made to induce
them to do so.   The obvious lesson of these failures
is that trade unions had best confine themselves to
the objects for which they are formed, and to
business which they understand.

About the same time began the movement for the
adoption of sliding scales to regulate wages.   From
1869 onwards the miners' unions were at last being

---

[1] *History of the Durham Miners' Association*, Ald. J. Wilson, M.P.,
1907, p. 110.

definitely recognised by the employers, and in Yorkshire, Staffordshire, Durham, and Northumberland
there began to take place joint meetings of employers and workmen's representatives, which soon
formed themselves into more or less permanent
Joint Boards or Committees having representatives
appointed by the owners' association on the one side
and representatives appointed by the council, or
the delegate conference, of the miners' union on the
other.   As pointed out by Mr. and Mrs. Webb, the
first result was that the men's leaders, in meeting
and discussing wages with the employers, insensibly
began to take their intellectual position, and a
general agreement ensued that the selling price of
coal was what mathematicians call the " independent
variable," and that wages must vary with it.   The
view that, if wages were taken as a fixed item in the
cost of production, the price of coal would adjust
itself to the wages, was scouted by the coal-owners,
and the miners generally had no confidence in it.
They were glad enough to have their organisation
recognised by the employers, and to have a formula
which would at least save them from the necessity of
fighting arbitrary reductions, and would automatically ensure them increased wages whenever the price
of coal advanced.   Sliding scales appear to have
been adopted in isolated collieries in the sixties ;
but the first to be applied to a district under a joint
agreement was that of South Staffordshire in 1874.
This was followed by one in South Wales in 1875,
after a great strike ; and the first scale in Durham

was adopted in 1877, the South Staffordshire scale being revised in the same year. Scales were soon adopted in every coalfield, but I need not describe any scale in detail here, leaving that to the next chapter.

Although most of the miners' leaders believed in the scales, there were a few like Alexander Mac-- donald who opposed them from the start, holding that the miner ought never to place his wages and his standard of life so completely at the disposal of the colliery proprietors, to be altered as they should find it convenient to vary the price of coal. It is true the first scale in Durham had a minimum ; but the second one had nòt, nor in most cases had those of Yorkshire and the Midlands. Experience taught the men in various coalfields the necessity of having a minimum, and their efforts became directed to forcing up the minimum when obtained.

At the same time there began to grow up during the eighties a formidable movement of opposition to sliding scales. In 1881 the Yorkshire Miners' Association, newly formed by amalgamating the South and the West Yorkshire Associations, formally terminated the then existing scale, and refused several propositions by the owners for new scales ; and Lancashire soon followed suit. The policy in regard to sliding scales was one of the most burning questions at the annual National Conference of Miners. The legislative Eight Hours Day question was also coming to the front. In 1885 a Midland Federation was formed of districts pledged to abolish the sliding scale, and to support the Eight Hours Day agitation.

In 1888 a split in the miners' ranks definitely developed and there was formed by a conference at Manchester the Miners' Federation of Great Britain by the adhesion to the Midlands, Yorkshire, and Lancashire of other districts, notably Fifeshire and part of South Wales. The miners of the country were now ranged in two national organisations : the old National Union, embracing all the districts still adhering to the principle of the sliding scale ; and the Miners' Federation, embracing those which had thrown it over, or were pledged to do so on the expiration of their existing agreements. It is not surprising that the National Union soon met with the same fate as the old Amalgamated Association, and rapidly decayed. I say " not surprising," because the only considerable districts still championing the scale were Durham and Northumberland, South Wales, and a small relic of the Amalgamated Association in South Staffordshire ; and in South Wales there was no real or solid union like that of Durham. The so-called unions of South Wales since 1875, when the men lost heart, were mostly mere formal associations for the purpose of operating the sliding scale agreement, the men's small contributions (2d. per month) being actually deducted by the colliery owners from the men's pay tickets and handed over in bulk to the miners' agent, who thus became very largely a servant of two masters. Some loose associations

in the different South Wales valleys, which did
collect voluntary contributions and support an
agent, had no lodge organisation. In effect, therefore, the National Union was reduced to Durham
and Northumberland, which counties in fact appointed practically the whole of the executive and
controlled the policy.

The last great service rendered to the miners'
cause by the National Union was the support of
the agitation for the strengthening of the law in
regard to safety in mines, and for improving the
position of the checkweigher, which were both
attained by the Coal Mines Regulation Act of 1887.
Whilst previous Acts had given the men power to
appoint a checkweigher at their own expense, the
contribution of the miners who benefited was
optional. This Act gave power to any checkweigher
appointed by a majority of the workmen in any
pit who were paid by the weight of coal gotten, to
recover from every such workman the due proportion of his wages as checkweigher. The appointment had to be made by ballot of the workmen
entitled to appoint. In practice, from this time
onwards, we find a checkweigher at every colliery
of any importance throughout the country, the
colliery company in most cases arranging to deduct
each miner's contribution at the colliery office, as
was expressly allowed by the Act. There must be
a committee of the workmen to supervise the election, or the removal, by ballot of the checkweigher,
and to fix his remuneration ; and this committee

is usually identical with the committee of the miners' lodge relating to that colliery. " It would be interesting," say Mr. and Mrs. Webb, " to trace to what extent the special characteristics of the miners' organisations " and, one may add, their rapid growth from this date, " are due to the influence of this one legislative reform. Its recognition and promotion of collective action by the men has been a direct incitement to combination. The compulsory levy, upon the whole pit, of the cost of maintaining the agent whom a bare majority could decide to appoint, has practically found, for each colliery, a branch secretary free of expense to the Union. But the result upon the character of the officials has been even more important. The checkweigher has to be a man of character, insensible to the bullying or blandishments of managers or employers. He must be of strictly regular habits, accurate and business-like in mind, and quick at figures. The ranks of the checkweighers serve thus as an admirable recruiting ground from which a practically inexhaustible supply of efficient Trade Union secretaries or labour representatives can be drawn."

Fourth Period—1888–1912

With the foundation of the Miners' Federation of Great Britain in 1888 we enter upon the fourth period in the growth of miners' trade unions, which closes with the permanent welding of the whole of the miners of the country in the Federation, which

2 G

may be said to have been achieved in 1912. This short space of twenty-four years has sufficed for a surprising improvement in the local organisation of the miners in the various coal-fields ; but even more remarkable has been the progress of the Miners' Federation of Great Britain, its steady growth in number and influence, the general acceptance of its policy, and its substantial achievements in numerous important Acts of Parliament, and in unification of the trade policy of different districts.

The Miners' Federation of Great Britain has always maintained from its inception an "advanced" policy. During the eighties the trade union world was riven by the rise of the New Unionism which the disciples of Karl Marx and Henry George had brought into being, assisted by the discontent with existing methods of a large body of unionists, particularly of the younger generation. The latter complained of the ineffectiveness both in trade disputes and in legislation of the old friendly society type of union, whose officials were mainly of a conservative and individualistic mind. The New Unionism stood especially for maintenance of the worker's standard of life—if necessary, by restriction of output—for the legal eight-hour day, and for nationalisation of land and the means of transport and production. Most, though not all, of the new unionists were avowed socialists ; but more important, perhaps, than their political opinions was their belief in aggressive tactics in trade matters.

All of these objects and ideals were adopted by

the Miners' Federation of Great Britain from the commencement, and have been steadfastly adhered to by the delegate conferences ever since, even if some of the prominent members of the Executive Council have not been sympathetic with all of them. In politics some of the leaders and officers of the Federation have been liberals rather than socialists ; but the general policy of this great union has from the first been remarkably consistent, being aggressive and progressive, with a distinct socialistic bias of a practical kind.

The membership of the Miners' Federation of Great Britain increased very rapidly in the early years, and has continued to increase from year to year, with slight fluctuations, ever since. In 1888 it was 36,000, next year 96,000, and in 1893 it had reached over 200,000. By 1900 it was over 363,000, but then fell off slightly with the depression of trade to 326,000 in 1904. Then by rapid strides it reached 458,000 in 1907, and with the adhesion of Durham and Northumberland, totalled 590,000 in 1908. It is now (end of 1913) over 670,000, and still growing rapidly. This remarkable growth is accounted for on the one hand by the successive inclusion of new districts, and on the other hand, by an increase of membership in most of the districts, which arose in two ways. In the first place, the total number of workmen employed in collieries continually increased ; and in the second place there came to be a higher percentage of unionists amongst the total employed, as a result of more

thorough local organisation. In 1889 the Federation consisted only of Yorkshire and Lancashire,
certain districts in the Midlands which had united
in the Midlands Federation, a district in South Wales
and Monmouthshire (8,000 members), and Fifeshire.
During the succeeding five years several small
districts joined ; but there remained outside the
Federation the whole of Durham and Northumberland, five-sixths of South Wales, and about the
same proportion of the Scotch miners, organisation
being still very weak in the last named two coalfields.
In South Wales there were in 1893 at least seven
independent miners' associations ; and these were
formed into a loose federation in that year under
the title South Wales Miners' Federation, largely
by the efforts of the well-known leader, Mr. W.
Abraham,[1] better known to the miners as " Mabon."
In 1898 the South Wales Miners' Federation was
consolidated and registered as a trade union, and
early in the following year it decided to join the
Miners' Federation of Great Britain. This involved
throwing over the sliding scale when the agreement
then running should terminate at the end of 1902,
and the concurrence of South Wales in the Eight
Hours Day propaganda and other objects of the
national Federation. The South Wales Miners'
Federation made clear the conditions on which it
was joining, namely, that it pledged itself to uniformity of action with the Federation on all questions
except wages, and that during the currency of the

[1] Now the Rt. Hon. William Abraham, P.C., M.P.

then existing sliding scale and conciliation agreements of South Wales, and the Federated Districts
of England respectively, any financial assistance
rendered, which would be mainly in respect of strikes,
should be of a voluntary character. Careful definition of the implications of membership of the
Miners' Federation of Great Britain was, indeed,
necessary, for the Durham Miners Association had
twice decided, once in 1892 and again towards the
end of 1896, to join the Miners' Federation of Great
Britain. In 1893 Durham refused to join in the
great strike, and was expelled from the Federation.
On joining for the second time correspondence arose
as to whether Object 5 of the Federation, " To seek
and obtain an eight hours day from bank to bank
in all mines for all persons working underground,"
bound Durham to approval of seeking the object by
legislation rather than trade agreements in the
various districts. On learning that Durham would
be bound by the majority, and that the majority of
the Federation favoured legislative action, the
Durham Association decided it could not alter its
attitude of opposition to the legal eight hour day.
After further correspondence in which Durham
affirmed its desire to be a member for wage questions,
the Federation rejected the county on June 30th,
1897, and returned the contributions.[1] Durham
and Northumberland joined the Federation in 1908,
when further opposition to the legal eight hours

[1] *History of the Durham Miners' Association*, Ald. J. Wilson,
pp. 277–80.

day had become futile.   A federation of the Scotch
Miners' unions, of which the principal were the
Ayrshire and the Fife and Kinross, was formed in
1894, and was soon joined by the Lanarkshire
miners, whose present union was formed in 1896.
Thus whilst Fife was an original member, Ayrshire
was added to the Miners' Federation of Great Britain
in 1894 and Lanarkshire in 1896.

The year 1893 was one of great unrest in the coal
trade, owing to the employers insisting on drastic
reductions of wages on account of the continued fall
of the price of coal.   A demand for a 25 per cent.
reduction in the federated districts of Yorkshire,
Lancashire and the Midlands was strenuously
resisted by a great strike involving 300,000 men,
which lasted from late in July to the middle of
November.   There were at the same time strikes of
shorter duration in South Wales and Scotland, a
total of 400,000 men being out at one time.   As
pointed out by Mr. and Mrs. Webb, the great
principle of maintaining the workers' standard of
life was at stake in this struggle ; and whilst the
actual settlement was a compromise, the miners
gained a very important point in the definite
recognition by the employers of a minimum percentage below which wages should not fall whatever
the price of coal.   The attitude always taken up by
the Miners' Federation of Great Britain has been
that a certain standard of wages should be just as
much a first charge upon the coal industry as rents
or royalties, and that any increase of wages when

trade is good is simply a legitimate share of profits and is not to be compensated by an indefinitely great fall of wages if trade becomes very bad. The Miners' Federation of Great Britain has recently [1] passed a resolution pledging all the districts of the country to support financially any district threatened with a percentage reduction which would bring the hewers' day wage-rate below 7s. per day. The Scotch Coal Owners Association has just (July, 1914) given notice of a reduction bringing the day-wage down to 6s. or thereabouts in most districts, and the dispute will therefore involve the Miners' Federation of Great Britain as a whole, and will possibly result in the first national struggle on the living wage question. The Federation has been, so to speak, clearing the decks for action by pursuing a policy which will enable it to terminate agreements with the owners simultaneously in all the coalfields of the country. At the conference at Southport in 1911 the following resolution was passed : "That no district within the Federation having agreements with the employers under their conciliation boards should renew them or sign any new agreement that would carry them beyond the expiration of the one having the longest period to run."

Existing Organisation

It will now be instructive to review the existing organisation of the miners, the growth of which we have traced through its various phases. There are

---

[1] Scarborough Conference, October, 1913.

certain features common to the miners' unions in all
the coalfields ; and there are certain other features,
particularly the degree of centralisation, in which
they differ considerably.

In miners' unions each colliery forms a natural
unit of organisation ; for not only do most of the
men live in the immediate locality but the disputes
and negotiations of that branch or " lodge," as it
is usually called, will be all with one employer.
Now that large mines employing some hundreds or
thousands of men are becoming the rule, the physical
condition of employment tends largely to favour a
strong organisation amongst the workmen, as we
shall easily see if we contrast such conditions with
those under which the printers or the carpenters
have to organise. Any one of their branches will
deal with a large number of small employers, and
their members are scattered throughout a number
of small workshops where the conditions of the
work differ considerably.

Every miners' lodge has its own committee with a
secretary, a treasurer, and a chairman. The secretary and treasurer are very busy officers, usually
devoting nearly all their spare time to the work,
whilst the secretary, if he keeps on good terms with
the management, often takes a day or two off from
work at the pit in order to cope with his secretarial
work, if he is in charge of a considerable lodge. The
secretary is usually paid for his work, or at least
for attending committee meetings and delegate
meetings. He is also paid a full day's wage when

obliged to miss a shift on miners' business. The treasurer and chairman are not usually paid unless they have to lose time at the pit through attending distant meetings. The committee meets weekly to consider a variety of business concerning members in arrear, disputes at the pit, claims under Workmen's Compensation Act, claims under the Minimum Wage Act, and other matters of routine business ; whilst from time to time they have to deal with special business such as the negotiation of a new price list or questions remitted to them by the executive council of the union or federation to which they belong. Most important discussions arise upon the instructions to be given to delegates attending a general conference of the coalfield in regard to some special matter. Generally there will be two delegates appointed from each lodge ; and often it is considered necessary to call a mass meeting of all the members of the lodge or to take a vote, sometimes by ballot, upon the instructions to be given.

Many problems and apparent inconsistencies in the proceedings of the miners' trade unions will be rendered clear when it is remembered how essentially democratic is their government. As time goes on the general level of education and capacity for independent thinking increases amongst the miners. They are becoming more and more jealous of handing over any considerable powers, especially in matters of policy, to their elected representatives, and this applies also to their permanent and highly placed officials, whose advice they may listen to

but do not always accept. Important questions
such as whether a strike shall be declared or terminated, what instructions shall be given to delegates
at a conference, what alterations in contributions or
official duties shall be made, are constantly being
referred either to votes of mass meetings, which are
attended by the lodge committee and often also by
the miners' agent, or to ballot.

There can be little strength in a miners' union
which is not based upon a lodge organisation.
Historically it is pretty clear that the large ephemeral unions, such as the Miners' National Association, died out as rapidly as they came into existence
simply because they had no effective lodge organisation.   An organiser was sent round the coalfield
addressing men at meetings near the pit-head at
colliery after colliery.   Many of the men paid contributions and thus became members, and a local
secretary was appointed to continue receiving the
contributions and to run the business of the union
in the locality.   But one man with no secretarial or
organising experience can be of little effective
service to some hundreds of workmen, however
enthusiastic he may be.   Consequently membership
of these large, quickly-formed unions gave the
members little material benefit, except in special
cases where a local dispute became sufficiently
important to be taken up by the central committee.

The stable form of the older unions appears to
have been the quite small unions which flourished
from the 'fifties and 'sixties onwards for many years

in different places, often with less than 1000 members in each. These were independent unions either of the workmen at a single colliery or of the workmen in two or three neighbouring collieries. Whenever they developed an effective organisation with one or more capable officers, they became fairly permanently established. Their weakness was in dealing with such a small section of the trade in their coalfield ; so that they had little influence on general trade questions affecting advances or reductions of wages, terms of contracts, and so forth. Such local unions gained strength when they became sufficiently numerous to form a federation, and this was very largely the course of events in the Midlands, Lancashire, and Scotland.

In South Wales the local organisation did not become strong, except in a very few places, until district unions were formed, each embracing a number of collieries in two or three connected valleys. In South Wales, and also in Durham and Yorkshire, the policy appears to have been to perfect the lodge organisation after a district or county organisation had been set up, it being clearly recognised how indispensable to stability is an efficient local organisation. Although the lodge committees deal mainly with routine matters of purely local importance, it has to be remembered that local disputes or encroachments by the management may become very numerous in a single colliery, and protection in these may be of greater importance to the workman than general trade matters. A central organisation covering a large area cannot deal promptly with a multitude of local matters. It must have efficient local committees under it to whom a good deal of discretion is given.

The standard type of trade union is one in which the lodges are part of a higher organisation, which maintains one or more permanent paid officials, besides having the usual officers and generally a Council of Delegates and an Executive Committee. The contributions of lodge members, which usually vary from 3d. to 6d. per week, may be either remitted in bulk to the central executive, which then allows a certain proportion of the funds of the union to be expended by each lodge and closely scrutinises its accounts ; or the lodge may retain, say, 1d. per member for its own expenses and forward the balance to the union treasurer. The last is, on the whole, the commonest arrangement.

The organisation in the different coalfields differs very much as to the size of the unions, the closeness of the control exercised over the lodges, and the organisation of different classes of workers. For example, in South Wales, the district unions of lodges are twenty in number, and these again are united in the South Wales Miners' Federation, of which they are now termed " districts." Each of these district unions has its own executive committee and delegate conference, has its own funds both for administrative purposes and strike pay, nearly every one employs a full time miners' agent, and generally conducts most of its business without interference from the Federation. The latter is a looser organisation, with an Executive Council composed mostly of the miners' agents of the various districts, and a Delegate Conference which meets annually and whenever necessary for the decision of specially important questions. It is supported by the districts remitting to the Federation Treasurer 6d. per member per month.[1]

On the other hand, the Yorkshire Miners' Association is a compact union of all the lodges, being organised for the whole of the West and South Yorkshire coalfields just like one district of the South Wales Miners' Federation.

In Durham the policy has been to form separate unions of the different classes of workmen such as colliery enginemen, coke furnacemen, etc. Each of these unions is parallel, so to speak, with the Durham Miners' Association, and each has a separate joint committee with the employers for dealing with trade matters affecting its own class of workers. For general trade purposes, however, all these unions of

---

[1] Payments by the Districts to the South Wales Miners' Federation are determined by Rule 4, which reads as follows :

Contributions.

" 4.—' Each member of this Federation shall pay not less than one shilling per lunar month or sixpence per fortnight. One-half of all contributions paid shall be forwarded by the District Treasurer to the General Treasurer every four weeks."

This amount is considerably less than that paid by miners in other coalfields, and the men's leaders are now urging that the amount should be increased.

the special trades are united with the Miners'
Association in the Durham Coal Trade Federation,
and the latter appoints the workmen's representatives on the Joint Conciliation Board which settles
county questions and particularly the county wage
percentage.

In South Wales, the Miners' Federation admits as
members all persons employed in or about collieries,
and has no separate organisation for workers of
different classes. They become members of the
lodges, and in turn of the districts, just like the
coal-getters. It is, perhaps, for this reason that the
independent union, the South Wales Colliery Enginemen and Mechanics', has for so long stood outside
the Federation. After long negotiations it has
at last decided to join the South Wales Miners'
Federation, feeling, I suppose, that that organisation
is now strong enough to give its artisan members
adequate attention and support. Mr. and Mrs.
Sidney Webb have come to the conclusion that an
organisation which provides separate unions for
different classes of workers engaged upon different
employments, and joins them in a federation, is far
more efficient and stable than one embracing all
classes of workers in each locality in the same union,
and merely federating the different localities. This
opinion is based upon their studies of trade unions
in all trades throughout Great Britain. I think
their opinion is supported by the organisation of the
Durham colliery workmen being in many ways more
efficient than that of South Wales. The independ-

ent unions for special classes can undoubtedly give greater attention to the particular requirements of the special trades.

The Lancashire and Cheshire Miners' Federation is organised much on the same lines as South Wales, but is somewhat more centralised, all the funds being in the hands of the Federation and not mainly with the districts. It is interesting to note that a recent proposal to centralise the funds of the South Wales Miners' Federation, and to centralise much of the office work and negotiation of important disputes in Cardiff, was decisively rejected by a ballot of the coalfield. The miners are very jealous of the independence of the districts, liking to manage their own affairs and distrusting a bureaucracy of paid officials, such as they imagined it was intended to create.

The Scotch Federation (formed in 1894) on the other hand is not a single union with districts, but a federation of distinct unions, such as the Fife and Kinross Miners' Association, the Ayrshire Miners' Federal Union, the Clackmannanshire Miners' Association, and others. It is paralleled in England by the Midland Counties' Federation, formed in 1886, which embraces seven distinct unions. The Scotch miners have just (1914) agreed by ballot to form a single centralised union for the whole of Scotland.

The Miners' Federation of Great Britain consists of the following fifteen unions and two federations :

UNIONS—	Membership on March 31, 1914.
South Wales Miners' Federation . .	134,190
Durham Miners' Association . . .	120,000
Yorkshire Miners' Association . .	107,224
Lancashire and Cheshire Miners' Federation	72,714
Northumberland Miners' Association .	40,786
Derbyshire Miners' Association . .	36,923
Nottinghamshire Miners' Association .	31,154
North Wales Miners' Association . .	12,000
Cleveland Miners' Association . .	9,295
Cumberland Miners' Association . .	8,525
Leicestershire Miners' Association . .	6,436
South Derbyshire Miners' Association .	4,700
Somerset Miners' Association. . .	3,893
Forest of Dean Miners' Association .	2,731
Bristol Miners' Association . . .	2,138
[Kent Miners' Association] . . .	150
FEDERATIONS—	
Midland Counties Miners' Federation .	54,000
Scottish Miners' Federation . . .	90,000
	736,850

The Miners' Federation of Great Britain occupies
a peculiar position in one way, that it not only acts
for the whole of Great Britain as regards legislation
and general trade matters, but also deals with
the wage percentage changes in the federated districts of England and North Wales, appointing representatives on the Joint Conciliation Board for the
federated districts, and regulating their policy.
The unions included in the federated districts are
indicated in the above tables. In this respect it

exercises the functions which some of its constituent
unions like the South Wales Miners' Federation, the
Durham Miners' Federation, or the Scotch Federation
exercise for their own districts.[1]

Miners' National Conferences

The annual National Conference of the miners
has always been a striking feature of their trade
union activity, even in far away times when organisation
was almost non-existent. The Miners'
National Association was formed at a conference at
Wakefield in 1841, and national conferences have
been held almost every year since that date in one or
other of the larger provincial towns of a coal mining
district, or at some convenient seaside resort. They
have been held under the auspices of the successive
national associations, there being, however, two
independent conferences during a few years when
both the National Union and the Amalgamated
Association co-existed. The important Conference
at Leeds in 1863 has been already referred to.[2]
At that time, and for many years afterwards, attendance
at the Conference was not limited to members
of miners' trade unions ; and social reformers or
propagandists who had interested themselves in the
miners' welfare were not only welcomed in the audience,
but permitted to speak, though apparently since
1863 they have not generally been accorded votes.

With the growth of organised trade unions, semi-
philanthropic leaders like Alexander Macdonald and

[1] The Rules of the M.F.G.B. are given in Appendix 6.   [2] See p. 454.

2 H

W. P. Roberts came to be replaced by professional
miners' agents, most of whom, like John Normansell,
had started work for their fellow-miners by being
elected checkweighers.   From about 1872 onwards
there grew up quite a body of paid officials of miners'
unions, and the Conferences came to be more and
more confined to delegates appointed by the different unions throughout the country, amongst whom
would be practically all the full time union officials,
though, of course, not all the checkweighers.   Hence,
during the past forty years the national Conferences
have been confined almost entirely to accredited
delegates, the meetings being sometimes public to
the extent of admitting newspaper reporters, but more
often being held *in camera*, a brief official report
of proceedings being afterwards issued to the press.

Whilst the Conferences have always been summoned and arranged by a national association or
federation, if one existed at the time, attendance
at the Conference was not confined to delegates
representing the unions which composed the national
organisation.   Membership of the Conference has
always been freely extended to the accredited representatives of miners' unions throughout the country,
even though they refused to join or support the
national union or federation, and this tolerance, and
the free interchange of opinion which it permitted,
have undoubtedly contributed to the strength of the
miners' movement.

There is no space for me to attempt a general review of the work of the annual Conferences ; but it

will be interesting to glance at some of the principal subjects of discussion in recent years. From the establishment of the Miners' Federation of Great Britain onwards until the passing of the Act, the Eight Hours Day question was the subject of resolutions every year, which as time passed became more and more precise in their instructions to miners' members in Parliament to take definite action. Only Durham and Northumberland were continually in opposition. Another constant subject of discussion and resolutions is the safety of mines, as to the necessity both for further legislation and better inspection. The Conference has been unanimous year after year in calling for the appointment of working colliers of at least five years' practical experience to be permanent assistant inspectors paid by the State, but as yet without success. Workmen's compensation and, until 1905, the position of checkweighers, have been frequent subjects of resolutions. Wages questions of a general character naturally come up for discussion occasionally ; but mainly on questions of principle, because the various district unions and joint conciliation boards provide the machinery for dealing with disputes and changes of wages. Uniform action by all coalfields of the country at the same time is most desirable where any new principle is involved. The Conference of 1908 resolved that " miners should be paid for long travelling underground," and reaffirmed that when the then current wages agreements terminated, no further agreement should be entered into unless a higher *minimum* wage should be inserted as part of the conditions. The 1915 demand is for the minimum wage for piece-workers of all coalfields in the Federation area to be raised to 8s. per day, and the Executive Committee is instructed to devise the best means of effecting this object, which is one still (1914) unattained by the Federation. In 1910 and 1911 the Conference gave much attention to the other and separate question of the individual minimum wage, and of course in 1912, the strike year, the subject was dealt with by special Conferences of the Miners' Federation of Great Britain, which then first came to represent all the miners of the country.

The Conference of 1910 was noteworthy for the number and range of the important questions handled. In spite of the time taken up by resolutions on abnormal places and demanding an individual minimum wage, Mr. Smillie had time again to deny that the selling price of coal should regulate the wages of miners, and to reaffirm that the fair day's wages of the miners should be taken as an unalterable factor of the cost of production, and should regulate the selling price of coal. It was then resolved by 114 to 3 that every effort be put forth by the Federation to secure *one* Conciliation Board to deal with miners' wages *throughout Great Britain*, and to have a uniform standard and agreement for all districts. Minor matters, such as extra payment for Sunday work, payment for long distance underground travelling, and weekly payment of wages, were rapidly passed over without dissentients.

In other spheres than wages and trade questions the Conference was active in 1910, as before and since. Northumberland moved that the basis of the International Miners' Union be broadened so that the International Committee be enabled at once to deal with national strikes and threatened international war in such a way as to prevent the latter and make the former successful. Lancashire and Cheshire moved that surplus funds from mining disasters be inquired into by the Executive Committee with a view to urging the Government to collect them all into one national fund for the relief of dependants of all mining accidents. There was the usual strong resolution calling for workmen inspectors, and one demanding a better standard of ventilation and improved facilities for ingress and egress, especially in the case of undersea mines. The housing question also received attention, not only in a resolution calling for legislation to prevent eviction from the employer's cottages during a trade dispute, which is a " hardy annual " at the conference, but also by Scotland's motion that every effort be put forth to secure by legislation proper housing accommodation for the working classes of the country.

In the foregoing resolutions the Conference is laying down a programme of national legislation in the interests of miners, which it wishes the Labour Party to adopt, and which it directs the members maintained by the Miners' Federation of Great Britain to support. Certain general resolutions are even more political. One, which has been frequently passed, demands the nationalisation of mines and of mineral royalties ; and the following even more sweeping resolution was adopted without comment as on former occasions : " That all land, minerals, mines, and railways, be nationalised in the interests of the industries of this country." Another resolution urges the Government to appoint a Minister of Labour ; and a special conference demanded the reversal of the Osborne judgment, whereby the use of the general funds of trade unions for political purposes was made illegal. The Labour Party was requested to regard this as the most important issue before the House of Commons.

The resolutions on nationalisation are little more than a pious registration of opinion, for even the Bill to nationalise the whole of the mines of the country, prepared and introduced in the House of Commons, cannot be said to have secured the serious support of the Labour Party. Yet the Miners' Federation of Great Britain has a way of sticking to things until it gets them, as witness the Eight Hours Day ; and its unanimity as to the desirability of nationalising the railways and mines means that as soon as the political kaleidoscope has turned to a favourable conjunction the united miners will bring all possible pressure to bear on the Government of the day to bring in a Government measure of nationalisation. The movement may take years to fructify ; but nationalisation of the coal mines is undoubtedly coming to the front as a practical policy amongst the rank and file of the thoughtful miners, every great colliery disaster giving it a big impetus. The Miners' Federation of Great Britain conferences accurately reflect the thoughts and wishes of the advanced and active trade unionists of the many hundreds of lodges throughout the country. When the men drive and the leaders are nothing loath, opinions may soon be translated into facts, even with our present slow moving parliamentary machine.

In Conferences subsequent to 1910, a noteworthy feature has been a growing readiness to pledge the support of the whole Federation of Great Britain in support of any district involved in a struggle through following a policy approved by the Federation. There is no question but that the great and successful strike of 1912 welded the Federation into a compact body which does not seem likely ever to be shattered. Northumberland and Durham have been promised support when they take action against the three-shift system, and each coalfield is to be supported against an attack on the wages question and the conditions of labour.

At the Conferences of 1912 and 1913 the most prominent trade questions were: (1) the establishment of a minimum day-wage of 7s. for the whole country at all times for all miners employed at the face or in ripping ; (2) the raising of the wages of surface workers about mines, which are now very low, to a minimum of 5s. per day throughout the country ; (3) the tendency to reductions of wages observable under the working of the Minimum

Wage Act, which it was decided to meet by insisting on a definite piece-rate of payment in respect of difficulties of working in abnormal places. The now familiar resolutions on eviction and housing were passed, as also the usual resolutions in favour of nationalising mines, minerals and railways. The position of aged and injured workmen also received attention ; but perhaps the most significant and important decision taken in October, 1913, is that which empowered the Executive Committee to enter into negotiations with other large trade unions with a view to co-operative action being taken in support of each other's demands. It would be difficult to exaggerate the importance of this decision of the miners. They have hitherto, even as lately as 1912, shown a disposition to rely on their own resources and combination, and to hedge themselves off somewhat in a world apart from the rest of the labour movement. The new policy, and a very bold and wise one, is that of the younger but experienced leaders now coming to the front ; and it has immediately borne fruit in the alliance already formed by the Miners' Federation of Great Britain with the National Union of Railwaymen and the Transport Workers' Federation, there being a standing Joint Committee appointed from their executives. All the tendencies indicate that the miners are likely to become both industrially and politically a stronger and stronger force in the life of this country, and that their growing power will be used for good.

CHAPTER XVIII

SLIDING SCALES AND CONCILIATION BOARDS

THERE are extraordinary fluctuations in the price of coal from year to year in accordance with the general cycle of trade, as is shown in the following table giving the average annual prices of exported coal from Cardiff.

AVERAGE MARKET PRICES OF STEAM COAL F.O.B. CARDIFF

		LARGE—Colliery Screened.				SMALLS.			
		Best.		Seconds.		Best.		Ordinary.	
1901	. .	18	1	16	11	9	5	7	2
1902	. .	15	4	14	5	8	4	6	9
1903	. .	14	6	13	8	7	11	6	3
1904	. .	14	9	13	8	7	3	5	7
1905	. .	13	0	12	0	8	4	6	6
1906	. .	15	2	14	4	9	8	7	9
1907	. .	19	0	17	1	11	10	9	4
1908	. .	16	1	14	4	9	3	7	1
1909	. .	16	0	14	3	9	2	7	2
1910	. .	16	5	15	2	8	8	7	4
1911	. .	17	8	16	0	9	3	8	0
1912	. .	18	1	16	2	11	3	9	11
1913	. .	20	3	18	9	12	6	10	3

The average price at the pit's mouth varies very closely with this.

489

Supposing that a colliery produces regularly a million tons annually whether the price is high or low, its revenue will fluctuate in exact proportions with the selling price, so that the revenue may in the course of two or three years increase by as much as 50 per cent. or decrease by as much as 30 per cent. Naturally, the colliery proprietors do not object to the former, but they are sometimes put into a difficult position by the latter event. In any case it is obviously difficult for a colliery owner to continue paying uniform rates of wages year by year unless his rates of wages are low compared with his total revenue, leaving a large margin of profit. Other expenses, such as rents, royalties, and interest upon debenture and preference capital are each in themselves small compared with the total revenue, and usually remain unchanged from year to year whatever the price of coal ; but there is greater difficulty in continuing wages at a uniform level, because they constitute in most collieries as much as 70 per cent. of the total cost of production.

During the nineteenth century, whenever a slump of trade set in, coal owners found the margin of profit being turned into a loss, and were forced to try and economise by reducing wages. More often than not this led to a strike, in which the men usually had to give way, because it was literally impossible for the owners to continue paying the wages which they paid easily a year or two previously. The men, finding that they had to accept a reduction of wages from time to time, took care to agitate

for an increase when trade improved, with the
result that there were sometimes also strikes when
trade was improving as well as when it was collap-
sing. More often, however, when trade was improving, a colliery owner would give way after
resistance because with increasing prices there was
ample room for higher wages.

The frequency with which labour disturbances
arose solely from the rise and fall of the market, led
to a general desire to make some arrangement
between masters and men which would allow for,
and regulate, changes of wages in accordance with
the state of trade.

After a good deal of local discussion, the idea of
the *Sliding Scale,* in which wages varied automatically with the price of coal, took definite shape ;
and a trial was made in South Staffordshire in 1874
and in South Wales in 1875. In 1879 and 1880, a
number of sliding scales were adopted, chiefly in
the coalfields of the North of England and of South
Wales, and these were revised from time to time
during the next fifteen years, and in some cases
abandoned.

As has already been explained in the foregoing
chapter the colliery price lists are merely standard
prices to which a percentage is added in order to
arrive at the actual money wage. When negotiating
new price-lists, in order that there may be uniformity in the percentage above standard, the new
rates are fixed on the basis of those paid in the datal
year in the particular coalfield in which the mine is

situated. The sliding scale itself simply consists of the figures of the percentages above the standard price lists which are to be in force when the price of coal is ascertained to be at the corresponding figure upon the scale. As an example, I may quote the beginning of the South Wales scale of 1892, which was the last actually in force.

SOUTH WALES SLIDING SCALE OF 1892

Price of Coal over—		and under—		Percentage of Wages above Standard.
s.	d.	s.	d.	
7	10·25	8	0·00	0 (at standard)
8	0·00	8	1·71	$1\frac{1}{4}$
8	1·71	8	3·43	$2\frac{1}{2}$
8	3·43	8	5·14	$3\frac{3}{4}$
8	5·14	8	6·86	5
8	6·86	8	8·57	$6\frac{1}{4}$
8	8·57	8	10·29	$7\frac{1}{2}$
8	10·29	9	0·00	$8\frac{3}{4}$
9	0·00	9	1·71	10
9	1·71	9	3·43	$11\frac{1}{4}$
	etc.		etc.	etc.

It will be observed that the scale was so constructed that a rise in the price of coal equal to one-seventh of a shilling, or 1·74d., was equivalent to a rise of wages by $1\frac{1}{4}$ per cent. of the standard wage. The standard price of coal was eight shillings ; and this meant that if the average price of coal was anything below eight shillings the men would be paid the standard rates of wages as in the price-lists without any percentage added. When the average price had risen above 8s. 1·71d., the percentage above standard was $1\frac{1}{4}$, and so on at the same rate, namely $8\frac{3}{4}$ per cent. for every rise of a shilling in the price of coal. The measure of the price of coal adopted in the South Wales agreement was a two months average of the net selling price per ton of large screened coal " free on board " in the nearest port. In the Durham and Northumberland agreements, the price at the pit's mouth has always been used, which would appear the better arrangement, as eliminating the cost of transport. Practically, however, there is little difference, as changes in railway rates are small in degree and seldom made, whilst f.o.b. prices are somewhat easier to obtain and average. The average price was ascertained by two accountants, one chosen by the masters and the other by the men, and each sworn to secrecy. Ascertainment was made in South Wales every two months and in the north of England every three months. Thus, in South Wales the average price ascertained for, say, January and February, would rule wages during the ensuing two months, and so on.

An important factor of the sliding scale is the selling price of coal, which is taken as equivalent to the standard wage. In the above example of a sliding scale the equivalent to the standard is 8s. If a lower price were taken as equivalent it would be to the advantage of the workmen, whilst a higher equivalent price would favour the owners.

In order to establish the sliding scale, and to give
effect to the changes of wages upon the reports of
the accountants, there was established in each
district a joint committee of the representatives of
the owners and the men, the former to be appointed
by the Coal Owners' Association and the latter by
the Miners' Federation of the district.   There was
little for such committees to do when once the scale
was established, as the fixing of the percentage was
purely a statistical operation ; but such a joint
committee sometimes undertook the settlement of
disputes upon price-lists, and other matters referring
to particular collieries which the local miners'
representatives had failed to settle with the colliery
proprietors.

The sliding scale did not survive in complete
form in any important district except South Wales
beyond the   nineties.   In Northumberland it was
terminated in 1887, in Durham and in the Lanark
coalfield, in 1889.   In Yorkshire, Staffordshire and
North Wales, there were several different sliding
scales in use for various periods, which were terminated at different dates before 1895 ; but in South
Wales the scale remained in operation until 1902.
Only in the iron trade, for iron-ore quarry-men and
blast-furnace men, are hard and fast sliding scales
still in use.   In Scotland and the Forest of Dean it
is still, however, virtually in force, but subject to
revision at short intervals.

Reasons for Abandoning the Sliding Scale

The movement to abolish the sliding scale came from the workmen, who were dissatisfied with it on several different grounds. It was argued that the selling price of coal was by no means the only indication of the coal owners' ability to pay wages. An important factor is the volume of trade—that is the quantity of coal being sold, which often increases at the same time as the price. And another point to be considered, the workmen held, particularly in South Wales, was the increasing price which coal owners were able to get for small coal. In the early nineties small coal was practically given away, but ten years later it was fetching a much better price owing to the increasing demand both for furnaces with mechanical stokers and for manufacture of patent fuel. Again, in many pits, economies in working were introduced, such as reducing the supply of pit props and limiting the frequency of haulage from the face, which, although beneficial to the owners, meant on the average a reduction of earning time to the men. It was urged, therefore, by the latter that the sliding scale, by having regard only to the price of coal, was against their interests.

One section of the men went further and averred that the sliding scale was a direct incentive to the colliery owners to sell at a low price whenever trade was depressed. So much of what they lost in price was made up by a reduction of wages, that the colliery owner's advantage lay in keeping the pit going at its full capacity without much regard to price, so that by keeping up a large volume of trade he would have a sufficient revenue to more than meet all his standing charges in addition to wages. The workmen would have preferred to work short time, thus keeping down the supply in order to keep up the market price. It was observed that they would suffer little or no reduction of total earnings by working systematic short time in a trade depression, and any slight reduction of earnings would be much more than compensated by the additional leisure.

A further objection of the men undoubtedly arose from the inconvenience of the wide fluctuations of wages which resulted. For example, after the periods of prosperity in 1891 and 1901, when wages stood at $57\frac{1}{2}$ per cent. and $78\frac{3}{4}$ per cent. above the standard of 1879, they fell in 18 or 20 months to 10 per cent. and $47\frac{1}{2}$ per cent. above the standard respectively. This meant a reduction of the actual rate of earnings by 31 and $17\frac{1}{2}$ per cent. respectively, due to these depressions of trade. The uneducated miner tersely summed up his objections to the sliding scale by remarking " the —— thing has no bottom ! "

An interesting table, showing the rise and fall of the percentage additions to standard wages in the different districts, is published by the Labour Department of the Board of Trade, and I reproduce it in Appendix VII. The following table shows the percentages above standard as they stood at certain of the highest years and at the lowest year (1896) since 1887.

PERCENTAGE ABOVE STANDARD OF COAL
HEWERS' WAGES[1]

Percentage above Standards of Hewers' Wages at end of—	Northumberland.	Durham.	Cumberland.	Federated Districts.	Forest of Dean.	Somerset.	S. Wales and Monmouth.	W. Scotland.	Mid and East Lothian.	Fife and Clackmannan.
1890.......	31¼	30	40	40	30	30	52½	50	50	50
1896.......	3¾	15	30	30	15	15	10	12½	10	*
1900.......	61¼	65	60	50	50	42½	73¾	100	100	97½
1912.......	38¾	46¼	52½	55	40	45	57½	68¾	68¾	68¾
Net percentage Increase between Oct., 1886, & 31st Dec., 1911.	38¾	42½	50	55	37½	45	50	75	73¾	68¾

* At Standard.

There is no denying that a sliding scale in which wages vary directly with the price of large coal is a very crude instrument of arbitration between the opposing interests, at any rate in the coal trade. Besides the objections above referred to, most of which have more or less force, there are always far-reaching economic forces at work tending to upset the balance of supply and demand, such as the influx of further labour or even a general rise of the cost of living, due to decreasing purchasing power of money. The growth of new industries, which

[1] Taken from Fifteenth Abstract of Labour Statistics (*Cd.* 6228—1912), p. 72, and Sixteenth ditto (Cd. 7131—1913), p. 86.

2 I

create a competing demand for labour, and the spread of education, which raises the workman's standard of living, are further disturbing factors which operate over any long period of years. Economic theory shows, indeed, that there are fundamental difficulties in the way of devising a sliding scale which could be satisfactory to both parties for more than a short period of seven or eight years, after which a revision of the nature of a new bargain would be necessary. This would probably remain true even if the sliding scale were not of the usual crude form, but varied the wage rate not only with price but also in proportion to other important factors such as the volume of trade and the price of small coal.

Conciliation Boards

The sliding scale period in the history of the coalfields has served as a most useful introduction to what is probably a more permanent method of regulating the rate of wages in accordance with the state of trade. The joint committees for the sliding scale accustomed both parties to their representatives meeting from time to time in formal conference to settle the rate of wages for the whole coalfield ; and such a committee passed naturally into what is now called a Conciliation Board. Again, the sliding scale accustomed both parties to regard the average selling price of coal ascertained by independent accountants as a principal basis of the wage rate, and this principle has also been main-

tained. It is curious to note a further survival of
the sliding scale period in the fact that wages are
generally still made to advance or decrease by a
percentage which is a multiple of $1\frac{1}{4}$, which, of course,
is the fourth part of the number five. Thus if the
men demand an increase of 5 per cent. and the
owners offer $1\frac{1}{4}$ per cent., it is probable that the
actual increase of percentage will be either $2\frac{1}{2}$ or
$3\frac{3}{4}$ per cent., not 3 or 4 per cent.

A Conciliation Board usually consists of an equal
number of representatives of the workmen and of
the employers. The owners' representatives elect
their own secretary and president and the men's
representatives do the same. Each section of the
Board meets frequently in private before, or in the
course of, negotiations ; and at a joint meeting of
both sides the president of each section acts as the
principal spokesman for his side. In order to avoid
a loss of voting power by either party through the
absence of any member from the joint meeting, it is
provided that each party shall have an equal number
of votes at every joint meeting. The agreement
states :—

" When at any meeting of the Board the parties
    entitled to vote are unequal in number, all
    shall have the right of fully entering into the
    discussion of any matters brought before them,
    but only an equal number of each shall vote.
    The withdrawal of the members of whichever
    body may be in excess to be by lot, unless
    otherwise arranged."

It is apparent that no important question can be settled by the votes of a joint meeting, for each side invariably votes solidly if there is a serious difference of opinion between the two parties. It is, therefore, the usual custom if the two parties fail to come to an agreement by mutual argument and discussion to make use of an independent chairman. He is really an umpire, and is an essential and vital part of the Conciliation Board machinery. The independent chairman does not usually attend except at an adjourned meeting after the parties have failed to agree. They then each state their case before him, and he gives his ruling either at the meeting or within a few days afterwards. It is specifically provided that the outside chairman shall not be financially interested in any coal mine in the United Kingdom.

The functions of conciliation boards differ a good deal in the different districts. All of them deal with the general rates of wages—that is to say the percentage addition to the standard, and some of them are confined to that purpose (Federated Districts, Northumberland and Scotland). In Durham and Cumberland, and in South Wales, general questions affecting the working of collieries other than wages (e.g. hours and number of shifts worked, methods of weighing coal, etc.) may be brought before the Board, so long as the question affects all, or a large number, of the collieries in the district. Disputes at individual collieries are not generally dealt with, but the exceptions to this are South

Wales and the small districts of Forest of Dean and
Radstock (Somerset).   When the conciliation board
deals only with the general wages level, there are
usually one or more *joint committees* to deal with
local disputes.   They are constituted in much the
same way as the conciliation board, only they are
smaller.   In Durham and Northumberland these
joint committees each cover the whole county, like
the conciliation board, and they have developed not
only an elaborate system of settling local disputes
by conference of the committee, and finally by
arbitration of an umpire, but also the system of
regulating colliery price-lists by the county average
referred to in the preceding chapter.

In the Federated Districts there are several joint
committees dealing each with one of the coalfields,
but in some areas of the Federated Districts there is
no machinery for settling local disputes other than
under agreements entered into between the Miners'
Association of the district and the large colliery
companies, separately or in groups, the smaller
collieries generally falling in with the larger as to
the terms of agreements.   The joint committees,
and the machinery in general for settling local
disputes, can be broadly separated from that for
settling the general level of wages which is the work
of the Conciliation Boards.   In what follows the
Conciliation Boards will be dealt with first, after
which the settlement of local disputes will be
discussed.

The procedure of Conciliation Boards is of a special

character required by their task, not, so far as I am
aware, developed in any previous institution. Most
of the Conciliation Boards meet quarterly upon a
regular day of the month for the purpose of regulating wages ; but the Board for South Wales meets
also at least once in each month for dealing with
disputes arising at various collieries. There is an
agreement in each district between the Owners'
Association and the Miners' Federation of the
district under which the Conciliation Board works,
such agreements being made to run over various
periods, usually three, four or five years. One
clause in the agreement usually states that it is the
desire and intention of the parties to settle any
difficulty or difference which may arise by friendly
conference if possible ; and in pursuance of this
resolution it is provided that the parties shall meet
to consider any application for an increase or
decrease of wages without the independent chairman
being present. If at such usual quarterly meetings
the parties fail to agree, and one or other of them
desires to persist in its claim, the meeting is adjourned for a few days to a date convenient for the
independent chairman to attend. The arguments
of both sides are then repeated before him, each
side votes solidly for its own view, and the independent chairman has to give the casting vote. He
may do so at the meeting ; but very frequently
reserves his decision for a few days until he has
studied notes made at the meeting and the written
evidence put in by both parties. He is usually at

liberty to call for any further evidence which he
considers essential, either at the meeting or soon
after it ; but his decision usually has to be given
within some specified period of five, seven or ten
days after the meeting. He gives such deferred
decision by communicating it in writing to the
secretary of each side at the same time. In some
agreements the chairman has also power to refer
the question back to a further meeting of the
Board.

An important provision of most of the agreements
is that the question submitted to the Board must be
stated in writing, and that it may be supported
by verbal, documentary or other evidence and
explanation. There is usually a chairman of the
Board, who presides in the absence of the umpire
(or independent chairman), and also a vice-chairman
who presides over ordinary meetings in the absence
of the umpire and of the chairman. The usual
custom is that the chairman is elected from amongst
the coal owners, whilst the vice-chairman is one of
the workmen's representatives. In South Wales,
however, there appears to have been a desire to put
both sides upon an absolute equality ; for instead of
a chairman and vice-chairman of the Board, each
party has its *president*. It is provided in the agreement that both presidents shall preside at all
meetings other than those at which the independent
chairman attends. By this peculiar arrangement
there are really two meetings conferring with one
another in the same room. The conference is

opened by the party making the claim and thereafter the right to speak falls to each party
alternately.

There is one very important rule limiting the
action of the independent chairman or umpire in
his decision in regard to an increase or decrease of
wages. It is that he has power only to decide for
or against the proposals previously made in writing
by one side or the other side. If he does not think
either of them justified there is no change of wages
for the ensuing quarter. Thus, if the workmen
make an application for a 5 per cent. increase of
wages, and the owners reply offering an increase of
$1\frac{1}{4}$ per cent., the chairman has no power to split the
difference but must decide on one of the other. In
1912, considerable feeling was aroused amongst the
miners in South Wales because they felt that their
Executive Council had been injudicious in asking
too much. It was generally felt that some rise of
wages was justified owing to the high prices which
the collieries were obtaining with a large volume of
orders immediately after the National Coal Strike.
The workmen's side applied for an increase of
$6\frac{1}{4}$ per cent., but the owners made no offer and the
independent chairman did not find $6\frac{1}{4}$ per cent.
justified ; consequently there was no increase at
all. For three months, therefore, miners throughout
South Wales lost 5 per cent. of wages, which would
almost certainly have been granted.

Peculiar as this procedure seems, there are in
reality distinct advantages in it ; the most important

being that any demand made by either side must be strictly reasonable and capable of being supported by good evidence so that it has at least a good chance of being accepted. Since each party before lodging an application for an increase or decrease has to make up its mind whether it has good evidence to support it, most of the work is really done before the parties meet. This procedure avoids also the difficulty of the independent chairman being confronted with two widely differing claims between which he might have a good deal of latitude as well as a good deal of difficulty in fixing the proper rate. By the present method, the Board meetings are carried through swiftly and harmoniously ; but if each side were able to put forward a claim in a bargaining spirit, with the object of having something knocked off it, there is no doubt that the Board meetings would be lengthy. Misunderstandings and contradictions would be abundant owing to the two sides being unable properly to support their claim, and the result would be acrimony and a generation of ill-feeling, whilst both sides would probably be dissatisfied with the decision of the independent chairman. At present one side is always satisfied.

In the Federated Districts there is a further restriction, for the agreement provides that no alteration shall at any one time (that is, before three months' interval) exceed 5 per cent. of the standard of 1888.

Conciliation Board Districts

The districts covered by the Conciliation Boards are shown in the following table, and the particulars are given of the area covered by each, and the number of men whom their awards affect :—

District.	Counties Embraced.	Approximate no. of men governed by decisions
Federated Districts	Yorkshire, Lancashire, Cheshire, Flint, Denbigh, Shropshire, Derbyshire (South Derbyshire Two separate unions), Notts, Lincolnshire, Staffordshire, Worcestershire, Warwickshire, Leicestershire . . .	589,900
Scotland .	Lanarkshire, Fife, Linlithgow, Lothians, Ayrshire . . .	140,000
Durham . .	Durham	160,000
Northumberland	Northumberland . . .	60,000
Cumberland .	Cumberland	
South Wales .	Monmouthshire, Glamorgan, Brecknock, Carmarthen, Pembroke	230,000
Forest of Dean.	Gloucestershire . . .	8,900
Radstock. .	Somerset (follow Fed. Dists. Conc. Bd.)	6,700
Bristol . .	Somerset and S. Gloucester .	
Kent . .	Kent	1,100

One point of considerable importance is the tendency towards centralisation, which has been adopted as a policy of recent years by the Miners' Federation of Great Britain.[1] One result of this

---

[1] See Chapter XVII.

policy is that the agreements in all the important districts end simultaneously on March 31st, 1915, when it is probable the negotiations for new agreements will be carried out upon uniform lines in all districts. It is possible that at no very distant date, the regulation of the percentage increase of wages may be made for the whole of Great Britain by one Conciliation Board, composed of representatives of the Coal Owners' Associations of Great Britain on the one hand, and representatives of the Miners' Federation of Great Britain on the other hand.

A few special features of the agreements in different districts require notice. In Scotland there is no permanent umpire, but an independent person is called in as a neutral chairman if the parties fail to agree at a second meeting. In South Wales and Monmouthshire a special arrangement is made for the anthracite district, in the far west. Shipments of anthracite coal are excluded in ascertaining the average selling price f.o.b. at the Bristol Channel ports, which is one of the factors regulating wages ; and the percentage above standard is always less in the anthracite district by 5 per cent. of the standard. In South Wales, as in Scotland, the 1879 standard is general, except that in a few South Wales collieries the 1877 standard is used ; and the percentage above that is found by subtracting 15 from the percentage above the 1879 standard.

There are some interesting clauses in the South Wales agreement designed to drive well home in the

minds of the miners the fact that the price paid for
large coal includes payment for all small coal which
may be gotten with it ; and also specifying very
exactly the method of weighing the coal so as to
find the amount of large coal. The South Wales
Conciliation Board Agreement is reprinted in
Appendix VIII.

The Conciliation Board of the Federated Districts
is interesting because of the very wide area, and
the number of different coalfields with varying conditions which it covers. The permanent offices are
in Manchester ; but the usual meeting-place is in
London, far away from any of the coalfields. It is
the success of this widely spread Board which leads
to the expectation that a National Conciliation
Board to arrange the changes of wage level for the
whole country at the same time is quite a possibility.

The world-wide fluctuations of trade make the
changes of selling price of coal simultaneous, or
nearly so, in all the coalfields of the country.

The Forest of Dean is a tiny coalfield with a peculiar and a special agreement. It really maintains
the principle of the sliding scale ; as a definite rise
and fall of the percentage with the price of coal is
provided in the following section :—

" It is agreed that from the 1st day of October,
1910, until the 30th day of September, 1913,
the rate of wages shall be advanced or reduced
$2\frac{1}{2}$ per cent. for each advance or reduction of
6d. per ton in the average price of the five classes
of coals at the five collieries above mentioned,

which has been ascertained now to be 14s. 6½d.
per ton ; but the minimum rate of percentage
shall at no period covered by this arrangement
fall below 27½ per cent. at the collieries now
paying 35 per cent., nor below 22½ per cent. at
the collieries now paying 30 per cent."

An important feature of agreements made in
recent years has been the limitation of the range of
fluctuation of wages. The men have naturally been
desirous to prevent their earnings falling below a
certain amount ; and the employers have naturally
demanded that if they thus bear a great part of the
burden of bad times, they shall have an opportunity
of making up in good times. Thus, in the Federated
Districts the minimum is 50 per cent. above standard
and the maximum 65 per cent. In South Wales
the minimum is 35 per cent. and the maximum 60
per cent., whilst in Scotland the corresponding
figures are respectively 50 and 100 per cent. above
standard.

Secondary in importance only to the minimum
percentage above standard is the selling price of coal
specified in the agreement as equivalent to the
minimum, and the price, if any, specified as
equivalent to the maximum, or other intermediate
percentage. Fixing such equivalents is a relic of the
ideas underlying the sliding scale,[1] and shows that
the price of coal is still considered the chief factor in
regulating miners' wages. The equivalent selling price
is thus specified in the South Wales agreement :—

[1] See *ante*, p. 492.

10.   (c) ...... The minimum of 35 per cent. above
the December, 1879, Standard of wages
shall, subject to sub-section (d) hereof,
be paid when the average nett selling price
of large coal is at or below 12s. 5d. per ton
f.o.b.   When the nett selling price of large
coal reaches 14s. and does not exceed 14s.
9d. per ton f.o.b., the rate of wages shall,
subject to sub-section (d) hereof, be 50 per
cent. above the rates paid under the Standard of December, 1879, and when the nett
selling price exceeds 14s. 9d. per ton f.o.b.
the workmen shall be entitled to claim
advances in the general rate of wages in
excess of the 50 per cent., and up to the
said maximum of 60 per cent., but in cases
of claims to advances above 50 per cent.,
50 per cent. shall be taken to be the equivalent of 14s. 9d. per ton f.o.b., and in the case
of claims to reductions 50 per cent. shall be
taken to be the equivalent of 14s. per ton
f.o.b.   The average nett selling prices shall
be taken as for large colliery-screened coal
delivered f.o.b. at Cardiff, Barry, Newport,
Swansea, Port Talbot, and Llanelly.

(d) At collieries where the Standard or basis
upon which wages are now regulated is the
rate of wages paid in the year 1877, the
percentage payable thereat shall be 15 per
cent. less than at the collieries where the
1879 Standard prevails.

When once the minimum is fixed, much bargaining may centre about the equivalent selling price, for it is to the advantage of the owners to raise the equivalent selling price and of the workmen to keep it as low as possible.   One of the concessions made by the workmen's side in negotiating the new agreement of 1910 in South Wales was to raise the equivalent selling price from 11s. 10d. to 12s. 5d.

The only criticism to be made of this policy is that the general trend of prices due to secular changes of the value of money interferes with the regular fluctuations of prices in the cycle of trade, which is usually 7 or 10 to 11 years in duration.   If the cycle of trade acted alone, the price of coal would fluctuate up and down between certain limits about a mean price ; and the percentage minimum and maximum of wages would be at an equal distance from the corresponding mean percentage. Since 1895, however, prices in general have been constantly on the upgrade, so that if any one in July, 1915, when the present agreements will probably have been terminated, will add up the total number of months for which the minima and maxima respectively have been in force in all districts during the past twenty years, he will probably find that the maxima have been in force for a longer total time than the minima, in spite of the fact that new agreements have from time to time raised both the minima and maxima.

The principal agreements in all the districts, and the principal duties of the Conciliation Boards, relate to the wages of the coal getters, whether working by time, piece or contract.   In nearly all cases, however, the wages of all other grades, such as repairers and hauliers, are governed by the same percentage addition to their standard wage—indeed, it usually applies to all underground workers and to most surface workers, except those who are members of other trades, such as mechanics, enginemen, smiths, carpenters, etc., who belong to their own trade unions, and are mostly paid a non-fluctuating time wage.   In some of these districts it has been found desirable, usually after much agitation by the men, to make subsidiary or separate agreements with classes of workers other than hewers.   There is, for example, a special agreement with the hauliers in South Wales ; and separate agreements have been made in Durham and Northumberland with the mechanics, and with the colliery enginemen, and with the cokemen.   These are only supplementary agreements, the interests of the parties being represented on the general Conciliation Board of the district for the purpose of fixing the general wage level by which they are bound equally with the coal-getters.

Local Disputes

The machinery for settling local disputes is far less efficient than that for dealing with the general level of wages.   There was a period, during the fifties, sixties, and early seventies of last century, when disastrous general strikes of whole coalfields occurred every few years, being often prolonged for several months.   The cause then was usually either a demand by the employers for a decrease of the general level of wages, or a demand by the workmen, who were just beginning to feel the strength of organisation, for an increase. Most of these strikes were the direct result of the fluctuation of the price of coal due to the cycle of trade ; and the sliding scales with their joint committees, and later the Conciliation Boards, have for a generation past entirely prevented strikes due to this cause, during the currency of an agreement. The only serious general strikes have been at the termination of an agreement, when one side or other desired to gain an advantage or remove a grievance in the next agreement ; and the National Strike of 1912, which was on the abnormal places question.

There is every hope, therefore, that continual effort will in time lead to the creation of an efficient machinery for settling local disputes. The existing machinery for local disputes is chaotic, differing in every coalfield, both in structure, power and efficiency. The Federated Districts are not federated at all for this purpose, and each has its own way of dealing with disputes—in some areas by leaving each colliery owner and the local or district Miners' Association to fight it out to the bitter end ; in others by reference to a joint Committee for the coalfield concerned, which may have great or small powers. Serious local disputes have from time to time been referred by agreement to the Conciliation Board of the Federated Districts ; but it is not part of their constitution or regular work. In South

2 K

Wales the local disputes are regularly dealt with by the Conciliation Board at its monthly meetings ; but so complex are the questions involved, that usually the Board appoints a committee, small or large, as seems necessary, sometimes with plenary powers to settle the dispute, but more often to report to the next meeting of the Board.   This is not always satisfactory, for the committee will consist of one or two members of the owners' side of the Board, and one or two members of the workmen's side. They are all very busy men occupied with their affairs in their own collieries or districts, and often they can ill afford the time to go minutely into all the details of a dispute which has probably been brewing for months ; and they find it difficult to meet often.   Hence there is a tendency for the proceedings of such committees to drag on for many weeks, during which the men lose patience and perhaps come out on strike, against the advice of their leaders.   The matter in dispute must, of course, be dealt with first between the local lodge and the colliery direct ; and when, after some weeks, they have failed, it is taken up by the district, and the miners' agent gives more attention to the dispute and negotiates under directions of the district committee.   When the latter comes to the conclusion it has failed, it refers the matter to the Executive Conciliation Board.   Meanwhile, the grievance— such as, perhaps, the suspension of the free delivery of the house-coal to which the miners are entitled at a low price, or the abolition of some other favourable

custom—has continued for several months, there is
very free grumbling, not only against the employers
in question and others in general, but also against
the seeming inefficiency of the Miners' Federation
officials.   A classic case of the failure of the Conciliation Board to settle a local dispute is that of the Cambrian Combine strike, 1910–1911, which arose over
the price list in a new seam of the Ely pit of the
Naval Colliery Company, a constituent of the
Combine.   The Board duly referred the dispute to
a committee of two with power to settle.   They
were unable to agree, however ; and after some
weeks of negotiations of the parties through the
respective members of their committee, the colliery
company, which was losing money by the delay,
locked out all its workmen.   Then another committee
of the Board was appointed of two on each side ;
but as described in Chapter XIX the workmen threw
over the recommendation of their own members of
this committee, which was ultimately accepted only
after a ten months' strike of 12,000 men.

Contrary to the advantage in the case of settling
the general level of wages, I think that local disputes
are best settled by joint committees dealing with a
small area, so that some at least of the members
will know intimately the local conditions at first
hand, and so that some of them may be personally
known to the workmen.   In Durham, and several
other districts, the joint committee dealing with
local matters has an independent chairman ; but
in South Wales no independent chairman is used

for local disputes, which seems a great mistake. In cases of disputed claims under the Minimum Wage Act[1] the matter is referred to an umpire chosen from a permanent panel, and the colliery representative and miners' agent state their cases before him. Such claims are of minor importance as compared with local disputes which cause serious strikes ; but if the workmen were satisfied of the impartiality of the umpire, and especially if there were the opportunity of the matter being re-opened by way of appeal after, say, two years' trial of a new price list, or a new working condition, I believe it would prove a highly successful method to have an independent umpire arbitrating as chairman of a small committee. There is so much work of this kind to be done that a permanent umpire might with great advantage be appointed for each coalfield, or section of it, as a State officer—probably under the Board of Trade, in the Industrial Conciliation and Arbitration branch ; but the use of his services should not be made compulsory.

Success of Conciliation Boards

Looking back over the past few decades of the history of the coal trade, one is forced to the conclusion that the Conciliation Boards have done a great work and are likely to remain as an effective bulwark against the disturbance of industrial peace from trade fluctuations. They have adopted all that was good of the old sliding scale system, in

[1] See Chapter XX.

principle, methods and organisation, and have added such improvements as experience suggested. Above all, they are elastic, and consideration and due weight can be given to all the changing circumstances of trade. At present their only failing is that a new agreement is necessary every three or five years and a rupture may occur in negotiating this.

Much of the success of Conciliation Boards in the coal trade must be attributed to the self-denying labours of the independent chairmen or umpires ; men who, without remuneration, have given much of their time and mental energy to the understanding, and to a just decision on the complicated evidence submitted to them, of questions affecting the interests of many thousands of persons. Men like Lord James of Hereford, for the Federated Districts, Lord St. Aldwyn for South Wales, Mr. Robert Romer for Durham, Lord Mersey for Northumberland, and others in other districts, deserve the deep gratitude not only of coal owners and miners but also of the whole country.

The Conciliation Boards have reached their present position by a process of evolution, which is always the safest and most permanent way for an institution to be established. Gradually they will gain prestige from long continued existence, and this may tend to enhance their authority in directions where, as yet, they have little power. As already indicated, I think there will be still greater centralisation ; whilst gradually, it is to be hoped, both sides will avail themselves of the Conciliation Boards on the one hand and of local committees of similar constitution, to be established everywhere by the Conciliation Boards, for every question that is of the nature of a bargain or involves standard conditions or rules of working.

It is far better to let this machinery gradually grow up in the course of many years than to try to establish it for flourishing and highly organised industries by legislation, as was done in Australia. In this country, at any rate where the protection of low paid workers is not in question, it would seem that matters of wages and of conditions of working, other than safety, should be left so far as possible to the parties concerned to settle between themselves by their voluntary organisations and by quasi-judicial bodies of their own establishing. The proper and most useful function of the State would seem to be to disseminate information upon the working of such institutions and methods in various districts at home and abroad ; so that by learning easily what is successful and what has failed elsewhere, both employers and workmen may come all the faster to the most permanent and successful method of settling their differences.

Some very thorny and difficult problems are aroused when it is considered how far the State shall give legal sanction to the findings of such quasi-judicial bodies as Conciliation Boards and local joint committees. This again is a question which they themselves can settle better than a hundred persons learned in the law, or in the arts and sciences ; but I think the tendency will be towards seeking and accepting more and more of the status and authority of a court of law.   The more this happens, however, the more certain it is that there will be new and great questions of principle arising from time to time which no board, committee, or court could possibly settle.   The minimum wage dispute was a case in point, and other questions must certainly arise from time to time.   For example, how could a Conciliation Board deal with a threatened strike of men who demanded nothing more nor less than that they should appoint one-third of the directors of the colliery company employing them.   In the evolution of industry, problems involving new principles will always be arising, and strikes will probably for long be the unavoidable preliminary to progress upon new lines beneficial to the workers. The purely wasteful strikes will be eliminated more and more by Conciliation Boards and joint committees ; and the growing organisation of labour will enable the workmen more and more to control the conditions, if not the remuneration, of their work to their own satisfaction.

CHAPTER XIX

ABNORMAL PLACES AND THE NATIONAL STRIKE

THE National Coal Strike of 1912 is a landmark in the history of the coal trade, not only because it led to the passing of the Minimum Wage Act, but also because it welded together the Miners' Federation of Great Britain into a compact trade union for the miners of the whole country, a result which cannot but have far-reaching influence in the future. It will be of interest therefore to trace the origin and growth of the movement which culminated in the National Strike, and to do so we must first fully understand the meaning of abnormal places and other abnormal conditions of work, to which a proportion of the hewers were subjected, often with hardship and injustice. The history will take us back to the early years of the present century, chiefly in South Wales, where the movement originated and gained its driving force.

In hewing coal the collier proceeds to make a cut at the foot of the seam, and the coal may then fall down or it may have to be driven down either by a wedge driven in at the top or by blasting when this is safe. The skill on the part of the collier comes in knowing just how to make his cut so as to bring down the most coal with the least labour, breaking up the coal as little as possible. In South Wales, particularly, the coal seams vary much from point to point in the mine. Sometimes the coal is crushed and produces a great deal of small coal for which, in South Wales, the collier is paid nothing. The coal seams are also liable to vary in thickness from place to place, which greatly affects the ease or difficulty with which the coal can be cut. Sometimes the roof is hard and firm, the coal detaching itself easily and making the work easy ; at other times it is so loose that the collier must be always on the watch for a fall or he must spend much time in protecting himself by putting up props to support the roof and generally by keeping the place in order. Sometimes stone occurs in the coal and has to be carefully sorted from it, or places may be wet, making it difficult and unpleasant to work in. Whenever these vagaries occur, the collier, who is paid so much per ton of large coal hewn, is likely to find his earnings short. A skilled man who can hew three tons of coal a day in a normal place may well find that, though he works harder, he can hew but one ton in such an abnormal place.

It is not only variations in the coal and roof and other natural conditions which affect the hewers' earnings on piece-work ; there are also many ways in which a hewer on piece-work may be handicapped by insufficient or faulty services of the kind for which he is dependent upon the management. The hewer, for example, must have trams into which to fill his coal sent regularly to his working-place, or he has to waste time sitting down doing nothing, as there is not space to accumulate much of a stock. The supply of timber for props or sleepers or the supply of rails, may run short. If he cannot get rails to bring his tram nearer than 30 yards he will have to shovel his coal the whole of this distance before filling into the tram, which means a great waste of time. Insufficient ventilation, which makes a man easily get hot and tired, besides adding to the danger, is another way in which the management may be responsible for reducing a man's output. Before the passing of the Minimum Wage Act it was not the colliery proprietors or the managers who suffered as a result of inefficiency but rather the hewer, and he was entirely dependent upon the arbitrary decision of the management from time to time as to whether he got any compensation for a shortage of earnings not caused by his own fault.

I do not wish to minimise the difficulties under which the management of a mine must labour in regard to this matter. No one can deny that the Minimum Wage Act has introduced new and serious problems of management which fall heavily upon the present managers and are in some mines tending considerably to increase the cost of production. At the same time the assurance of a minimum day wage to underground piece-workers was undoubtedly a just and necessary measure. The subject is one of such great practical importance that it seems hardly beyond the scope of this book to attempt a full statement of the causes which may lead to the earnings of hewers on piece-work falling below the normal, including therein the faults of the hewer himself. The following list applies to steam coal collieries where the large coal is screened from the small and the hewer paid for large coal only. If the hewer is paid on the total weight of " through and through coal " some of the following causes of short earnings would be eliminated.

Natural Conditions

1. Seam becomes thin, requiring the removal of much stone.
2. Seam exceptionally thick, requiring scaffolding.
3. Loose joints in the coal, which means that it is partly crushed and produces 50 to 70 per cent. of small.
4. Soft or broken roof, requiring continual attention in setting props.
5. Bands of stone (clod) in or over coal.
6. Sporadic concretions of stone in coal.
7. Working place becomes dripping wet.
8. Blower of gas.

Every one of these conditions hinders the hewer in producing coal, but No. 3 applies only when the collier is paid solely for large coal sent out, as in the South Wales steam coal pits. Nos. 5 and 6 delay the collier, because he must pick out the stone from the coal or incur the expense of employing a boy.

Conditions the Fault of Management

1. Putting too many men in mine, whereby it is impossible to wind out of the pit all the coal which the men could produce, so that they are kept waiting for trams to fill.
2. Insufficient stock of trams.
3. Inefficient, or old-fashioned, winding machinery.
4. Bad organisation of underground traffic.
5. Frequent delays of traffic through falls, " journeys " running off lines, etc., through keeping main roads in bad state of repair.
6. Insufficient ventilation.   (There are complaints that parts of many mines are not kept up to standard, the collier thereby becoming more easily exhausted.)
7. Short supply of timber (props, cogs, sleepers, etc.).
8. Short supply of rails for collier to get tram near to the face so that he must shovel or carry coal 20 or 30 yards.
9. Bad condition of main haulage ways, also careless handling of trams, leading to unnecessary jolting of coal. Thus, the collier may have 70 per cent. of large coal and have it reduced to from 40 to 50 per cent. by the time it is weighed. This is most unfair where he is paid nothing for the small.
10. Turning a new stall. There is no room to store any coal, and special arrangements ought to be made to supply trams promptly.

Insufficient output from hewer's fault

1. Want of knowledge of coal seams and of skill in hewing.
2. Want of physical strength and ill-health.
3. Previous neglect to keep place in proper order.
4. Laziness or satisfaction with low earnings. (This applies particularly to newly-arrived agricultural labourers.)

The Consideration System

The system adopted for very many years to meet shortages of earnings due to natural conditions, and to some extent where the management was at fault, was the granting of an allowance or " consideration " in respect of the difficulty of working. Wherever possible the management would make the allowance a tonnage rate at, say, 3d. per ton of coal for a soft roof, or say, 2d. per yard for ripping of roof, required owing to an inequality of the coal. The allowances for " clod " (that is loose stone just above or in the coal) are sometimes stated in the price list[1] ; and the more of such items are included in the price list the better. It is impracticable, however, to meet half the possible contingencies without a very lengthy and elaborate price list ; and in the older price lists there are fewer fixed allowances than in more recent lists. The rule, therefore, was that for cases not provided for in the price list a hewer must make a claim, and arrange the allowance with the overman

[1] See page 341.

or undermanager. A bargain was struck, and the
allowance was generally put upon a piece-rate basis.
If no piece-rate could be agreed, it was customary
to grant a consideration of from 6d. to 3s. or 4s.
per day for working in a poor place.

If this system had really been worked as a system,
the allowance being properly adjusted to the need
by a highly skilled man devoting, perhaps, in a
large mine, the whole of his time to the inspection
of places, and assisted by two or three clerks, probably
it would have been satisfactory, and there would
generally have been no real injustices. In practice
the allowances were granted sometimes too freely,
and often unequally, and sometimes withheld when
they should have been given, much depending upon
the personal equations of the manager and under-
manager and the supervision by the directors, and
also upon the persistence of the men and their agents.

In the last quarter of the nineteenth century there
were many mines in South Wales in which consideration money was undoubtedly too freely paid.
In one or two cases which have come under my
notice, and probably in several others, it became the
custom for the management to make up the men's
earnings to 4s. 9d. plus percentage whenever they
were short, with practically no investigation beyond
the knowledge that the man was, generally speaking,
a skilled workman. Whilst very happily situated
mines could afford to be generous, others could not,
and yet were ; and this explains the lamentable
history of financial troubles of some of the former

colliery companies in South Wales. Little or nothing was paid by these companies in dividends, and large private fortunes were lost. The fact is that the generous colliery proprietors were undercut by their rivals, there being no legal or trade union minimum standard of average daily earnings.

The increasing burdens imposed by Parliament about the close of the nineteenth century in the shape of the Coal Mines Act (1896), the Workmen's Compensation Act (1897) and the export tax on coal (1901), and by the new regulations issued by the Home Office (1902) to secure greater safety in timbering, haulage, etc., all added considerably to the cost of production, and mine owners naturally cast about them for sources of economy. Wages forming about 60 per cent. of the cost of production, owners naturally turned, as they always have done, to this class of expenditure as the most likely on which to save. The sliding scale being in operation, however, till 1902, they could not attack the percentage of wages above standard, and the miners were equally able by their strong organisation to resist any appreciable decrease of the percentage, except in proportion to prices, when the conciliation board was established to replace the sliding scale. Any proposed reduction of the cutting price of coal was always met by the miners with united front and the support of other districts in the inevitable strike. Hence several owners turned to the allowances in which they had to deal with the men only as individuals.

Cost accounting, already well established in some industries, was beginning to be applied to coal mining by progressive directors and managers in the late nineties, and this revealed precisely the average expenditure per ton on allowances and dead-work generally. The custom was adopted by one after another of the bigger companies of keeping detailed cost accounts of each district of a mine, and thus putting different overmen and under-managers in competition for the lowest costs on dead-work. Success meant a substantial cash prize in one case which came to my notice ; but in any case first promotion. In some companies, severe pressure was put upon the managers to cut down costs, and those who failed had to make room for others.

It is not, therefore, surprising that some of the managers showed no fine discrimination in carrying out the new policy of drastic reduction of the cost of dead-work. With no intention whatever to be unfair to individual miners, such unfairness arose, and chiefly because the policy of reducing the cost of dead-work was undertaken without any well considered or carefully thought-out scheme. In many cases the managers, already overworked, adopted very rough and ready methods in cutting down allowances. The pays were made fortnightly until the Coal Mines Act (1911) came into operation ; and it was the custom for the under-manager to go round the working places with the overman of the district, and also with a clerk, to make a general tour of the working places and arrange the rate of

allowances with the colliers. The days of such
inspection are called " measuring-up days," and
they used to occur once a fortnight. " Measuring-
up " throughout a large mine would take two or
three days. Properly speaking, if the under-
manager had time it ought to take longer. The
kind of abuses which arose and of which numerous
specific instances have been quoted to me came
either from (1) a man failing to get any promise of
an adequate allowance on " measuring-up day," or
(2) finding that the allowance promised to him
verbally was not forthcoming on pay day.

As to the first class of grievance, I was told that
in some mines if the measuring-up party could
possibly pass by a working-place they would be
only too glad to do so, and that the collier had to
keep sharply on the look-out to see that he got their
attention and made his claim, or he would be too
late. Whenever a workman made a claim the
procedure was for the officials to discuss it with him,
the argument sometimes becoming very hot. In
most cases an agreement would be reached, and a
verbal arrangement made, which the clerk at the
manager's request would take down as best he could
in the dim light underground. Sometimes no
agreement was reached and the under-manager and
his party would leave with the dispute open. It
was stated to me by miners who worked in one of
the pits concerned that the manager would make
fairly reasonable allowances to three-fourths or
more of the men, namely those to whom he happened

2 L

to come first on his round, but the rest had to go
without anything.   He would say quite frankly :
" Sorry, boys, but I have not got a penny left for
allowances."   It appears that a lump sum was set
aside each fortnight for the granting of allowances
but the manager failed to make an equitable
distribution of it.

As regards the second class of grievance, where
allowances were promised but not paid, this also
generally arose from the fact that a manager found
afterwards that he had promised more than he could
afford to pay.   He would then strike out, or reduce
by half, the allowances in what he considered the
less deserving cases, or sometimes it was the men
who he thought would give least trouble or could
be most easily intimidated who would go short.
One manager's method, if he found he had promised
more than he could pay, was to reduce all the allowances promised in the same proportion by 10, 15 or
20 per cent. as might be necessary.

It will easily be understood that very great sore-
ness was caused by these arbitrary reductions of
allowances.   The workmen's Lodge Committee
always had to take up every case separately with
the managers ; and in a mine employing a thousand
colliers it would be no uncommon matter for the
committee to bring sixty or seventy cases before the
manager for reconsideration after the committee
itself had most carefully sifted the complaints made
to it.   Individuals who endeavoured to see the
manager and obtain redress personally, as they

were generally required to do before the committee
would investigate their case, had the greatest
difficulty in getting a hearing. They would be kept
waiting about for an hour or two day after day at
the pit head when they came up from work in the
hope of an interview with the busy officials, and
when obtained the interview often resulted only in
heated argument.

The movement to cut down allowances began, as
already stated, in the first years of the present
century, and it received a great impetus during the
depression of trade in 1902–1904. The South
Wales Miners' Federation took the matter up upon
the request of certain of the districts, and some of
the cases were taken to the County Court. In
September, 1907, however, a case, regarded as a
test case, was brought by a skilled collier residing
at Ynysybwl against the Ocean Coal Company for
payment of an allowance on the ground that that
allowance was customary in the pit, and he was
therefore entitled to it.[1]   He was represented at the
expense of the Miners' Federation, but Judge Bryn
Roberts gave a decision adverse to his claim—a
decision which has had far-reaching results. The
judge's decision was that allowances which were not
specifically provided for in the price list were mere
gratuities given by grace of a colliery company and
were not recoverable at law, however long it may
have been the custom to pay such allowances. This
view was almost in direct conflict with that of Judge

[1] Walters *v.* The Ocean Coal Co., Ltd.

Bryn Roberts's predecessor, Judge G. Williams, who gave many decisions upholding allowances where it could be shown that they had the sanction of long custom although not included in the price list.

The way was now open for colliery proprietors anxious to cut down costs to pursue the policy of reducing allowances to what they considered it was possible to pay without fear of being troubled in courts of law, and the practice of thus reducing costs spread practically to all the larger companies.

The usual result was that a man finding himself in a bad working place worked much harder in order to try and make up his short earnings, which generally he was able to do until the passing of the Eight Hours Act. The colliers in South Wales had been accustomed to work 10 or $10\frac{1}{2}$ hours from bank to bank, but from the 1st of July, 1909, onwards, they were restricted to $8\frac{1}{2}$ hours. In the older mines a man lost much time walking a mile or two from the pit bottom to his working-place and back in the evening, added to which there would be a wait of from five to twenty minutes for his turn to ascend. Although the meal time was reduced from about three-quarters of an hour to twenty minutes there was a reduction of over 20 per cent. in the net working time. It became practically impossible for a man having to work in a poor place to make up his earnings by extra exertion. Under the 10 hours a day *régime*, a man in a normal place took the work fairly easily and put forth his utmost exertion only when in a bad place and without hope of an allowance. Under such circumstances men occasionally seriously overworked themselves. With the eight hours working-day in force, a man's working speed was to some extent increased in a normal place so far as he could get trams to clear the coal ; but in a bad place he had so much the less time in which to work at full pressure and the monetary loss was severely felt.

It was this reduction of earnings, and particularly the arbitrary manner in which the men working in abnormal places were being dealt with, which led to the strike of the autumn of 1910, when nearly 30,000 men were out. This struggle originated at the Cambrian Combine Collieries and persisted there long after the men had returned in the Ogmore and Aberdare Valleys, so that it is generally known as the Cambrian Combine Strike.

The history of this strike is worth relating as it was a direct antecedent of the agitation which led up to the National Strike. The Ely pit, belonging to the Naval Colliery Company, which was controlled by the Cambrian Collieries, Ltd., under Mr. D. A. Thomas's chairmanship, was the scene of the dispute which led technically to the outbreak of the strike. One of the seams approaching exhaustion and proving no longer profitable, the management decided to open the Upper Five Foot seam, and a price list had to be fixed. The usual practice was followed of engaging about 70 to 80 men to work on day wages at the consideration rate of 5s. plus percentage on the 1877 basis (then making about

6s. 9d. per day). A record was kept both by the owners and by the men as to the output from each stall. It was asserted that the workmen, knowing that the average output over a few weeks would be taken as the only available basis upon which to fix a cutting price of the coal, did not exert themselves as much as they would have done upon a piece-rate payment, and to some extent this was probably true. Negotiations went on for a long time. On December 16th, 1909, the dispute had got to the stage of being referred to the Conciliation Board, and one representative on behalf of the owners and one on behalf of the workmen were appointed to settle it. On June 8th, 1910, a formal offer was made by the owners' representative of 1s. 9d. per ton, 1d. extra for stone irrespective of thickness, most of the dead-work to be paid for as in the price list for the Five-Foot seam of the same pit which had long been worked. The workmen, however, continued to demand the cutting price of 2s. 6d. per ton, alleging the stone in the coal was more trouble-some than elsewhere and the coal not so easy to get down.

As the dispute appeared to be getting no nearer settlement, notices were given on August 1st, 1910, to the whole of the workmen employed in the Ely pit, about 950 in number. The owners' sugges-tion to refer the matter to arbitration was received with hesitation by the workmen and rejected by the Coal Owners' Association, which did not like the introduction of this new principle in the

fixing of prices.   Consequently the lock-out notices
took effect on September 1st.   The Naval Lodge of
the South Wales Miners' Federation appealed to
their fellow-workmen throughout South Wales and
Monmouthshire, and issued a long statement of
their case, in which they pointed out that four other
collieries of the Combine were ready to work the
same new seam, and that the price list fixed for the
Ely pit would hold for these others, which contained
many millions of tons of coal yet to be worked.
The Naval men were, however, in no temper to
await the decision of a conference of the coalfield,
and they struck without notice on September 5th.
The remaining workmen of the Cambrian Combine
came out without notice on September 19th, and
this was in defiance of the South Wales miners'
delegates' conference held at Cardiff on the 17th,
which by a small majority voted against the handing
in of notices by the Cambrian Combine workmen,
chiefly on the ground that sectional strikes weaken
the Federation without any corresponding gain.
There was an extraordinary acerbity and excitement
in the temper of the workmen, who refused to listen
to the moderate councils of their leaders.   However,
on the urgent advice of their leaders, who objected
to an unauthorised strike without notice, the workmen resolved to resume work the next day, and this
was permitted by the owners at all except the
Naval pits.

A ballot of the coalfield now gave a substantial
majority in favour of a stoppage of the Cambrian

Combine Colliery, and a levy in favour of the workmen employed there, and notices were served on
October 1st, some negotiations in the meantime
having proved ineffective. The position throughout
the central part of the coalfield was one of great
unrest, notices having been handed in at a large
number of collieries upon various questions, to a
large extent on account of the employment of non-
unionists. During the month when notices were
running, Mr. F. L. Davis, on behalf of the owners, and
" Mabon " on behalf of the workmen, who had been
appointed by the Conciliation Board as a second
sub-committee, arrived at an agreement on the basis
of a cutting price of 2s. 1.3d., this price to include
dealing with all stone up to 12 inches in thickness.
This settlement, although approved by some of
the miners' leaders, was rejected by the whole of the
Cambrian Combine workmen, and the strike of the
12,000 men employed by the Combine began on
November 1st. Meanwhile, a strike had begun
without notice in the Aberdare valley upon such an
apparently small item as the management prohibiting the men carrying off broken props for firewood
without making a small payment. Soon all the
pits of the Aberdare valley were idle, and the workmen tabulated a long and formidable list of grievances in which the abnormal places figured pro-
minently. With strikes also in the Ogmore valley
there were about 30,000 men out.

Very great bitterness of feeling developed both
in the Rhondda and Aberdare valleys. The miners

insisted on the colliery enginemen coming out, and the owners had to take immediate measures to rescue the ponies from the pits before they became flooded.   Serious and regrettable violence developed, mainly through attempts to prevent the introduction of " blackleg " labour to keep the pumps and ventilation going ; and very large forces of police were drafted in from the surrounding neighbourhood, and from the Metropolis.   After violent riots, in which not only colliery property but many shops were destroyed, both the Rhondda and Aberdare valleys were occupied by the military, who handled the situation with the greatest restraint and success.

The riot fever was over by the end of November, and the workmen settled down with grim determination to a long struggle.   The Cambrian workmen were being supported by the Federation, but the Aberdare miners, who had struck without the authority of the Federation, were not, and they were obliged to return to work after many weeks without having gained anything.   As the strike continued for month after month in the Rhondda, a settled gloom fell over the district ; but the workmen were all the time very active in the discussion of their policy, the unofficial local workmen's leaders going far beyond and differing from the officials of the Miners' Federation.   Negotiations proceeded from time to time, pamphlets and fly sheets were issued by each side to explain their position, the Board of Trade intervened, and the Miners' Federation of Great Britain gave financial support to the South

Wales Miners' Federation in respect of the strike, and then came into the negotiations. Much discussion centred round an assurance given by Mr. D. A. Thomas that earnings of the men on piecework would be made up to 6s. 9d. per day if they were short through difficulties met with in the seam of coal. Great controversy raged about the value of this assurance, as to whether it meant a definite and certain minimum wage in abnormal places or not. Apparently Mr. Thomas meant that the ordinary practice of making up by an allowance to this consideration rate would be continued exactly as heretofore, and would be paid to every man considered to be a good workman, and who had, in the opinion of the management, met with a difficulty which would make it impossible to earn 6s. 9d. a day upon the price list. Such an assurance the Cambrian Combine workmen considered valueless in the light of their experience in the past two years, but leaders of the M.F.G.B. thought otherwise. On May 15th a settlement was made by the M.F.G.B. with the owners upon the basis of the 2s. 1·3d. per ton, and the assurances of the allowances being made, with a proviso for a reference of any dispute as to an allowance to a committee of the South Wales Conciliation Board.

The struggle then entered upon a new and interesting stage ; for the strikers threw over the M.F.G.B. settlement, and refused to return to work. Instead of confessing themselves beaten they started a violent agitation for a minimum wage of 8s. a day for all colliers, and 5s. a day for all unskilled workers, demands which had been privately discussed and proposed for some months and they demanded that national action under Rule 21 of the M.F.G.B. be taken in support of this policy, which meant that the M.F.G.B. should ballot its members throughout the country and call a strike if it obtained a two-thirds majority.

A South Wales Conference held at the end of May, 1911, resolved that, failing a decision by the M.F.G.B. at the forthcoming conference on June 14th to ballot for a national stoppage with a view to securing for all colliery workmen a definite guaranteed minimum wage, the South Wales Federation should declare a stoppage to secure this end. There followed in South Wales much discussion and criticism of the policy of the coalfield strike, and at another South Wales conference on June 12 it was dropped, the general opinion being that national action would be more effective, and that the M.F.G.B. should be got to go as far as possible. It was, therefore, resolved that failing national action to secure a definite guaranteed minimum wage, the South Wales delegates should support national action upon the question of abnormal places and lower paid workmen.

It will be seen that the Cambrian Combine struggle —still dragging on—was now fading into the background. The group of Rhondda enthusiasts who had stiffened the dispute by their insistence on a guaranteed minimum of 6s. 9d. per day as a condi-

tion of accepting the owners' revised offer of a
cutting price of 2s. 1·3d., had during the period of
the strike converted the South Wales coalfield to
their policy of the minimum wage. In a few months
they were to convert the whole country ; but for
the moment they seemed to be defeated when on
June 13th the M.F.G.B., at a conference held in
London at the Westminster Palace Hotel, decided
that the terms of May 15th secured the object sought
by the last conference, and that the M.F.G.B. would
no longer take any responsibility in reference to the
Cambrian dispute. The conference also postponed
the consideration of the abnormal places question
until another meeting to be held on July 28th, by
which time the districts were to report on the
matter.

A South Wales conference on July 1st again
rejected the May settlement, and drew up a manifesto to be sent out to the English and Scotch coalfields, and undertook to send speakers to the various
coalfields in support of national action on the question of a guaranteed minimum wage. On July 7th
the M.F.G.B.'s contribution of £3,000 per week
towards the support of the 12,000 Cambrian workmen still out, came to an end ; and the South
Wales levy was increased so as partially to fill the gap.
But, although the Cambrian workmen heroically
decided to continue the fight on a much reduced
strike pay, the strike had already done its work in
the national sense ; and there is no permanent
interest in the final phases through which it dragged

on for a few more weeks, until the men were forced to accept the terms which had been offered the previous October providing for 2s. 1·3d. cutting price, and at last resumed work on September 1st, 1911.

The movement for dealing with the abnormal places question by national action was now fairly under way. At the M.F.G.B. conference of July 29th, held in London, the abnormal places question was considered, and the Federation definitely committed itself to a settlement of the question upon a national basis by adopting the following resolution :

" That the officials of the Federation be instructed to arrange with the coalowners of the United Kingdom for a joint meeting to consider the question of paying the district minimum rate of wages for working abnormal places ; that failing to get satisfaction on this question, a conference be called to decide upon a ballot of the members of the Federation to ascertain if they are in favour of ceasing work until the district minimum wage is obtained."

It will be observed that the idea underlying this resolution was to ask solely a minimum wage for abnormal places, and it did not cover cases of men being unable to earn the minimum rate for all causes other than their own negligence or want of skill, as does the legal minimum wage which was ultimately obtained. The district minimum rate of wages was intended to be the daily rate which is recognised in each coalfield or part of a coalfield, or in classes of collieries in part of a coalfield, as being the rate to be paid to hewers when put upon day work. These day rates vary very much from coalfield to coalfield, and by as much as 8d. or 9d. per day in the standard rates within a coalfield.

There was some difficulty in acting upon this resolution, for the M.F.G.B. was taking the initiative in negotiating nationally for the first time in history. There was no federation of the employers' associations corresponding with the M.F.G.B., and the secretary of the latter wrote to Sir Thomas Ratcliffe Ellis, secretary of the Mining Association of Great Britain, suggesting a joint meeting of representatives of his Association and of the M.F.G.B. in order " to consider the question of a rate of payment for working in abnormal places, and also to consider the low wages paid to day rate men." Sir T. R. Ellis replied, informed Mr. Ashton that the Mining Association of Great Britain did not deal with questions of wages, and that if there was to be a meeting it must be with representative coalowners appointed by the different districts.

Such a meeting of coalowners was held in London at the Westminster Palace Hotel on September 19th to consider the M.F.G.B. application ; and whilst there was a general readiness expressed to meet the workmen's representatives, the proposal to pay a fixed minimum wage for abnormal places was viewed with great disfavour in all quarters, excepting Yorkshire, where the workmen and the owners had come to an agreement, and in Lancashire, where an

agreement would probably have been reached except
that the miners had now made the matter a national
one. In the Midlands the owners were also likely
to come to a settlement on the basis of payment for
abnormal places, but in the other districts, for one
reason or another, no agreement had been reached
in the negotiations which had been going on for
some months. These were undertaken, it will be
remembered, as a result of the resolution moved from
South Wales at the annual conference in October,
1910, and confirmed at the special conference held
on January 24th, 1911, when the districts were
instructed to press for the average or minimum rate
of wages, whichever is in operation, to apply to all
workmen engaged in working in abnormal places
or under abnormal conditions. The strongest opposition came apparently from the coalowners of South
Wales and of Scotland.

The joint national meeting of the representatives
of the coalowners' associations and of the M.F.G.B.
was held upon September 29th. The workmen
asked that there should be a definite understanding
that whatever was the agreed and understood wage
paid when a man was on day-work should be paid
to the workmen whilst working in abnormal places.
The owners replied that they were, as heretofore,
willing to give " consideration " wages and to take
into account not only the character of the abnormal
place but also the labour efficiency of the workman
employed there. They did not consider it right
that a skilled workman normally earning, say, 7s. a

day, should be brought down to the ordinary average day wage of the colliery in which he was working and which would have to be paid to a bad or indifferent workman who happened to be employed in an abnormal place. They adhered to the view that the relative efficiency of the workmen must be taken into account and that the fixing of any particular amount of wage could only be done on a purely local basis. They desired the negotiations to be referred back to the districts, as it was impossible to fix any particular amount of wage which a man should earn throughout the whole country.

The last contention of the owners was conceded by the M.F.G.B., which itself found that the wage-rates recognised in the different districts varied so widely as to present great difficulties in negotiations on a national basis. The other main contention of the owners, that payment for working in abnormal places should be dependent upon the man's known efficiency as a worker, could not for a moment be admitted by a responsible trade union, for it contravenes the essential doctrine of the " common rule " which is the very foundation of trade union policy : that for doing the same kind of work all men shall be paid the same rate. This is maintained whether the payment is by piece-rate or by time wage. The trade unions continually insist that the responsibility must be upon the owners of rejecting workmen who are not efficient enough to earn their day's wage and this responsibility the coalowners have persisted in trying to evade. Whereas this

battle was fought and won by the trade unions of the
engineering and building and many other great
trades half a century ago, it was only in this abnormal
places dispute that it became a dominant question
in the coal trade.   Much of the struggle in the coal
trade of the last few years becomes clear when the
fundamental character of the common rule in trade
unionism and its absence in the mining industry
until the Minimum Wage Act was passed, is borne
in mind.

The joint meeting failed to reach an agreement,
and this was the end of the attempts to settle the
question of abnormal places by itself.   The advanced
section of the M.F.G.B., who favoured the policy of
an unconditional minimum wage, now obtained
control of the councils of the Federation.   On
October 3rd, the M.F.G.B. annual conference opened
at Southport.   On October 6th the following
important resolutions were carried :—

" That the Federation take immediate steps to
secure an individual district minimum wage for all
men and boys working in mines in the area of the
Federation without any reference to the places being
abnormal.   In the event of the employers refusing
to agree to this, then the 21st rule be put into
operation to demand assent."

" That a conference be called on November 14th
for the purpose of taking action under rule 21." [1]

---

[1] The 21st rule of the M.F.G.B. used to read as follows :—
" That whenever any Federation or District is attacked on a
general wage question, all members connected with the Society

2 M

" That in order to put the preceding resolutions into operation. districts are hereby instructed to meet their employers on the questions contained therein as early as possible, and that all delegates from the various counties, districts and federations come prepared to report upon these to the special conference on November 14th."

The various districts now proceeded to formulate their demands and to present them to the various coalowners' associations. Nearly all the districts prepared a full schedule of minimum rates for each of the various grades of labour. Some districts, like Yorkshire and Nottingham, joined with South Wales in asking 8s. for hewers, Lancashire and Durham asked 7s. and Scotland 7s. For the hauliers and other low-paid day-wage men, all the districts adopted the minimum of 5s. and for boys 2s. for the youngest age. These demands were rejected by the owners in all the districts ; in South Wales the ground stated being that the demands violated the wages agreement of 1910, and in Scotland that they violated the Board of Trade agreement of 1909, whilst various reasons were given for their rejection in other districts. It was only in the Federated Districts of England and North Wales, where the question came before the

shall tender a notice to terminate their contracts, if approved of by a Conference called to consider the advisability of such joint action being taken."

It was altered at the Conference on July, 1911, so as to be applicable to aggressive as well as defensive action to the present form, See Appendix VI, p. 824.

Conciliation Board on November 10th, 1911, that some progress towards a settlement was made. The owners agreed that they could recommend the adoption of the principle of the minimum wage, but it was impossible to agree as to the rate, though the margin of difference was not large.

At the special conference of the M.F.G.B., held in London on the 14th and 15th November to receive the reports of the district, great satisfaction was expressed at the admission by the owners' side of the English Conciliation Board of the principle of the minimum wage. As the English Conciliation Board was to consider the matter further on the 6th December, it was decided to adjourn the conference until the 20th December to receive the final reports, and the intended resolution to ballot the whole of the members of the Federation in favour of a national stoppage was postponed. At the same time, however, the conference resolved that in view of the reports from the districts, the best course to pursue to obtain their object with the least delay, would be to negotiate nationally, and the executive committee of the M.F.G.B. was therefore instructed to formulate the claim of each district, and, with some additional representatives, to meet the colliery owners of Great Britain at the earliest possible date and report immediately thereafter to a national conference. The resolution provided that these negotiations on a national footing were not to interfere with any already proceeding locally in the districts. At this period the coalowners of nearly

all the pits of the country expressed themselves
perfectly ready to consider an equitable arrangement for the payment of a guaranteed wage to men
working in abnormal places if the miners would
agree to certain rules or safeguards to secure that a
man should do a full day's work for the payment.
The workmen, however, had now got quite beyond
being satisfied with a settlement of the abnormal
places question only.   They had now definitely made
up their minds to obtain a guaranteed minimum
wage, and all their demands were uncompromising
in their adherence to this principle.

At the further meeting of the English Conciliation
Board on December 6th, no great progress was
made towards a settlement.   The miners' representatives could not yield on any points as they did not
wish to compromise the negotiations on a national
basis which were also proceeding.   At the national
meeting of owners and men's representatives on
December 18th, the conditions which the miners
would agree should be associated with the minimum
wage were discussed, and the owners agreed that
the matter should be again submitted to the districts
for further consideration.   The miners, however,
were to some extent losing patience, as the owners
in nearly all the districts had not given any hope
that the acceptance of certain conditions by the
miners would lead to the owners' agreeing to a
guaranteed minimum wage.   The M.F.G.B. conference held on December 20th and 21st took
drastic action.   It resolved that the ballot vote of

all members be taken on January 10th to 12th, 1912, the result to be sent to the secretary not later than the 16th ; and that in case the ballot vote should result in a two-thirds majority in favour of a national stoppage, notices should be given in every district so as to terminate at the end of February, 1912. Each district was also instructed to send to Mr. Ashton, the secretary of the M.F.G.B., a tabular statement of what it desired to be its minimum rates of wage, and that the executive committee of the Federation should consider these statements and report to a national conference to be held at Birmingham on January 18th. It was decided that the form of ballot paper should be as follows :—

" Are you in favour of giving notice to establish the principle of a minimum wage for every man and boy working underground in the mines of Great Britain ? "

The voting duly took place and the situation was not affected by a meeting of the English Conciliation Board on January 15th, which rejected the schedule of minimum wage-rates presented by the English miners, as the other schedules had also just been rejected in the other districts. On January 18th, at the special conference at Birmingham, Mr. Enoch Edwards presiding, the results of the voting were announced. They were as follows :—

For giving notice . . .	445,801
Against ditto . . .	115,721
Majority . . .	330,080

The only district in which a majority voted against a national strike was Cleveland, and the explanation of this is that this district contains a large proportion of ironstone miners, who are admitted to the same union as the coal miners but to whom a guaranteed minimum wage had already been granted. The requisite two-thirds majority having been obtained, the conference resolved that notices should be tendered in every district to terminate at the end of February, but at the same time the employers should be notified that the workmen's representatives were still prepared to meet them and continue negotiations both in the districts and nationally, with a view to arriving at a satisfactory settlement.

The conference adjourned until February 1st, in London, when the executive committee submitted the schedule of claims which it had prepared from the materials submitted by the districts ; and on the following day they were adopted by the conference. The minimum day-wage rates claimed for piece-workers at the face of coal were as follows, the figures being stated in descending order :—

Notts	7s. 6d.
Yorkshire . .	7s. 6d.
Derbyshire . .	7s. 1½d. to 7s. 6d.
South Wales . .	7s. 1½d. to 7s. 6d.
Leicestershire .	7s. 2d.
Northumberland .	6s. 0d. to 7s. 2d.
Lancashire . .	7s. 0d.
Midland Federation .	6s. 0d. to 7s. 0d.
Cumberland . .	6s. 6d.

South Derbyshire	.	.	6s. 6d.
Durham	.	.	6s. 1½d.
North Wales	.	.	6s. 0d.
Scotland	.	.	6s. 0d.
Cleveland	.	.	5s. 10d.
Forest of Dean	.	.	5s. 10d.
Bristol	.	.	4s. 11d.
Somersetshire	.	.	4s. 11d.

It was also decided to maintain the claim that no
underground worker should receive a rate of wages
less than 5s. per shift, and no boy less than his
present wages, nor in any case less than 2s. a day.
From these minimum rates for day-wage men and
boys the small coalfields of Somersetshire, Forest of
Dean, and Bristol were to be excepted, those districts having lodged emphatic protests against the
originally proposed rates as being likely to throw a
very large number of men and boys out of work.
It was also resolved that the individual minimum
wages for all piece-workers, other than colliers,
should be arranged by the districts themselves and
be as near as possible to the present wages ; and,
as the rates paid to underground workers employed
on day-wage are so complex and difficult to deal
with, that these also be left to the districts with
instructions that they endeavour to arrange minimum rates for each class or grade locally in each
district. It was an unfortunate, but perhaps
inevitable, weakness in the men's case that the
M.F.G.B. was quite unable to formulate complete
schedules of claims for all the workers of different
grades in all the districts. It would have been

impossible, however, without a considerable office staff and the assistance of statistical experts besides persons with extensive local knowledge, to have prepared schedules which would not be open to serious criticism, and there was no time to organise such statistical work.

In studying the list of figures claimed for hewers and other piece-workers at the face, it will be seen that Notts and Yorkshire claimed the highest minima, South Wales and Derbyshire claiming the same figure for the highest of the classes of collieries into which it was anticipated these coalfields would be divided. There is a still greater disparity in this revision than in the original schedule between the lowest and the highest figures, Bristol and Somerset claiming only the very low figure of 4s. 11d.

The conference adjourned until February 13th, in the hope that some progress would have been made towards a settlement before the whole of the notices need be tendered. South Wales required to give a month's notice, and this was duly tendered by all workmen on the last day of January. Many of the districts had, under their agreements, a fort-nightly notice others a weekly, and a few had only to give a day's notice. At the conference it was reported that the South Wales notices had been duly given; and as no progress had been made with negotiations, other districts were instructed to hand in their notices in due course, all to terminate on February 28th. This conference passed the fol-lowing resolution, which explains itself: "That we

express our regret that the coalowners have refused
to accept the principle of an individual minimum
wage for all men and boys employed underground,
as we know that there can be no settlement of the
present dispute unless this principle is agreed to.
In view of the fact, however, that we have no desire
for a serious rupture in the coal trade of the country,
we are willing to meet the coalowners at any time
to further discuss the matter if the coalowners
express a desire to do so." After some further discussion the conference was adjourned until the
executive officials should think desirable.

By this time the strike was felt to be inevitable,
and the dispute was beginning to receive considerable attention in the press. A good deal of alarm
was felt ; but it was hardly thought seriously that
anything so tremendous as a simultaneous strike of
all the coal miners of the country could really occur.
Nevertheless, the price of coal was beginning to
rise in most of the principal markets. Sir George
Askwith, who had been keeping in touch with the
development of the dispute, now intervened on
behalf of the Board of Trade, and attempted to act
as conciliator, but without success, although negotiations were still proceeding in the federated
districts. On February 20th definite Government
intervention began with letters addressed by the
Prime Minister to the coalowners on the one hand,
and the miners on the other. He invited representatives of each side to meet him separately upon the
22nd. At these conferences Mr. Asquith was accom-

panied by Mr. Lloyd George, Sir Edward Grey, Mr. Sydney Buxton (President of the Board of Trade), and Sir H. Llewelyn Smith and Sir George Askwith as chief permanent officials of the Board of Trade. Members of the Government gained much information as to the views of each side and the grievances of the miners, but no progress was made towards a settlement.

On the same date there met in London the Executive Committee of the International Miners' Federation. French, Belgian, German and Austrian delegates met the British delegates in order to consider what action these countries should take to support the British miners. The main object was to prevent the export of coal from the Continent to Britain, a policy which was successfully carried out. On Saturday, the 24th, there was a further conference between the Government and the coalowners' representatives ; and Monday, the 26th, saw the beginning of the fateful week when notices would run out in all parts of the country. National alarm was widespread at the unyielding attitude of both owners and miners. Several questions were asked in Parliament.

Next day the *Times* made a feature of exposing the syndicalist policy of the miners by quoting at length the most revolutionary parts of a recently issued pamphlet called "The Miners' Next Step." This pamphlet was prepared by a group of active propagandists in the Rhondda Valley, and was very cleverly put together. Arguing that trade unionism of the existing type of organisation was played out, it advocated an aggressive policy with a view to getting the control, and ultimately the ownership, of the mines into the hands of the workers. The method was to be " the irritation strike," which meant every man reducing his output by one-half over a long period until the colliery had been rendered unprofitable, and could then be bought up at a nominal figure by the workers' new trade organisation, to be called into being for the purpose. It was, however, incorrect to suggest that this was a policy approved by more than a very small section of the miners. On the same day (the 27th) there was another national conference of miners which was addressed by Mr. Asquith. Members of the Govern ment subsequently met the Executive Committee of the M.F.G.B. The Government's policy was to help the owners and the miners to come to an agree-ment themselves on the difficult questions involved in the minimum wage dispute. Mr. Asquith announced, however, that the Government would be prepared to propose a scheme if necessary. At most of the collieries in Derbyshire notices expired on this and the previous day, and the men without hesitation brought their tools out. The miners in some parts of Leicestershire and Notts also fin-ished on Tuesday, the 27th, and a pessimistic feeling was becoming general.

On Thursday, February 29th, the newspapers reported that on the previous day 115,000 miners were already idle, and that Mr. Asquith had spent his time between conferences with the coalowners and miners, and visits to the King at Buckingham Palace. The Government issued the following important statement of their proposals for settling the dispute :—

(1) His Majesty's Government are satisfied, after careful consideration, that there are cases in which underground employees cannot earn a reasonable minimum wage, from causes over which they have no control.

(2) They are further satisfied that the power to earn such a wage should be secured by arrangements suitable to the special circumstances of each district. Adequate safeguards to be provided to protect the employers against abuse.

(3) His Majesty's Government are prepared to confer with the parties as to the best method of giving practical effect to these conclusions, by means of district conferences between the parties, a representative appointed by the Government being present.

(4) In the event of any of the conferences failing to arrive at a complete settlement within a reasonable time, the representatives appointed by His Majesty's Government to decide jointly any outstanding points for the purpose of giving effect in that district to the above principles.

These proposals were accepted with good grace by the owners of the English federated districts, and reluctantly by the owners in Durham and Cumberland. The South Wales, Scotch and Northumberland owners rejected them. The miners gave a qualified acceptance of the proposals on the understanding that the principle of the guaranteed individual Minimum Wage was accepted, and not open for discussion, and on the understanding that they adhered to their schedules of minimum wage rates, which had already been presented to the Government and the owners. This was no more than a re-assertion of the attitude which they had consistently adopted for the past three months. On Friday, March 1st, reports from the districts showed that there were now 803,000 miners on strike. The last strike notices expired on the 2nd, so that on this last day of the week there were over a million men out on strike—practically all the miners of the country—the numbers employed on December 31st, 1911, being returned by the Home Office as follows :

Underground	.	.	.	878,759
Above ground	.	.	.	210,331
Total.	.	.	.	1,089,090

Of this number about 800,000 might be held to be concerned in the minimum wage demands of the Miners' Federation, though only about one-third of that number, or at most one-half, would be likely to obtain any benefit whatever by their being granted. Only a sufficient number remained at work to feed the ponies, keep the pumps going, and in some collieries to execute absolutely necessary repairs. The policy of the Miners' Federation was to allow the mines to be kept in working order

on condition that no coal was gotten or wound to the
surface.   Further   negotiations   were   proceeding
between the Government and both parties.   The coal
owners of the federated districts expressed themselves as willing to grant a minimum wage of the
amount asked for by the men, but only when working
in abnormal places, and they offered a guaranteed
minimum wage under all conditions of 1s. per day
less.   Owners in the other districts continued to
offer only a minimum wage for abnormal places.
The Prime Minister announced that he was prepared
in the last resort to introduce legislation to secure a
minimum wage for miners throughout the country,
provided adequate safeguards were included and
reasonable latitude allowed in the fixing of the rates.

By March 5th the effects of the strike were beginning to be widely felt.   The price of coal was rising
considerably.   A few works and factories in various
parts of the country had already closed down, and
many more were put on short time, whilst a very
large number took the precaution of giving their
employees a week's or a fortnight's notice.   All
the great railway systems in the country took
measures of economy by considerably cutting down
the railway service.   Nearly all express trains were
suspended, and from one-fourth to one-third of the
slow and local trains, causing much inconvenience.
Meetings of the Industrial Council were held daily
from March 4th to March 7th ; but as they effected
no progress towards a settlement, the negotiations
were again taken up by the Government.   The

Government invited both parties to meet them
again ; and a national conference of the miners,
assembled for the purpose, agreed to a further
meeting on the understanding that the principle
of the Minimum Wage would be excluded from
the discussion. They did not, however, this time
exclude from the discussion the actual figures of the
minimum rates, and the mode of their determination. The joint meeting under the chairmanship of
the Prime Minister began on March 12th, and continued on the two following days. On March 15th
the Prime Minister and his colleagues had separate
conferences with the Coalowners' Consultative
Committee and the Miners' Executive. The Government was obliged to recognise now that their attempt
to settle the dispute by mutual consent had failed ;
and that there was no alternative but to introduce
legislation to meet the demands of the miners so far
as reasonable, if a most disastrous prolongation of
the strike were to be avoided. In a statement
issued by Mr. Asquith he indicated that the Bill to
be introduced by the Government would provide
for the fixing of district minima locally by Joint
Boards in each district, consisting of representatives
of the coalowners and the miners, and presided over
by an independent chairman, to be selected by the
parties themselves, or if necessary, by the Government. There would be provisions to secure prompti-
tude in the presentation of the cases of both parties,
and in the adjudication thereon.

In the meantime distress was beginning to be

felt in various parts of the country, and it was
estimated that nearly 400,000 persons were now
thrown out of employment in other industries
through the reduction of the railway staffs, paying
off steamship crews, and the closing of various works
and factories.   Fortunately the weather was mild,
but as there was great demand for house coal,
persons in the colliery districts occupied themselves
with digging for coal in disused levels, and in small
quarries on the outcrops, a course which was in
many cases charitably permitted by the landowners
without charge.   Very large quantities of coal were
also gathered off the spoil heaps, colliery sidings, etc.

On the continent of Europe, and also in Pennsylvania, there was at the same time great unrest
amongst the miners.   On March 13th began a strike
of about 210,000 men in the Ruhr district of the
Westphalian coalfield.   This was supposed to be
partly a sympathetic strike with the British miners ;
but the German miners had many distinct grievances
against which they were striking.   It was not by
any means a unanimous strike, and after a fortnight
began to collapse.   The French miners were agitating
for a general strike, but were held in check by their
leaders as the Government had promised legislation.
The issues there were mainly in regard to safety
provisions rather than wages.

On March 19th, when the price of coal in London
had reached £2 per ton, the Minimum Wage Bill
was introduced in the House of Commons.   The
Prime Minister, in his speech, made a point of the

special conditions of the particular trade and the special emergency with which the country was faced. The necessity of the Bill was pretty generally recognised, and controversy now turned on the inclusion in the Bill of the figures 5s. and 2s. as the minimum wages for adult day-wage men and boys respectively. Mr. Asquith announced that the Cabinet had decided against the inclusion of any figures in the Bill. Owing to a further conference with the owners and workmen, the Government did not take the second reading until the 22nd, when it was carried by 348 votes to 225. On the following days the committee stage was taken, and Mr. Enoch Edwards, President of the M.F.G.B., moved an amendment to include the schedule of January as the lowest figures at which the local Boards might fix the minimum wage rates for piece-workers at the face. This was rejected by a large majority ; but the controversy over the 5s. and the 2s. continued. On the 26th Mr. Brace moved an amendment to include these figures. He said they were reasonable because they were then paid in the inferior collieries, and it was only the wealthy colliery companies which paid lower rates. After a long discussion the amendment was rejected. It should be said that the Government held special conferences with both parties on the 25th and 26th to see if an agreement could be reached to include the figures of 5s. and 2s. ; but without result. The Government appeared, indeed, to be confirmed in their opinion that all figures must be left for settlement by the District

2 N

Boards. The Committee stage was concluded by the House sitting far into the night of the 26th ; and the Bill passed its third reading in the House of Commons at 2.30 a.m. on the 27th. A special sitting of the House of Lords was held at 3 a.m. of the same day, at which the Minimum Wage Bill was read a first time. It passed the House of Lords with only slight amendments, which, after a slight disagreement, were accepted by the House of Commons, and the Bill received the Royal assent on March 29th.

On the day that the Bill passed the House of Commons, meetings of the coalowners and of the Miners' Federation were held. The coalowners' representatives recommended that when the Bill became law every endeavour should be made to give effect to its provisions. The miners' conference decided to take a ballot of all members of the M.F.G.B. on the following question :—" Are you in favour of resuming work pending the settlement of the minimum rates of wages in the various grades of work by the District Boards, to be appointed under the Mines Minimum Wage Act ? " It was decided not to make any recommendation to the workmen as to how they should vote.

The result of this ballot was announced on April 4th as follows :—

For resumption of work .	201,013
Against	244,011
Majority against . .	42,998

The Executive Committee was embarrassed by this result, as the rules of the Federation contained no provision as to the majority to be required to continue or terminate a strike. They therefore summoned a national conference to meet two days later and submitted the following recommendation : " Seeing that there is no provision in the rules or regulations of the Federation to guide this Committee as to the majority required to continue the strike, except the resolution passed at the conference held on December 21st, 1911, that a two-thirds majority was required to declare a national strike, we agree that the same majority should be required to continue the strike, and seeing that a two-thirds majority is not in favour of the continuance of the strike and acting upon that vote, we advise the resumption of work." This the conference, after some discussion, accepted. It was also agreed that instructions should be sent out to the various districts that work should be resumed as soon as possible. As the Easter holidays were approaching, however, not many collieries could make arrangements to re-start until after Easter. As a matter of fact, during the past week or two several thousand men, particularly in districts like Warwickshire, where non-unionists were in large proportion, had resumed work.

In some of the coalfields, however, the workmen expressed extreme dissatisfaction at the decision of the conference, that they should return to work without the minimum wage rates having been actually fixed. As a matter of fact, all the important coalfields of the country, except South Wales, had voted by considerable majorities against the resumption of work. It was curious that South Wales, which had originated the strike, should be the most eager to end it, and there are several reasons which may be given for this. In the first place, as the conference had given no recommendation as to voting in the ballot, the miners' agents had felt themselves free to express their own personal opinions ; and in South Wales the majority, who realised that the Act gave them practically the power to obtain what they had been fighting for, advised a return to work. Further, the South Wales miners had exhausted the funds of their union, which were lower per man at the commencement of the strike than in any other coalfield. This was mainly due to the Cambrian Combine and other strikes in 1910–11. Also the South Wales miners were getting rather tired of strikes ; as probably about 60,000 of them had been out on strike at one time or another in 1910 or 1911. In Yorkshire and Lancashire there was considerable opposition to returning to work. On April 10th the Lancashire miners decided they could not return until the minimum rates were fixed, and rioting arose in connection with attempts to prevent resumption of work at a few collieries which were opened. A conference of the Lancashire Federation, however, ordered the men to return, and this they did in a few days. By the middle of April there was a fairly general resumption of work throughout the country, and by the end of April it may be said that all the collieries were again open, with the exception of just a few small and old pits, which were permanently closed. The rank and file of the miners expressed themselves as strongly dissatisfied with the Act as passed, and when the District Boards came to be constituted and awards were made by the independent chairmen, there was much grumbling because the minimum rates fixed in many of the districts not only fell below the M.F.G.B. demands, as regards the piece-workers at the face, but gave a lower minimum than the 5s. to the underground day-wage men. Nevertheless it may be said that the passage of the Act represented a distinct victory for the workers, who gained one great principle for which they were striving. By enquiries of the miners' agents I find it is generally admitted that the Act has been a great advantage to the workmen, now that they are beginning to understand how to make their claims. I find, too, that many of the colliery managers do not consider that they have lost much through being obliged to make up the wages of piece-workers. Certainly the burden has been very small in comparison with the great profits made during the boom of 1912 and 1913.

The miners did not find the national strike quite so effective a weapon as had been anticipated. In some districts the idea had been widely circulated that a national strike could not possibly last more than a week, and a fortnight was regarded as the utmost limit before the nation would be at the miners' feet. As a matter of fact, public alarm decreased considerably after the first week, for it was found that there were very large stocks of coal available. The Great Eastern Railway said that it could run its ordinary service for two or three months at least, and the other railways could keep going for a similar period on a reduced time-table. Some of the collieries, particularly in the North of England, had accumulated enormous stocks of coal during the two months preceding the strike. Some attempts were made by strikers to prevent this being put on the market; but the leaders discountenanced such interference, adhering to the principle that no coal must be wound. Such coalowners as had accumulated stocks undoubtedly profited greatly by the strike. The financial loss of other owners was not serious, for in their contracts for delivery they were protected by the strike clause; and the higher prices ruling for some weeks after the strike went a long way to recoup them for their loss in the heavy standing charges accruing during idleness.

Another source of the miscalculation to those who had predicted a short and sharp struggle, was the fact that they had forgotten what an enormous quantity of coal there is at any moment under way by sea to London, so that ships continued arriving and unloading in the Thames two or three weeks after the strike had commenced. The price of coal rose considerably in London, but there was never an actual coal famine, such as did develop in some of the northern and midland towns, where many of the

factories had kept going to the last moment.   Thus,
whilst the strike was less rapidly effective than had
been anticipated, the very serious effect upon their
fellow-workpeople in other trades had been to a
great extent overlooked by the miners.   In a very
great number of trades the workmen were obliged to
have the assistance of the out-of-work benefit of their
trade union, with the result that the funds of the
principal trade unions became very seriously
depleted.   Thus the miners' strike was not altogether popular with the rest of the labour world.

The cost of the strike to the country, as a whole,
is difficult to estimate, but it must have been enormous.   The actual loss of miners' wages was probably very nearly £10,000,000.   The total loss to
the community, taking into account all indirect losses
of income and profits in other trades, in addition to
the losses in the coal trade, can hardly have been less
than £50,000,000.   It may seem extraordinary that
the total effect of such a loss upon the course of the
country's trade during 1912 was not greater than it
appears to have been.   It must be remembered,
however, that 1912 was a year in which a great
boom of trade was beginning, and further, it is
doubtful whether the large figure I have suggested
for the total loss is more than about 2 per cent. of
the total of profits and incomes, including all wages
earned in the United Kingdom.   There is also the
interesting point that in coal mining it is possible
to some extent to make up the loss incurred through
so long a stoppage.   It was found that in the two

months succeeding the resumption of work the output of coal was quite abnormally high. It was said that the workmen must be working unusually hard to make up the losses of six weeks' idleness, and that it showed how far their usual rate of work fell below what was possible if they worked steadily. Whilst it is probably true that hewers did apply themselves with greater energy in the few weeks just succeeding the strike, the greater output was also due to the coal having become easier to work. The same phenomenon occurs after every strike, for the interval when no coal is removed gives the coal face a substantial " rest." There is time for the weight of the superincumbent strata gradually to bear down on and slightly crush the outer few yards of the unworked coal, which results in its being far more easily worked than usual. That is the chief reason why the total output for the year is by no means reduced by the proportion of time for which the mines were stopped.

CHAPTER XX

Principles of the Act

In the last chapter I brought the history of the abnormal places question and the National Strike to a close with the passage of the Minimum Wage Act and the gradual return of the men to work. In this chapter we shall see how far the Act realised the hopes of the miners, and then consider the working of the Act itself. We shall see how the District Boards are constituted, and how they carry out their duties. We may then turn to appraise the benefits conferred by the Act as administered by the District Boards and the umpires ; and finally we shall consider whether any important alterations would be fair and desirable.

We have already seen that the miners were extremely anxious to have the figures of the district minima prepared by the M.F.G.B. inserted in the Act, but that the Government had persuaded them that this was impossible, the coalowners having refused to discuss figures in London without the guidance of statistics relating to the various districts. Whilst obliged to give way upon this point, a strong stand was made by the miners' leaders to get the

minima of 5s. for day wage men and of 2s. for boys
included in the Act. In this again they were in
the end baulked, the Act being passed without the
concurrence of the miners. Subsequently, owing
to the inconclusive result of the general ballot,
the Executive Committee decided to accept the Act
as passed, and recommended a general return to
work.

It is interesting to notice that a new and very
important principle was involved in the strong stand
made by the miners for the 5s. minimum per shift.
They desired it to be considered as the lowest limit
of a reasonable living wage for an underground
worker in any part of the country ; and that this
should therefore be a bedrock minimum for adult
workers. English legislation always proceeds as
far as possible by compromise ; and where new
principles have to be introduced into legislation
they are adopted tentatively, and are not followed
to their logical conclusion, but only just so far as
may be necessary to remove a particular grievance.
There appear to have been two distinct principles
underlying the Minimum Wage agitation and
National Strike : in the first place the injustice
to a proportion of individuals in their being forced
from time to time by the circumstances of their
employment, and through no fault of their own,
to work for wholly inadequate remuneration. Here
was a particular grievance specially related to the
coal mining industry ; and having once realised the
nature of the grievance, and having been convinced

of its actuality, British parliamentarians, as typified by the Prime Minister and his colleagues, became perfectly ready to supply a legislative redress for this particular grievance. On the other hand, the claim for a reasonable living wage for the day-wage men stood upon quite a different footing, these men being paid a time wage similar to that customary in so many industries. The Government hesitated to give legislative sanction to the principle that what is a reasonable living wage, could or should, be determined apart from an examination of the rates of wages actually paid for such labour in every district. They felt it impossible to ask Parliament to commit itself to the task of itself fixing a reasonable living wage, especially in the hurry of what was avowedly panic legislation. The easiest course appeared to be to leave the Minimum Wage rates for all classes of workers to be settled by the Joint District Boards ; and it was therefore in this form that the Minimum Wage Bill became law.

The result is that the Minimum Wage Act establishes one great principle in British legislation, but does not establish another which it was urgently desired by many Trade Unionists that it should. It does establish the principle that the State may in justice to individuals in certain employments find it necessary to fix certain minimum rates of wages, below which it shall not be legal for employers to pay their workmen. The criterion of whether such legislative interference was necessary appears to have been solely the degree of hardship inflicted upon individuals and the number of men affected by the absence of such legislative protection. The principle which the Act does not establish is that of the State taking responsibility for insuring to every class of worker a wage rate which shall be adequate to provide them with a recognised minimum of the necessaries of a healthy, cleanly and respectable life. The independent chairmen have power to take this principle into consideration if they so wish ; but are not required to do so.

Provisions of the Act [1]

The Act which Parliament asked the miners to accept is a mere skeleton, practically all the detailed provisions necessary for achieving its objects being left for settlement by District Boards. If these could not have been constituted, or had failed to perform their work, whether through negligence or deadlock, the Act would have become a dead letter. Furthermore, there was the possibility, which has to a slight extent been realised, of the Boards in different districts, or their independent chairmen, deciding upon rules so different in different districts that the law would be essentially different in one part of the country from another.

Glancing rapidly through the clauses of the Act we notice that the first states that " It shall be an implied term of the contract for employment of a workman underground in a coal mine, that the employer shall pay to that workman wages at not

---

[1] See Appendix IX for extracts from the Act.

less than the minimum rate settled under this Act and applicable to that workman, unless it is certified in manner provided by the district rules that the workman is a person excluded under the district rules from the operation of this provision, or that the workman has forfeited the right to wages at the minimum rate by reason of his failure to comply with the conditions with respect to the regularity or efficiency of the work to be performed by workmen laid down by those rules ; and any agreement for the payment of wages in so far as it is in contravention of this provision shall be void."

This defines clearly the nature of the Act. Workmen are only entitled to the minimum if they comply with conditions as to regularity or efficiency of the work, and certain workmen are exempted altogether. In the second clause it is provided that the district rules shall lay down the conditions respecting the exclusion from a right to the minimum rate of aged, infirm, or disabled workmen, and that the district rules shall lay down conditions with respect to the regularity and efficiency of the work, and the time for which a workman is to be paid in the event of any interruption of work. It also provides that in case of failure to comply with these conditions, except from a cause over which he has no control, the workman shall forfeit his right to the minimum wage. The rules are also to decide who is to settle whether a workman is one to whom the minimum rate is applicable or not, or whether a workman has complied with the conditions. The third clause makes the Act retrospective in its action to the date of passing, granting from that date the minimum rate of wages afterwards to be settled.

There then follows an important safeguard to the effect that nothing in the Act shall prejudice the operation of any agreement entered into, or custom existing, before the passing of the Act for the payment of wages at a higher rate than the minimum rate settled under the Act. Had it been possible for the coalowners to claim that all price lists or Conciliation Board agreements were rendered void by the passing of the Act, as happened in some districts after the passing of the Eight Hours Act, there would have been complete chaos in the coal trade. The Minimum Wage Act is entirely supplementary to all the previously existing and complicated machinery for adjusting and paying wages, and the Act was required solely to rectify certain anomalies in that system.

It is also provided that in settling any minimum rate of wages the Joint District Board shall have regard to the average daily rate of wages paid to the workmen of the class for which the minimum rate is to be settled. The words "have regard to" are delightfully vague, and rather savour of the lawyer throwing dust in the eyes of the anxious trade unionist who wanted to bind the independent chairmen of the District Boards to fix the minimum nearly as high as the average for each class of workmen. I am not saying that the minimum ought to be put nearly as high as the average; but am merely pointing out that this provision only means that the Board is to have before it for guidance the average daily wages of each class of workmen. Then follows an important clause allowing the Board of Trade to recognise as the Joint District Board for any district, any body of persons fairly representing the workmen and employers, and the chairman of which is an independent person appointed by agreement between two sides, or in default of agreement by the Board of Trade. The object of this clause is to enable the Joint Conciliation Boards already existing in the various districts to act also as Joint District Boards under the Act. It is secured that there must be equality of voting power between the workmen's representatives on the one side, and the employers' representatives on the other side, the chairman having a casting vote.

The Joint District Board is required to settle the general minimum rates of wages and the general district rules for its district ; but it is given power to make special minimum rates either higher or lower than the general district rate, and special district rules either more or less stringent than the general rules, for any group or class of mines for which it is shown that the general rates or rules are, for any reason, not applicable. As a matter of fact there are many cases in which this power of fixing special minimum rates has been taken advantage of wherever there is a group of mines working a special class of coal, or lying in a remote part of the district, or for any other reason paying a different average rate of wages. The Joint District Boards are also given power to subdivide their district into two parts, or, if both sides agree, into more than two parts; and different districts may be united for the purpose of settling rules, a provision which I do not think has been made use of. There are several cases in which District Boards have divided their districts into two or more parts.

As regards alterations of the minimum rate of wages from time to time, it is provided that they may be varied (a) at any time by agreement of both sides, (b) after one year has elapsed since the rates or rules were last settled or varied, on an application made (with three months' notice given after the expiration of the year) by any number of workmen or employers, which appears to the Joint District Board to represent any considerable body of opinion amongst either the workmen or the employers concerned. It will be observed that there is no provision that the minimum rate of wages shall rise or fall together with the percentage above standard fixed by the Conciliation Board; but apparently there is nothing to prevent such an arrangement being made so long as both sides are agreeable in conformity with provision (a).

The Act prescribed that the Board of Trade should recognise Joint District Boards within two weeks after it was passed or proceed itself to appoint a Board or person to act for any district, and that if the Joint District Board had not, within three weeks, or within such extension of time as the Chairman decided, fixed the minimum rates and rules, the district chairman should himself proceed to settle the general district minimum rates, and the general district rules, and any special rates or rules required.

These are all the effective provisions of the Act, all the rest being merely formal. It will be realised at once that a very great deal of responsibility and labour was thrust upon the independent chairmen of the District Boards ; and it is immensely to their credit that, with a certain amount of assistance from the Board of Trade, the Act was within a few months got into full working order, in spite of a tendency to obstructiveness or obstinacy on both sides in certain districts.

The Joint District Boards

In a schedule of the Act is given the following list of twenty-two districts for which Boards were to be constituted :—

Northumberland.
Durham.
Cumberland.
Lancashire and Cheshire.
South Yorkshire.
West Yorkshire.
Cleveland.
Derbyshire (exclusive of South Derbyshire).
South Derbyshire.
Nottinghamshire.
Leicestershire.
Shropshire.

North Staffordshire.
South Stafford (exclusive of Cannock Chase) and East Worcestershire.
Cannock Chase.
Warwickshire.
Forest of Dean.
Bristol.
Somerset.
North Wales.
South Wales (including Monmouth).
The Mainland of Scotland.

2 o

In practically all of these there existed previously a Conciliation Board or Joint Committee for the regulation of wages, and in almost every case this became the Joint District Board under the Act, without modification, except in some cases by reducing the number of members. In two or three districts, the Joint District Board succeeded in settling both the district rules and general minimum rates of wages by agreement within the extended time decided on by the chairman. In some other cases, for example, Cannock Chase, the rules were agreed within the time limit, but not rates of wages. In most of the districts, however, the time limit expired after an extension of three to seven weeks, making six or ten weeks in all, without an agreement having been reached on any of the district rules or rates of wages excepting some few of minor importance, so that the burden of arbitration fell upon the independent chairman, and he had to decide all the important rules and all the rates of wages. The Board of Trade kept in touch with the different District Boards and by this means secured some degree of uniformity in the important rules, especially as regards principles ; but there is a decided divergence of practice on many points between different Boards. Some of the rules are brief and avoid details, and others are much more lengthy and endeavour to provide for all possible contingencies. In every case there are many matters which must be referred from time to time to the decision of umpires ; as for example, whether a workman is entitled to the benefit of the Act or not, and as to whether he has disqualified himself or not under the rules for claiming a minimum wage for any particular period. Almost every Board appointed a panel of umpires to whom was delegated the decision of these points ; and these umpires are now constantly deciding cases under the Act. The umpires have generally been selected from amongst barristers and solicitors, or well known public men, usually Justices of the Peace. The smooth working of the Act depends very much upon the careful attention and impartiality with which these attend to the hearing of numerous cases, which often bristle with technical points.

The General District Rules

It will be of interest to examine the rules of one or two districts closely in order to understand precisely what the legal minimum wage amounts to in practice. We may take the South Yorkshire District first, and compare some others with it. Sir Edward Clarke was the independent chairman, and he substituted two months for the limit of three weeks granted by the Act for the Board's decisions ; but the Board failed to agree on the principal rules and rates, and they were left for his decision.

The rules deal first with the aged and infirm. Workmen over 65 are not entitled to the minimum rate of wages, and the same limitation applies to others over 60 years of age who in the opinion of the sub-committee are by reason of age unable to do a fair day's work, and to younger workmen who from infirmity, illness, accident or disease are unable to perform their tasks satisfactorily. A man returning after illness is to be regarded as infirm for the first week.

The next section cannot be expressed better than in the original : " Any man who through his own default fails to do a fair day's work, or to work his place to the best advantage, or who refuses or neglects to carry out any reasonable order given him by the deputy or other superior official to ensure him working his place to the best advantage, or who without good cause delays in going to his work, or who ceases work before the customary time at the pit, unless there is no work for him to do, shall forfeit his right to the benefit of the minimum rate of wages."

The next rule says that a workman who in any week fails to attend and work 80 per cent. of the possible number of shifts he might have attended, shall forfeit the right to receive payment at the minimum rate. Saturday afternoons, from the termination of the ordinary morning shift, and Sundays are excepted. This means that if the pit works six days in the week, the miner must attend five days, and if it works five days, he must attend four days ; but if the pit works only four days, as sometimes happens, the workmen must attend each day, or lose the benefit of the minimum rate. It is well known that miners are irregular at their work, particularly so in Durham and Northumberland, where few work more than five days, and four days has long been customary with many of the men, particularly when wages are high. In all of the coalfields the owners have a perpetual grievance against the men for lost time, which renders so much of the capital expended in opening the mine unremunerative ; and it is natural that the owners, having been forced to grant the minimum wage, sought this reasonable concession in return. It is fair to add, though, that men who have to walk two or three miles to the pit-head often unwillingly miss a day's work in very wet weather because they would arrive soaked through at the pit-head, and would have to remain in their wet clothes all day, as the colliery provides no place where they can change and leave clothes or even leave a coat.

The fourth rule deals with the application of the minimum to cases where the hewer employs one or more helpers, as for example, trammers and fillers. They are usually paid a fixed daily wage by the hewer in charge of the stall, or by the colliery company on his behalf, he being paid the balance. In other cases men working under the man in charge of the stall agree to take a certain proportion of the gross earnings of the group. In order to avoid the hewer claiming the minimum rate through paying too much to his assistants, it is provided that " in ascertaining the earnings of coal getters, or their workmen paid by the piece, there shall not be deducted for their trammers, fillers, or others working under them, a wage more than 1s. in excess of the minimum rate fixed for those classes of workmen respectively."

The fifth rule provides that in any shift or shifts in which a stoppage occurs by reason of strikes of any men or boys employed at the colliery, a workman shall forfeit his right to receive payment at the minimum rate of wages. This does not apply, however, when he is ordered to perform, and does perform, some other work in lieu thereof. Cases have come before the umpires in different districts as to whether, when a man is thus ordered to perform other work and does so, the minimum rate applicable is that of his own work or of that which he is put to do, the colliery usually claiming the latter when it is lower. In some cases, and I believe in several districts, it has been decided that the man is entitled to whichever is the higher minimum rate, for his own work, or for that which he is put to do—which is obviously fair.

The sixth rule deals at length with the question of stoppage of the pit due to accident or various causes, such as shortage of wagons, which may prevent the company from continuing work except at abnormal cost. The rules provide that the collier must take the risk of such stoppages occurring. If, when he presents himself at the pit bottom, or at the lamp station near to it where his lamp is examined, he is informed that something has happened in or about the mine to prevent his working, he is not entitled to claim any wage at the minimum rate in respect of that day. In fact, he gets no compensation for having got up very early and travelled perhaps several miles and descended the pit. In the building trades a man under such circumstances would always be entitled to payment for his " walking time," or else to two hours' wages. The workman is also only allowed the proportion of the day for which the pit is actually working, or for which his district or his own place can remain open for working, if anything whatever prevents the pit, district, or stall continuing to be worked. He may thus only get one-quarter, one-half, or three-quarters of the minimum wage for the shift according to the time when it became necessary to " knock off." It is provided, however, that when a workman is thus prevented from working he shall be allowed to come out of the mine without any unreasonable delay.

The last rule provides that the interpretation of the rules shall be referred to the chairman of the District Board, whose decision shall be final ; and it also lays down the mode of settling disputes as to the application of the minimum rates in particular cases. After discussion between the workman or workmen concerned and the officials of the mine, and after they have failed to agree, the matter in dispute is to be submitted to the manager of the mine and some person working in or about the mine nominated by the workmen ; and if they fail to agree the dispute is then to be referred to a joint committee. Every such local committee is to be appointed by the two secretaries of the Joint District Board, or by the District Board itself, sitting without, or if necessary with, the independent chairman. In the event of such local committee failing to agree, it is to refer the matter to a neutral chairman, whose decision shall be final. Pending any decision, which is to be given as promptly as possible, " the workman shall continue at work, his right to receive the benefit of the minimum wage as from the date of the complaint being reserved until the decision shall be given."

In these rules, that is, for South Yorkshire, and in some others, the miners' agent is ignored, for the dispute is referred to the manager and " some person working in or about the mine," who would usually be an official of the colliery lodge, or a member of the committee, but could not be the agent. In most of the districts it is the miners' agent who has to confer with the manager ; and there is in South Wales a preliminary stage before the dispute reaches these busy men. Possibly the intention in some of the districts was to keep the miners' agent and the manager out of the dispute until it comes before an umpire, when the manager and miners' agent would each state his case before him, like counsel.

An abnormally long time was taken in settling the district rules for South Wales ; and though they need not be any the better for that, they are certainly more detailed than the rules for most of the districts. They regulate the work of 200,000 men, a larger number than are included in any other district under the Act ; and for this reason have an importance which has led me to reproduce

them in full in Appendix X. On looking through
them it will be observed that an aged workman is
defined as one who has reached the age of sixty-three.
Rule 5 requires that a workman shall work five-
sixths of the possible shifts, which is more stringent
than in South Yorkshire and most of the other
districts, which require only four-fifths. It is now
modified by the succeeding sentence, which permits
a man to miss one day, even if the colliery works
only five days, provided he has not missed a day
during the previous week. This modification was
added in 1913, and during the first twelve months in
working the Act there was a good deal of hardship
in the rule, which often meant the loss of a man's
right to his minimum through a single day's absence.

There is a greater difference than appears at first
sight between requiring a man to work five-sixths
and four-fifths of the possible time because a man
cannot split a shift. The length of a " pay," which
used to be fourteen days, became, since the Coal
Mines Act (1911) came into force, generally seven
days, that is six weekdays. As the rule was, if the
mine stopped for one day in the week through
shortage of wagons, or of orders, the collier in South
Wales had to attend every day the mine was working ; for if he missed one out of five, he failed to
work five-sixths of the possible time. Thus, if he
overslept himself one morning, or thought it too wet
to go to work one day, he risked losing the benefit of
the minimum wage should the colliery subsequently
be stopped a day that week—only twenty-four

hours' notice of the stop being usually given. Considerable friction was caused by this rule when it first came into force, because some considerable classes of night workers in South Wales normally work only five shifts a week ; and for some months, until they were expected and permitted to work only 25 nights in every six weeks, they lost the minimum wage if absent a single shift.

It is further to be noticed in Rule 5 that the collier or other piece-worker is not entitled to the minimum wage unless he claims it from the overman in charge of his district, whom he must find and convince that his working place or other circumstances are such as to prevent his earning the minimum wage on the price list. If the official does not agree, a technical dispute arises, and this is to be settled in a manner which it is interesting to compare with the rule for South Yorkshire. The rule (No. 8) provides that the workman must first try to settle direct with the official in charge of the mine, meaning generally the under-manager. Failing agreement, the dispute is to come before two officials of the colliery representing the employers, and two members of the committee of the local lodge of the Miners' Federation, or not more than two representatives appointed by them. This is an easily constituted local court of persons likely to have full knowledge of all the local circumstances. Its success depends very much, however, on the degree to which its members can detach themselves from mere partisanship ; and it is necessary for every member to be fair and judicial, for, if one member exhibits partisanship all the rest feel immediately forced into the same attitude, and deadlock ensues.

If these four persons fail to settle the dispute, it is to be referred to the manager of the mine and the miners' agent for the district. If they fail to settle the dispute, as happens not infrequently, it is to be referred to an umpire to be selected by them, or if they disagree, by lot, from a panel of persons chosen to act as umpires by the two chairmen of the employers' and workmen's representatives on the Joint District Board. The employers and the workmen may call or submit such evidence as they think proper, and the umpire, or the referees in the earlier stages, may make such inspection of the workings as they think necessary. In practice the umpire is sometimes confronted with rival plans or diagrams showing an abnormality in a seam. Cross-questioning the witnesses will usually determine how far each diagram is correct ; but in some cases the umpire may feel obliged to go and see for himself. The rule further wisely provides a time limit within which decisions must be made ; and expressly permits the attendance of the colliery representative and miners' agent before the umpire.

A great bone of contention in South Wales has been the period of averaging the earnings in order to determine whether the earnings under the price list fall short of the minimum wage. Rule 7 provides that " the earnings of two consecutive weeks shall be divided by the number of shifts and parts of

shifts he has worked during such two weeks." The
result is that if a man has had his wages made up
to the minimum as the result of an average of the
past fortnight, and if his working place improves,
or he is moved to a new place, or even if he works
harder, and so earns more than the minimum in
the ensuing week, that excess goes first of all to
pay off the deficit of the previous week before he
can touch any of it. An amount so " debited shall
be deemed to be a payment on account of wages
to become subsequently due to him," says the
rule.

There is marked disparity between the periods
adopted for averaging in the various districts
throughout the country. In a few (*e.g.* West
Yorkshire, Cumberland) the period is one week.
In the majority it is two weeks, and in some of the
less important districts it is longer—*e.g.* Cannock
Chase, three; South Staffordshire, four weeks.

The Miners' Federation of Great Britain claims
that in making any rule requiring a longer period
than one pay, *i.e.* one week, for averaging, every
District Board or independent chairman who did
so acted *ultra vires*, and beyond the authority of the
Minimum Wage Act, which nowhere says that
the period of averaging shall be the subject of a
rule. The Act says nothing about average earnings
and only gives power to make rules regarding the
conditions of efficiency of work, so that there is
nothing to over-rule the Coal Mines Act (1911)
requiring weekly pays, which presumably are to be

self-contained.   A test case [1] was fought in the courts
by the South Wales Miners' Federation, and given
against the colliery owners by Mr. Justice Pickford.
On appeal, Lords Justices Vaughan Williams, Kent,
Buckley, and Kennedy were unanimous in holding
that there was "no power and no authority in the
Joint District Board to make any such rule as that."
At the same time the Court declared that it could
not endorse Mr. Justice Pickford's declaration that
the Joint District Board could settle the period of
averaging at one week.   The Court of Appeal
declared that it expressed no opinion as to how the
rate is to be ascertained.   In default of any agreement between the two sides of the Joint District
Board as to what should be done, the rule requiring
the two weeks average has remained in force for a
further six months.   As I write, the matter is still
unsettled, although it is two years since the offending
rule was made.

The opposition of the coalowners arises from a
fear that some workmen may take advantage of a
short-period average to get paid for work twice over.
This a man could do by spending most of his time
one week in repairing his place, and making extensive
preparations for the getting of coal, but sending
very little out.   He would thus perhaps get twenty
shillings or so added to his price list earnings through
the operation of the minimum, and the following
week he would work and send out as much coal as
would normally take a week and a half.   This

---

[1] Davies and others v. Glamorgan Coal Co.

practice cannot be carried on from fortnight to
fortnight, but can be from week to week. The
owners, therefore, insist that the workmen's
representatives on the District Board shall devise
some satisfactory rule which may prevent such
malpractices, without too great cost in supervision.

General District Rates

The Joint District Boards were less able to agree
as to the minimum rates of wages than as to the
district rules, and in the great majority of the
districts practically the whole of every schedule
of rates was settled by the independent chairman. [1]
The organisation of the coal-mining operations
differs much in the mines of different districts,
and consequently the classes of workmen employed
are different both in number and duties. In every
district the most important and difficult rates to
settle were those of the hewers on the one hand,
and those of the lower-paid day-wage men on the
other hand. Those are the two most numerous
classes of employees in a mine ; and also the
minimum would be generally speaking more costly
per man in their case than for other grades, because
in many mines a fair proportion of the hewers when
working in abnormal places had undoubtedly had
to put up with wages less than a fair minimum ; and
many of the lower-paid day-wage men in most districts
were evidently entitled to a substantial increase.

[1] Exceptions were South Derbyshire, Warwickshire, and the
district of Lancashire and Cheshire.

NORTH STAFFORDSHIRE

Schedule of Minimum Rates Fixed

	Present Average Wage.		East and West District.		Cheadle District.		Ironstone Mines Special Rates.	
	s.	d.	s.	d.	s.	d.	s.	d.
1. Contract Colliers	—		7	0				
2. Other Colliers	6	7·08	6	6	6	0	6	3
3. Loaders	{ 6	0·29E	6	0 }	5	6	5	9
	{ W5	7 W	5	6 }				
4. Crutters	6	6·73	6	6	—			
5. Assistant Crutters	5	8·61	5	9	—			
6. Datallers	5	8·83	5	6	5	3		
7. Assistant Datallers	4	6·45	5	0	4	6		
8. Packers working under supervision	4	8·11	5	0	—			
9. Coal Cutter Attendants	5	10·03	5	9	5	6		
10. Assistant Coal Cutter Attendants	4	9·6	5	0	4	6		
11. Onsetters	5	8·08	5	6	4	6		
12. Motor and Enginemen	4	4·75	5	0	4	6		
13. Roadmen	5	5·49	5	0	5	0		
14. Boys	1/10½ to 4/- at 20		2/- to 4/6 at 21		2/- to 4/6 at 21			

The rates for colliers are in all cases net rates, that is, free from any deduction for explosives or tools.

There were also a number of other important rates to settle. In districts where the hewer does not do his own filling of coal into the tram, the "fillers" are a numerous class; and there were also the timbermen, the roof-rippers, the packers or wallers, the various classes of enginemen, and the attendants on coal-cutting machines. One anomaly is that in some of the districts shotfirers and horse-keepers have not been considered to come within the Minimum Wage Act because they have statutory duties, and in other districts they have been granted a minimum wage.

The award made by Judge J. K. Bradbury for North Staffordshire is of interest because he states that he adopted the rule of fixing the minimum for a class at about the average daily earnings, and gives the figures of these averages on which he based each rate. The table (page 591), which I quote in full, also shows which classes of workers had their wages substantially raised.

I explained fully in the former chapter on Conciliation Boards, how wages are varied from time to time in every district by the Board altering the percentage to be added to the standard wages. It is unfortunate that no provision was made in the Act defining whether or not the minimum should rise and fall automatically with the rise and fall of the district percentage above standard. It seems to have been generally thought immediately after the passing of the Act that the minimum wage was intended to be a fixed figure for each class of worker in each district, which would not vary with the state of trade. The awards in several districts, particularly the earlier ones, were given on that basis, it being assumed apparently that the Joint District Board would meet from time to time to alter the minimum wage, if it was considered by either side that an alteration of the percentage addition having been made by the Conciliation Board warranted reconsideration of the minimum wage.

It is natural that workmen do not care to have a minimum which remains stationary when the percentage addition to wages is raised. In districts with a fixed minimum the workmen have generally in such circumstances applied for an increase, and have obtained it, wages generally having risen during the past two years (1912–14). It is, however, a very cumbrous arrangement for the Joint District Board to have to meet and do all over again each time the work done by the Conciliation Board for the district. Much the better arrangement is that adopted by South Wales and Monmouthshire, and also by the Bristol, Somerset, and Warwickshire districts, by which the minimum wage fixed is a standard rate which varies automatically with the percentage addition fixed by the Conciliation Board.

It was the failure to have such a convenient arrangement which led to the disastrous strike throughout the South Yorkshire district in the spring of 1914. There had been a fixed minimum established for each of the several grades of workers ; but this had been regarded as unsatisfactory by the Conciliation Board of the Federated Districts. The Conciliation Board therefore resolved that men earning less than the minimum wage (which for hewers was 6s. 9d.) should have the increased percentage added to the minimum. In the course of a few months the Conciliation Board added three 5 per cents. ; so that 15 per cent. above the standard of 1888 was to be added to the minimum wage. The standard day-wage was 4s. 6d. per day for some collieries and 5s. for others, making the 15 per cent. additions as generally understood approximately 8d. and 9d. respectively. The absurd result was that men earning just under 6s. 9d., say 6s. 8d., had their wages made up first to 6s. 9d. and then by the percentage to 7s. 5d. or 7s. 6d. ; but men earning just over, as for instance, 6s. 10d., received no addition at all.

Annoying as this was, there would have been no strike had not other troubles developed. It so happened that one of the largest colliery companies had not read the vaguely worded resolution of the Conciliation Board as meaning that the 15 per cent. additions were to be calculated on the standard day wage, and adopted the plan of calculating them on the man's actual earnings at standard rate, which would mean adding anything from 1d. to 7d., instead of 8d. or 9d. Some other collieries had misunderstood the resolution in the other direction, and assumed the minimum rate itself was to be taken as raised from 6s. 9d. by 15 per cent. to 7s. 10d., and actually paid on that basis, so that in those collieries men earning just over 6s. 9d. did get the addition, and everyone got more than in the other collieries. By degrees the owners paying most found out what interpretation others put on the Conciliation Board resolutions, and

naturally tried to adopt the interpretation most
favourable to themselves.

Meanwhile the workmen's side of the Joint
District Board had applied for an increase of the
minimum wage, and the independent chairman,
Sir Edward Clarke, gave as his award an increase
from 6s. 9d. to 7s. 3d.   Correspondence ensued as to
whether Sir Edward intended to include the 15
per cent. granted by the Conciliation Board in his
7s. 3d., making only a 10 per cent. increase on the
Conciliation Board basis, or was it to be added to
the 7s. 3d. ?   He very naturally and properly
replied that the 7s. 3d. corresponded with the 6s.
9d., and that he had nothing whatever to do with
any Conciliation Board agreement providing for
payment in addition to the minimum wage.

A strong section of the workmen now claimed
that they were entitled to have the 15 per cent.
added to the 7s. 3d. ; and in this they were supported
by a further passage in the independent chairman's
explanations to the effect that, so long as the Conciliation Board's resolutions remained in force, it
" appeared " to him that the 7s. 3d. would be supplemented by the percentage.

The unfortunate owners, confronted with the
prospect of a fall in the price of coal, refused to be
squeezed any more ;   and they mostly began to
insist that they were only legally bound to pay
Sir Edward Clarke's award of 7s. 3d., and proceeded
to do so.   It was then that the strike, which had
already begun in the neighbourhood of Rotherham,

began to spread. By this time the men were so much annoyed at not getting the 8s. (7s. 3d. plus 15 per cent. of 5s.) which had been practically held out to them, and I think, actually paid in some cases, that they went out on strike against the strong appeal of their leaders, who urged them to wait until the Conciliation Board could deal with the matter.

To cut a long story short, there was a strike of 150,000 men for about five weeks, in which the whole of the Federated Districts of North Wales, Lancashire, and the Midlands supported the Yorkshire miners. Meetings were held both of the Conciliation Board and of the Joint District Board for South Yorkshire ; and finally the owners agreed to a large extent to the men's claims. The chief source of difficulty in arriving at a settlement was the position of the owners of the older and smaller collieries prevailing in the eastern district, who declared that it was an absolute financial impossibility for them to pay the 15 per cent. on top of Sir Edward's 7s. 3d.

It was therefore agreed that the workmen should be balloted as to acceptance of a scheme by which certain of the older and smaller collieries which proved they were unable to pay on the higher basis should be constituted a special group or class of mines under the Minimum Wage Act, for which the Joint District Board should award the lower minimum of 6s. 9d. A majority voted in favour of accepting this proposal, and the men returned to work with two minimum wages, namely 7s. 3d. plus percentage (making 7s. 11d. or 8s.) for the majority of mines, and 6s. 9d. plus percentage (making 7s. 5d. or 7s. 6d.) for a special class of mines which could prove to the Joint District Board that they would be unprofitable at the higher rate. There ensued a most unusual eagerness on the part of many colliery managements to disclose the entire details of their working costs to the workmen's representatives.

Taken as a whole the outcome of the strike was a distinct victory for the men, a minimum wage of 8s. being the highest figure paid at any time in the whole country. In the differentiation of collieries there is a new and very interesting example of the application of the principle of "ability to pay" in determining wages. The tendency of competition, and the policy of trade unionism, has generally been to establish a uniform rate of wages for the same task throughout each district. But trade union conbination, or a minimum wage board, in forcing up wages must sooner or later get to a point where a further increase of the rate of wages, if applied to all firms, will close so many, that the loss in unemployment becomes so serious as to over-balance the advantage of increasing the wages. Under such circumstances with a uniform wage, the firms owning the most cheaply worked and best situated mines or plant make a large profit out of their advantage over the badly equipped or situated firms who cannot pay more. By differentiating

the wage-rate for the same task, according to the
employer's ability to pay, a part of that profit is
transferred to the workman. Theoretically, the
mines paying the lower rate would be left with the
less efficient workmen; but, practically, as they are
older mines in well settled and thickly populated
districts, this tendency, though acting, will be
modified accordingly, and only operate slowly over
a period of many years.

I have gone in detail into the causes of the
Yorkshire strike, because it is one of the strangest
and most serious muddles I have ever heard of in an
industrial dispute. There appears to be no blame
resting on anyone in particular. Sir Edward Clarke,
in awarding a fixed minimum wage of 6s. 9d., did
the same as most other independent chairmen of
the District Boards. The Conciliation Board
acted with the best intentions in adding the percentages. It would appear now, however, that it might
have been wiser of the Conciliation Board if it had
not acted directly in regard to the minimum wage,
but had merely issued an invitation and request
to the several joint district boards in its area for
them to meet, and by agreement of the two sides
to add the percentages recommended by the Conciliation Board. The lesson of this strike, however,
which I wish to enforce, is that it would be far
better for all the joint district boards to fix the
minimum wage as a standard wage and make it
subject to the Conciliation Board percentage,
as was done in South Wales. It would then form

part of the general machinery for adjusting the rise
and fall of wages with the state of the coal trade ;
many occasions of friction through the alteration
of the minimum wage not immediately following the
percentage would be avoided ; and the work of
Joint District Boards would be lightened, or reserved
for their most useful function of constantly over-
hauling and improving the District Rules, and
making relative re-adjustments of standard minimum
rates.

Success of the Minimum Wage Act

The Minimum Wage Act was an extraordinary
piece of hastily prepared experimental legislation,
rushed through Parliament in the shadow of an
unprecedented national calamity.　At the time both
the coalowners and the general body of miners
were intensely dissatisfied with it, and only some
of the miners' leaders agreed with the Government
and the general press that it ought to provide a
satisfactory solution of the abnormal place question.
The Act has now been in force for two years ; and
it is possible to say of it that in general it has proved
a great benefit to the miners without putting a
heavy burden on the mining industry.　The two
outstanding benefits are : (1) that it has distinctly
increased the wages of the unskilled underground
day-wage men ; and (2) that it has secured to hewers
a minimum day-wage when prevented by causes
over which they have no control from earning the
minimum by piece-work.

The claim of the M.F.G.B. for a 5s. per day
minimum for the latter class has not, it is true,
been realised in most of the districts, in spite of
Mr. Asquith's declaration in Parliament that he
considered it a reasonable demand, which statement many of the miners regarded as almost amounting to an instruction to chairmen of Joint District
Boards to grant it.   Some of the district chairmen
(*e.g.* Lancashire & Cheshire) have acted on the
suggestion ; but the great majority, and particularly in the larger districts, have fixed the minimum
for adults at 4s. 6d. or 4s. 9d. per shift, which
on the usual average of 11 shifts per fortnight
means earnings of 24s. 9d. and 26s. 1d. per week
respectively, which cannot be called satisfactory pay
for adult underground workmen.   The contention
of the owners, however, is that it is only the unskilled helpers who are paid rates as low as this,
and that these are almost entirely young men
learning the run of the mine and waiting their turn
for promotion to better paid work.   Whilst this
is true in some collieries, there are a great many
of which it is not true, practically all the unskilled
day-wage men being employed in such collieries
at the minimum, and there being amongst them
many men well on in life with families to support.
In South Wales the lowest minimum rate for day-
wage men is 3s. 2d., plus percentage, which, at the
present level of 60 per cent., is equal to 5s. 1d.   It
seems unlikely that wages will ever again for long
at a time fall below 50 per cent. above standard, which

is equal to 4s. 9d. per shift. For the majority it
has recently been raised from 3s. 2d. to 3s. 4d., which
at 50 per cent. is 5s. I am told that there were,
before the Act came into force, several hundred
different day-wage rates being paid in South Wales
and Monmouthshire, counting as different rates the
same figure paid to a different class of workman.
Only about 40 of these now remain in the unskilled
grades, all the rest having become merged in the
minimum. Of the total number of day-wage men
who were employed in the South Wales district
in February, 1912, just before the strike, 56 per cent.
have had their wages raised by the operation of
Lord St. Aldwyn's awards.

In South Wales, when the percentage stands
at its maximum of 60, the day-wage men have got
a minimum 1d. to 4d. higher than the 5s. which
was claimed for them by the M.F.G.B.; and in a
few other districts the 5s. or more is being actually
paid during the present good trade. To bring it
up in the other districts to this level would involve
no great burden on the industry, and would give
a large class of workmen the chance of leading
a decent life in respectable homes. It is this
section of the miners which is now compelled to
live either two families in a cottage, or to put up
with old and insanitary cottages or " flats," with
no proper sanitary arrangements. To have a large
class of workers so ill paid is a drag upon the
social advancement of large districts in our coalfields.

The hewers, colliers, timbermen, rippers, road repairers, and other skilled piece-workers are in an altogether different category, and the minimum rates settled for these classes in any one district are generally at approximately the same level. The hewers are, of course, the most important and numerous class, and it is interesting to compare the rates they have actually obtained in the principal districts with those demanded by the M.F.G.B. in their final schedule. To make a fair comparison, I have stated the rates first awarded by the district chairmen, as well as those now prevailing. The latter, after a period of remarkably good trade, in most districts show an increase ; but are likely to be subject to a reduction in the near future when the price of coal is falling.

	M.F.G.B. Final Schedule.	First Minimum Wage.	Present Minimum Wage. (June, 1914.)
South Wales	. 7s. 1½d. to 7s. 6d.	6s. 10½d.	7s. 4d.
South Yorkshire	. 7s. 6d.	6s. 9d.	7s. 4d. to 8s.
West Yorkshire	.		
Derbyshire .	. 7s. 1½d. to 7s. 6d.	6s.	—
Nottinghamshire	. 7s. 6d.	7s. 3d.	7s. 11d.
Scotland	. 6s.	5s. 10d.	—
Durham	. 6s. 1¼d.	5s. 6d. to 5s. 10d.	—
Northumberland	. 6s. to 7s. 2d.	5s. 6d.	—
Lancs. & Cheshire	. 7s.	6s. 6d.	—
Cumberland .	. 6s. 6d.	6s.	—

It will be seen that the districts which most nearly obtained what they had demanded were

Scotland, Nottingham, and South Wales. The rates awarded in Durham, Northumberland, and Scotland were particularly low, very nearly as low as in the typical low-wages districts like Bristol and the Forest of Dean coalfields. It is clear that a higher wage level rules in the newer coalfields for the same work, as for example Nottingham, South Yorkshire, and South Wales ; though it must be remembered that a large proportion of the miners of Durham, Northumberland, and Scotland have cottages provided for them free or at a nominal rent.

One serious defect of the Act is that it did not clearly determine the nature of the workman's right to receive a minimum wage, with the result that it has been differently interpreted by different independent chairmen. In certain districts [1] it is laid down in the rules that the workman is entitled to receive the minimum wage except when his failure to earn the minimum is due to causes over which he has any control, in other words—is due to his inefficiency or laziness. It is intended that, if the workman's earnings prove short of the minimum rate the colliery officials shall question him and investigate the cause. If they find they have good reason to believe he could have earned the minimum if he had worked the place properly as directed by the overman, or find evidence that he has been " resting " or otherwise neglecting his work, the manager will issue a certificate that the

[1] *e.g.*, West Yorkshire, South Yorkshire, Nottingham, Shropshire and others.

workman has forfeited his right to the minimum wage, and invite him to sign an acceptance. If the workman thinks he can prove he has not been inefficient or lazy, but ascribes the shortage to other causes, he will probably refuse to sign, and then a dispute arises and is to be settled as provided in the rules, and ultimately by arbitration of an umpire. The important point is that in rules of this class the burden of initiative, and largely also of proof, is put on the management.

In most of the districts, however, there is a provision that the workman shall forthwith give notice to the official in charge of his district of his inability to earn wages at piece-rates equal to the minimum rate and shall at the time state the cause or causes to which he attributes this inability to earn the minimum. He forfeits his right to the minimum wage, however short he may be, unless he has immediately given this notice.

There has been a good deal of friction over rules of this character made in the different districts, the chief objection of the miners being that it puts on them the onus of proof that they are unable to earn the minimum rate. The South Wales Miners' Federation sought a decision of the court that the rule was *ultra vires*, but the decision went in favour of the rule, and was upheld on appeal. In many collieries the workmen were very shy of giving the necessary notice, many fearing that if they were too persistent in making claims they would be marked for dismissal at a convenient opportunity.

The process of establishing a claim to the minimum wage when disputed by the management is highly technical, and requires ability in collecting and marshalling evidence which the ordinary collier does not possess. The South Wales miners are, therefore, finding that their ordinary agents have not time to handle the mass of claims which have arisen under the Act, and many district committees of the South Wales Federation have appointed special agents to deal with the minimum wage cases, a policy which has been found to pay well, as such agents have recovered considerable sums. In the smaller districts where the minimum wage work is not sufficient to occupy an agent's whole time he combines such work with that of acting as workmen's safety examiner.

I regard the necessity for the existence of district rules as a distinct advantage of the Minimum Wage Act. They have been looked upon merely as unpleasant but necessary adjuncts to the existence of a minimum wage, designed to prevent workmen from defrauding their employers by being slack or inefficient in their work. It is doubtless desirable to have rules with this object in view, otherwise the cost to the coal owners of supervision would be greatly increased, and their only remedy would be the dismissal of a man who was thought or proved to be slacking. Frequent dismissals would create great unrest amongst the workmen.

Although the foregoing was the specific object of the rules, it appears to me that they have a further and general advantage in tending to regularise the relations of the management and the miners. In most factories there is a well established discipline depending on certain clearly stated and well known rules as to conduct and the handling of machinery. The rules are not necessarily detailed, officious, or irksome ; but merely what is necessary to provide a basis for discipline, and to prevent constant disputes with the foremen.

In coal mines generally, probably owing to the former prevalence of the sub-contract system, there does not seem to be the same tradition of ordered working as is found in our great industrial establishments. The statutory rules as regards safety were pretty well enforced ; but in other respects the discipline of the mine depends very much on the personality of the manager. A change of manager means bringing new rules which he thinks necessary into operation, whilst others insisted on by the former manager are dropped. Customs and rules of working differ from colliery to colliery, and this means trouble with workmen who change from one mine to another. The skilled miner is in general a very independent person, much alive to his rights, and unwilling to submit to orders which, too often perhaps, he regards as merely arbitrary whims on the part of the official. Friction arising from these causes has greatly increased with the better education of the miners who can often argue on equal terms with their superiors, and with the growth in the size of mines in recent years. A mine employing two or three thousand men is a small community requiring a carefully thought-out code of laws, and their enforcement by adequate and reasonable authority. Too little attention has been given to thinking out such codes of rules ; and it is precisely because the district rules under the Minimum Wage Act supply a uniform basis for rules of working in all collieries in each district that I think they will gradually come to have a most salutary effect upon industrial relations in the coal trade.

I hope that Joint District Boards will give further careful attention to elaboration of their rules, and that each district will try to learn from the experience of others. Some central agency is needed for acquainting each of the Boards with the difficulties met by other Boards and how they were overcome ; and presumably this intelligence work could best be carried out in a strictly impartial manner by a department of the Board of Trade.

In the light of two years' experience the Minimum Wage Act can certainly be pronounced a success. It has remedied almost completely certain serious grievances ; and there is the prospect that when certain alterations in the policy and rules of the District Board have been effected many causes of friction will be removed, and a state of more efficient and harmonious working than has ever been known in our coal mines will be introduced.

CHAPTER XXI

The Collier's Daily Toil

THE collier's life is a monotonous one, and a description of one day's toil will serve as an example of the routine which he follows throughout the year. If the shift in which he works starts at 7 a.m., he has usually to be out of bed from one or two hours earlier, according to the distance between his home and the pit. We may therefore assume that the miner is astir at half-past four or five. While he dons his clothes and his heavy hob-nailed boots, his wife is busily engaged in preparing his morning meal and filling his box and billycan with food which he will require for the meal time underground. Immediately after breakfast he sets forth to catch the workmen's train, or to walk the whole distance to the pit. During dry weather this walk often proves a healthy and enjoyable one ; when, however, as often happens, the miner has to tramp through the cold drizzling rain or heavy snow, he reaches the pit-head drenched to the skin. Here he hides his wet overcoat, mackintosh, or umbrella in the weigh-house, engine shed, or the carpenters' shop, and his matches, cigarettes or pipe in some secret

608

place near the colliery entrance, for no provision is made for storing at the pit-head the clothes or smoking requisites of underground workers. It is a serious offence to take matches or anything smokable underground.

The lamp room is now visited and the miner is handed a numbered lamp and he then passes on to the pit-head. As the accommodation in the cages for lowering men into the mine is usually small, it often happens that the men have to form a queue, and the late arrivals have to wait ten to twenty minutes for their turn to be lowered.

Each time the cage comes to the surface the banksman lifts the bar and a prescribed number of miners step in and take hold of an iron rail round the inside of the cage. The banksman pushes a button to ring the electric bell in the engine-house, the cage rises an inch or two to allow the grips beneath to be withdrawn, and then down drops the cage with its human freight into the dark inky blackness of the pit shaft. When the cage has gone a certain distance, the engine-driver applies his brakes and the cage slackens speed gradually until stopped at the bottom.

The pit bottom is high, roomy and comparatively safe. Here the miner rests a moment until he gets his " pit eyes," for the dim light of his safety lamp is of little use until he has accustomed his sight to the darkness. He then starts on his way along one of the haulage roads to the district of the mine where his stall or working-place is situated. Before he is allowed to enter his district and proceed to his stall he must have first received the sanction of the overman or another responsible official. A careful miner is not content to commence on the assurance of an official that all is well ; he himself makes a careful examination of the roof and sides and tests for gas. Having satisfied himself that there is no immediate danger the miner commences his actual daily toil, in which he is assisted by a boy or adult helper who not infrequently is a member of his own family.

The methods employed in mining the coal vary in different coalfields and in different mines, according to the thickness and character of the seams and other circumstances. Usually the first thing done is to " hole " or cut away the shale from underneath the coal which is to be brought down. This is done by means of a pick, and in order to more effectually discharge this task, the collier is compelled, if the seam is not very thick, to adopt a very cramped and uncomfortable position—in many cases, indeed, he has to lie sideways at full length on the ground in order to be able to swing his pick in such a way as to win the coal satisfactorily. When it is necessary to support the weight of coal during the holing process, short wooden sprags or props are placed under the edge of the coal at frequent intervals, and when the holing work has been completed the sprags are withdrawn and the huge pressure of the overlying strata causes the mass of coal to fall forwards and downwards and be crushed. The collier and his assistant now set to work to fill the coal into the tubs or trams brought to their working-place by hauliers with the aid of ponies.   Great care has to be taken in loading the trams with clean coal only and no stone, and to pack the coal so that it cannot shake loose with jolting.   This is especially necessary in South Wales, where the small coal, as well as dirt and stone, is not credited to the collier for payment.   When the tram is loaded, the collier chalks on it the number of his stall so that the weigher and check-weigher at the surface will know to which man they must credit the weight of coal brought up in each tram, for it must not be forgotten that hewers—except when the conditions of work are abnormal—are paid on the basis of the amount of coal sent out by them.

The usual stone and other rubbish not loaded into the tram is thrown into the " goaf " or " gob," which is the area from which the coal has been worked and which in most mines is carefully walled off from the roadways so as to support the roof and preserve ventilation.   When the working space has been cleared, the collier sets props to support the roof, the roadways with their tram lines are brought forward nearer to the new coal face, and the preceding operations are repeated.   In South Wales, the collier (hewer) does all this deadwork, so far as is requisite for keeping his place in order ; but in the North of England there is a division of labour practised, special workmen being employed to do the filling and others the prop-setting and tram-line extension.   The South Wales collier is proud of

being able to perform all the various tasks connected with filling and keeping his working-place in
tip-top order, and prefers not to trust to any other
set of workmen.   As regards the setting of props,
and also drawing them, it is in South Wales desirable
that the collier shall be able to do this himself ; for
the strata, particularly the roof, are more treacherous
than in England, and may need attention at any
moment, whereas a few hours makes little difference
usually in the north of England and the setting of
props can be left for the special shift to come.
Naturally the price list includes payment to the
collier for this deadwork.

From the above description it would appear that
a collier's task is a simple one requiring little or no
skill.   This assumption, however, would be a great
mistake ; a novice at the task would no doubt be
able to cut the coal after a fashion, but it requires
skill and intelligence as well as muscle to extract
the coal in good condition and with a minimum of
accidents.   The task is an arduous one, and in
order to avoid the unseen perils that daily threaten
his life, the miner has constantly to keep a watchful
eye on likely sources of danger.

The amount of coal a miner and his assistant can
cut in a given time varies considerably on account
of the different conditions prevailing in different
mines and seams : from two to four tons a day is
about the average.   When, however, the miner is
working in disturbed or difficult ground—in abnormal places as they are called—his output will

be considerably less, and the legal minimum wage comes into operation if he takes the trouble to find the overman of his district and make his claim and it is not disputed.

During the course of the day the miner's stall is visited by different officials, whose duty it is to see that the working places are quite safe and that the methods employed for extracting the coal are such as to yield the best results to the mine-owners. In many mines, coal can be profitably won by ordinary methods, but if explosives require to be used the miner has to purchase them from the colliery company, who must sell at cost price. Only certain varieties are allowed to be used. The miner keeps his supply in a little tin can, and when he has by means of a drill driven a hole or holes into the coal, he very carefully inserts charges and then awaits the arrival of the shotman, who alone is empowered by law to fire the explosives. This official connects the charges by means of wires attached to detona-tors to an insulated cable many yards in length, which he fastens to an electric machine. The men leave the stall and proceed some distance away so as to be safe from possible injury from the flying pieces of coal and stone. The shotman now sends an electric current along the wire. Immediately the solid mass creaks and groans and subsides to the floor, breaking up into lumps which can easily be loaded into the waiting trams. Lumps which are too heavy for men to lift are broken, but otherwise the coal is loaded as large as possible. In the course of his work the miner has often to wait short periods for the shotman to make his round or for a haulier to bring an empty tram, and it is usually during such waits that he partakes of the meal which he brings with him from home in the morning. No provision is available to enable the miner to clean his hands of the coal dust with which they are soiled, but this is only one of many inconveniences with which the miner must put up.

After from seven to eight hours of almost incessant strenuous physical exertion the miner puts aside his tools—sometimes secured by a locked chain—and makes his way along the narrow and low-roofed gateways, crouching down as he walks and sometimes having to scramble over a fresh fall of roof until, after some minutes, he comes to a main haulage road where there are permanent electric lights, and a further walk of half a mile or so takes him to the shaft bottom. In the main road he has passed out of the dangers of gas and roof falls, but only to encounter a fresh danger, for it is like walking along a main line of railway. Journeys of trams hauled by ropes come rushing along at twenty miles an hour and the miners walking must flatten themselves against the wall or rush for the refuge holes when they hear a journey coming. It used to be the custom to stop hauling coal while the men were coming out of the mine, but since the Eight Hours' Act has limited the period of winding, this has been discontinued in most mines, with the result that there have been many fatal accidents.

Few people realise the amount of walking which falls to the miner's lot. The working-places are often a mile or so from the bottom; and in older mines, and particularly in under-sea mines, they may be two or three miles from the shaft. To get to his working-place the miner has to trudge through pools of black slush, water percolating from the roof often dripping on to his head and back and leading to great discomfort. Working continuously, as some must do, in a dripping wet place generally leads to diminished vitality and disease. Whilst some mines are very wet, others are too dry, so that there is constantly coal dust in the air which is somewhat unwholesome to breathe and highly dangerous as an explosive unless zones or sections of the mine are kept well watered. The high temperature of many mines must also not be forgotten. When the temperature is over 80° the miner generally strips naked to the waist. The high temperature is a serious hindrance when the mine is very wet and the ventilation very poor. If constantly fanned by a moderate breeze the miner can continue his work comfortably and safely; but if the ventilation of his particular place or district has failed or weakened he first feels the effect by serious fatigue, and this warns him to keep a good watch upon his lamp, for a sign of the blue cap of flame inside it, which is evidence that he is in air containing gas sufficient to form an explosive mixture. Under such circumstances it may require a man's utmost nerve and resource to extricate himself without fatal results to himself and his mate. This is particularly the case if a blower of gas happens to develop in a working-place, usually through a movement of the roof. There may be such a rush of gas that even the best ventilation is temporarily overcome. The blue cap appears and the miner must beat a hasty retreat and warn the men in neighbouring stalls to do likewise. He cannot rush away ; his lamp must be kept near the ground and be handled most gingerly, for if he wave it fast through the air or if he drop it ever such a short distance, the dreaded flame may come through the gauze and all will be lost.

Our miner, forbidden to ride on a journey of trams, has safely walked homewards along the main roadway. After waiting his turn at the bottom of the shaft he is wound to the surface, and generally shivers from the rapid change of temperature. He returns his lamp to the lamp room, and it is the lamp man's immediate duty to report if any of the lamps are not returned, as this means that the men are for some cause lost in the mine. Men have not infrequently been known to fall asleep in their working-places, overcome by fumes from a fire originating by spontaneous combustion in the waste slack or dirty coal thrown into the goaf. After collecting his hidden treasures, such as pipe, tobacco and matches, and overcoat or mackintosh, if the weather is cool or wet, the miner starts for home, which usually means a walk of one or two miles and sometimes longer or a journey by train. It is

during these homeward tramps, or when waiting in
cold and draughty shelters for the trains or at the
colliery offices for "pay" or to see an official, that
many miners contract the diseases which sap their
strength and lead to early death.

When home the miner usually performs his ablu-
tions in a tub placed in front of the fire, while the
other members of the family pursue their ordinary
avocations. Whilst the men of the household bath
in turns the house-wife prepares their meal, and
arranges the wet clothing on the backs of chairs or
on the oven door. In some colliery areas where the
houses are small and consequently overcrowded,
considerable inconvenience is occasioned in miners'
families by this undesirable arrangement; especially
so is this the case when, as often happens, four or
more are awaiting their turns for the tub, there being
perhaps a father and two of his sons, and one or two
lodgers, all working in the same pit.

Pit-head Baths

In recent years the desirability of providing
accommodation for miners to wash and change their
clothes close to the pit-bank has been widely recognised. As already pointed out, the health of the
miner is seriously risked by his coming hot and
perspiring from the close air of the mine into a
biting winter wind. A place where he can cool
himself under shelter, and at the same time have the
very necessary bath would be the greatest advantage
from every point of view. It would give him the

opportunity of getting out of his wet and dirty working clothes, to come home clean and in neat clothes. Above all it would vastly improve conditions in the home. Nine-tenths of the miners' cottages have no fixed bath, not one per cent. have a separate bathroom and hot water. Consequently, for a couple of hours when the men return, the principal room of the cottage through which all members of the family, men or women, must pass, is converted into a bathroom, and the miner's wife and perhaps her elder daughter, if there are grown up children, has no end of trouble in fetching in the water, heating it on the fire, and in cleaning up after the men have had their baths, and their clothes have been dried. The wife of any workman has quite enough drudgery in the home without this daily addition ; and where the men of the household may be working on two different shifts, or even three as in Durham, the work of preparing the baths must practically destroy all possibility of having a decent home.

In every way, therefore, it would be an immense boon to the mining population if a pit-head bath installation were established at every colliery and its use made compulsory. The Royal Commission on Mines reported in favour of the provision at collieries of suitable washing and changing arrangements ; and they gathered much information as to the extensive arrangements of this kind made in many of the Continental mines, particularly in Germany and Belgium. A definite step forward was made by inserting a clause in the Coal Mines Act of 1911,

which enabled the workmen by ballot to demand
the provision by the employers of pit-head baths
provided the estimated cost of maintenance does
not exceed 3d. per week for each workman, of which
sum the workman is liable to contribute one half by
deduction from wages. There are, I believe, only
two collieries where pit-head baths have so far been
provided in this country, at Atherton in Lancashire,
and at one of the Ocean Coal Company's pits in South
Wales. Ballots have been held amongst the workmen of some other pits ; but so far they have been
inconclusive. This is partly the result of the faulty
wording of the Act itself, which reads as follows :—
" Where a majority, ascertained by ballot of two-
thirds of the workmen employed in any mine to
whom this section applies, represent to the owner of
the mine that they desire that accommodation and
facilities for taking baths and drying clothes should
be provided at the mine and undertake to pay half
the cost of maintenance of the accommodation and
facilities to be provided, the owner shall forthwith
provide sufficient and suitable accommodation and
facilities for such purposes as aforesaid." This can
be taken to mean that a majority out of two-thirds
of all the workmen employed, that is to say, slightly
over one third of the total employees of a mine,
can demand the provision of the accommodation.
The coalowners contend, however, that the intention was that the ballot must show a majority equal
to two-thirds of the workmen employed in the mine,
and the dispute it appears must go to a court of law

for decision.[1]   It is lamentable that Acts of Parliament are not drawn up by persons who know their business !

It may be thought strange that it is difficult to get two-thirds of the men employed in any mine to vote for having pit-head baths ; but miners are a conservative body, and it is very difficult to get them to favour changing the habits of a lifetime for baths of a kind which they cannot picture to themselves. They are shy as to leaving their good clothes in a public place whilst in the pit, as to whether they will have hot water, and whether they may not have to dress in a cold place rather than in front of a fire.   Such a variety of drawbacks present themselves to their minds that perhaps it is not surprising that the movement to secure the adoption of pit-head baths is making but slow progress in spite of a vigorous campaign undertaken by the wives of some of the leading miners' agents. The provision and use of pit-head baths should be made compulsory after some date three or four years hence.

The Mining Population

A mining community differs in several important respects from other industrial communities.   In manufacturing areas, as a rule, the population is

---

[1] The owners' view probably represents the intention of Parliament.   The other version apparently originated when a ballot taken at the Llanerch Colliery, Pontypool, failed to show that two-thirds of the total number of men employed at the colliery were in favour of the system.

made up largely of workers of different trades and classes, and there is considerable variety in their habits and outlook. In most of our coalfields, however, mining is the principal, and often the only industry. If we take, for example, an area like the Rhondda Valley in the South Wales coal-field, with a population at the last census date of more than 150,000, we find that about 95 per cent. belong to families engaged in, or dependent upon, the mining industry. There are few works or factories, or other employment except for the comparatively few openings for employment on the railways and in shops. There is, in most mining areas, a very limited choice of trades, and it is partly the lack of other opportunities that constitutes the chief reason why the sons of miners for the most part adopt their fathers' vocation ; besides which the mines can well absorb the labour of the youths of the district. More so than in many of the skilled trades, such as engineering, carpentry or commercial pursuits, labour in the mines is recruited mainly from the children of parents in the same industry. The bulk of the men now employed underground are the children and often the grandchildren of miners.

There are perhaps four reasons why boys in mining districts, after leaving school, take up mining as a means of livelihood : (1) The first is that already referred to, viz. :—that the mine is the nearest and often the only centre of employment. (2) The wages earned by boys and youths in collieries are considerably higher than those paid in most other branches of industry. For example, under the Minimum Wage Act in South Wales, boys under fifteen years of age receive 1s. 6d. as the standard rate, and as the percentage above the standard is usually over 50, the weekly amount earned will be on the average more than 13s. In each of the following years until the youth attains the age of 21, he receives an additional 3d. per day, so that in his twentieth year, the minimum amount earned is 27s. (3) The work of mining, although it often requires considerable skill for its efficient discharge, does not require a long apprenticeship to earn fairly good wages; the work is comparatively easily learnt. A young man possessed of a fair amount of intelligence, strength and skill, is able to earn in a few years after commencing work, as large a wage as a collier who has been engaged in mining for thirty or forty years. (4) There is also a charm about mining that constitutes a considerable attraction for boys. Underground work appeals to their love of mystery and adventure, and school-boys in mining villages look forward for many years with great eagerness to the time when they will be privileged to descend the pit shaft and experience for themselves the novelty and adventure of which their elder companions often speak. Contrast this with the dread of going underground which prevents so many workless men in the big cities near the coalfields from trying the mines.

Miners' families, as a rule, are fairly large, and this is perhaps one reason why the elder boys of the families commence work as miners immediately they pass through the elementary schools, the parents being anxious to have the earnings of the elder children to meet the growing expenses of their families. The better-class collier is desirous, when his means afford, of giving his children advantages which were denied to himself, and many send their sons to the University or technical schools. Whilst the majority of miners belong to the local population brought up in the industry from childhood, the newer coalfields contain a large proportion of men who have formerly been engaged in a variety of occupations. There has been a great drain of the younger agricultural labourers to the coalfields, and men of all trades, industries and races are met with. It is rarely the case, however, that there has been any organised effort to introduce " foreign " labour. The principal examples are in Scotland, where there are colonies of Irishmen, of Germans, and of Poles, working in pits largely apart from Scotch miners. Such colonisation was attempted to a small extent both in Durham and South Wales many years ago, so that there is in two or three localities in South Wales a considerable proportion of Irishmen and of Spaniards and Italians. Since the growth of trade unionism any organised introduction of foreign labour would not have been tolerated.

Permanence of Mining Communities

Just as residents in colliery districts are limited in the choice of vocation, so are they restricted in the change of vocation : " Once a collier, always a collier " is an epigram which has the merit of being true. Changes of employment usually arise, either from the desire to improve one's position, or because opportunities of employment in the original occupation have disappeared. Miners rarely adopt other industrial pursuits, because there is no occasion for them to do so. The wages they earn are usually more than those they can obtain in other spheres of employment, so there is nothing to attract them elsewhere. Moreover, the growth of coal production is so great that the supply of labour is usually below the demand, and unemployment is comparatively rare. There is the additional fact that miners on the whole are conservative in temperament, and do not readily favour changes in important matters.

There is, however, some degree of migration between different coalfields, the tendency being for workmen to go from the older coalfields to the new and rapidly developing fields—Fife, South Yorkshire, South Wales, Kent—where far higher wages are paid. Although naturally there are exceptions, it is, as a rule, true to say that the immigrants are usually men of the least desirable and responsible class. They have been weeded out from the mines of the older coalfield, perhaps during a period of depression, and they do not compare well with the indigenous population of the districts to which they come. Some of them settle down, however, and improve considerably with the high and steady wages which they can earn in their new home.

Character of Miners

Homogeneity of the population as regards employment is a factor which has a far-reaching effect on the character of the individuals of mining communities. There is not such diversity of aims, ambition and character amongst miners, as is found in towns where several kinds of industries are carried on. Whilst individual miners naturally vary much, these workers, as a class, become stamped with certain common features, due to the conditions of their employment. In industries in which small firms are the rule a considerable proportion of the workers hope eventually to become owners or masters ; and even though few achieve this ambition it distinctly influences their attitude both towards the work in which they are engaged and also towards the owners. Mining, however, is an industry which requires an extensive capital, highly technical knowledge, skill, and considerable business capacity. These essentials can, it is true, in exceptional instances, be acquired by the men working at the coal face ; the vast majority of workers, however, lack both the means and the opportunities, and can never hope to become the owners of mining undertakings. A small proportion can qualify to participate in the management of the mines, though they may never become owners. The ordinary miner, however, is tied to the coal seam. His industry is a highly specialised one

requiring intelligence, experience, courage and re-
sourcefulness, but requiring little or no capital, no
business capacity except to make a good bargain on
a piece-rate, and only an equipment in tools which
is more or less primitive in character. For the great
majority, therefore, it is a case of "once a miner
always a miner." There is an open road only in
the direction of becoming officials of the mine, on
the one hand, or of the trade union on the other, and
in both progress is difficult and the proportion of
openings to the total number of workmen engaged in
the industry is exceedingly small. Although a
number of miners have climbed to positions of
honour and influence in various spheres of life,
compared with the total mining population they
are few.

We observe, therefore, that the miner's work is of
such a character that it cannot absorb all the
energies of his mind, but tends only to exhaust his
physical strength. The intellectual and emotional
tendencies must find an outlet, and they do so in a
variety of ways, good and bad, at which it is worth
while for us to glance.

Combined with the monotonous and unexciting
character of the employment, working in dark and
grimy surroundings, and living in secluded districts
where all are of one occupation, has a depressing
effect upon the minds of the men and tends to drive
them to extremes in various directions. The miner
is practically always a very serious man. The large
proportion are of a sober and religious disposition,

but considerable sections are wild and reckless and flee for recreation to rabbit-coursing, gambling, intemperance, and other exciting forms of pleasure. A devotion to football as a sport is fortunately taking a strong hold of the miners given to this temperament.

It is probable that in few industries does religion exercise so great an influence on the minds of the believers. The better class of miner seems to realise the constant presence of the unseen dangers in the mine, and the sense of insecurity has made him turn to religious sources for help and guidance. This has been perhaps particularly marked in South Wales, where the miners have for generations been devoted to their chapels, and revivals have from time to time swept through the valleys. On the other hand, the younger generation in the valleys of Glamorganshire is breaking loose from the traditional religion ; and, indeed, to a large extent from any religion except that of the salvation of the workers by the establishment here and now of a just and liberal social order. For this younger element, consisting of earnest and devoted men, and active though unguided thinkers, socialism and syndicalism in their various forms are a real religion, so that there is a continual and tragic clash of thought and sympathies between the younger generation and the older—to whom the chapels and their religion are still the great inspiration of life. No one interested in social and industrial problems should fail to read the play called " Change," by

Mr. J. O. Francis, the scene of which is set in one of the Glamorganshire mining valleys, and portrays with an extraordinary fidelity the present character of the mining community, not even forgetting the English immigrant.

A large and growing body of miners in the older coalfields take keen interest in the more serious forms of recreation. In South Wales there is scarcely a mining village which does not boast of at least one choral or dramatic society, and the same is true to a large extent in the older coalfields of Yorkshire and Durham. Latterly, political and social clubs and institutes have been widely established. Nearly every village has its workmen's institute, or at least a reading-room, controlled by the miners themselves. The activity of the miners upon local authorities, particularly the County and Borough Councils, is reflected in the provision by the Councils of evening classes in a great number of mining villages, and the Workers' Educational Association and University Extension organisations are also active towards the same end. The most encouraging sign, indeed, is the ever increasing attendance recorded at evening schools and other educational institutions in the mining districts. All these educational facilities for adults are provided as the result of action taken by the miners themselves to secure them, and where they have not had the knowledge or desire requisite for taking the initiative the educational and recreative facilities of the best kind are still quite inadequate. There are many mining villages, miles away from the nearest town, with no library, no evening classes, and no institute where men can see newspapers and periodical literature, or meet for social and business purposes. Only the public-house is ubiquitous, though the cinema is becoming nearly so. This last, if in some degree controlled, might of course be a most useful educational instrument, and especially so in mining districts—which are naturally very largely cut off from the outer world, and from the broad flow of national life, so that the miners are particularly lacking in knowledge and experience of certain ordinary matters.

Miners' Societies

Few people have the faintest conception of the extent to which the miners are organised for various purposes, and of the amount of administrative experience which is obtained by those entrusted with the management of trade unions and societies, and with safeguarding the interests of their fellow-workmen in various directions. In many mining villages the men belong to Trade Unions, Friendly Societies, Co-operative Distributive Societies, Medical Clubs, and a large proportion also to building clubs, institutes or social clubs, churches, and a variety of other organisations of a political or social character. During recent years the number of such bodies has greatly increased, and various organisations now cover almost every interest, industrial and social, in the miner's life.

By far the most important, because it involves practically all the miners in a district, is the Trade Union. The Miners' Federation is perhaps the strongest trade union in the country, and in most districts includes all but a very few of the employees engaged in any colliery. Each colliery has its own special Trade Union Lodge, the management of which is entirely in the hands of the workmen themselves. It arranges terms and conditions of working with the management, and generally looks after the men's industrial interests. When a question of considerable importance arises, the men hold a general meeting for the purpose of discussing such questions, and then the decision of this meeting is carried into effect by a committee appointed by the men themselves. It is this committee which negotiates with the management. As a rule it is composed of the ablest and most active men in the colliery. Members of the committee are elected periodically, usually the majority retain office for many years, as their superior ability and great experience enable them to retain the confidence of the men. It is by service on these committees that young and aspiring miners obtain their business experience and qualify for positions of greater usefulness as miners' agents, members of Parliament, and for similar offices. The committees sometimes, of course, make mistakes, but on the whole they discharge their duties to the satisfaction of their constituencies.

The great orders of Friendly Societies have

numerous lodges in colliery districts, and in addition there are numbers of local friendly societies. In most cases the lodges consist almost entirely of colliers ; and here again the more intelligent have opportunities for the development of business skill. On the whole, these societies have been conspicuously successful, and have been productive of great advantage to the workers resident in mining communities. Where these societies have proved unsuccessful, this has been occasioned by a want of brotherly spirit amongst the men who claim on the society or club so often that they eat up its funds.

In some mining villages, especially in the North, the majority of the miners are also organised into co-operative distributive societies for the purpose of purchasing domestic commodities. Many of these societies have been in existence for some years and have large annual turnovers. Their " stores " are usually the largest and most imposing building in the district, and on account of their constantly increasing business, extensions of premises are often necessary. The management of each local society is vested in a committee elected quarterly, half-yearly, or annually, as the case may be ; and in this sphere also the intelligent miner has ample scope for the development of his business capacities. When it is considered that the committees are composed mainly of miners, the important results achieved are highly creditable. Most of the active members of co-operative societies have a clear understanding of retail business and of the principles underlying the co-operative movement; and, happily, the spirit of co-operation is rapidly extending.

The organisation of medical treatment has existed for many years in colliery districts, but it has often given rise to considerable difficulties. There has usually been a doctor connected with every colliery, and in most cases, before the National Health Insurance came into force, it was customary for each miner's contribution towards the cost of medical treatment to be deducted from his wages. In some cases the appointment of the medical practitioner has practically been made by the colliery owners, and the miners have little or no control; but in other localities the funds have for long been administered by medical committees consisting of representative miners. The workmen constituted practically a compulsory medical club, providing medical attendance and medicines and sometimes benefits in hospitals and convalescent homes for the miners and all their families. The British Medical Association was always strongly opposed to these medical clubs, and prevented its professional members being engaged by them, so that the various societies have had some difficulty in obtaining an adequate medical service. The introduction of the National Health Insurance brought a great deal of confusion into these medical arrangements, because the employers were bound to deduct the contribution from the workman's wages in addition to what he was paying to the colliery doctor. The colliery medical club could not be made an approved society, because it provided not only for the workmen but for every member of their families. The miners have, therefore, had to form special approved societies, and there has been a readjustment of the contribution for the colliery doctor to maintain his services for those of the miners' families not obtaining benefit under the Act.

Achievements of Miners

Large numbers of miners have climbed to positions of honour and influence, and have done much good work in various spheres of public life, political, industrial and municipal. Miners are largely represented upon Parish and District Councils of the coalfields, and to some extent upon the County Councils. They generally form a distinct labour party in municipal affairs ; and men who are good speakers and able to be of public service are officially recognised by the local miners' union and are recompensed for their time devoted to public work. It is significant that the local authorities on which the miners are strongly represented are far more progressive in their aims and achievements than public bodies recruited mainly from the business and professional classes. Large numbers of miners' agents and some working miners have also been admitted to the magistracy, and the manner in which they have carried out their judicial duties has frequently won for them well deserved praise.

In the House of Commons the miners are fairly

well represented. There have always been representatives of miners in Parliament ever since the veterans, Alderman John Wilson and Thomas Burt, representing Durham and Northumberland, were first elected. The latter was for many years " father " of the House of Commons. The mining representatives in Parliament have, generally speaking, been of the liberal-labour complexion, but in recent years they have tended to amalgamate more closely with other labour representatives in the formation of the official labour party. There are altogether nine Members of Parliament officially the candidates of the Miners' Federation of Great Britain, whilst there are four others directly representing District Federations or Associations. Mr. Keir Hardie, although himself originally a miner and elected by a miners' constituency, is a representative of the Independent Labour Party. These miners' members, besides having been able to secure and influence legislation to the advantage of the miners, have served upon Royal Commissions and Departmental Committees with such success as to reflect credit not only upon themselves but also upon the class which they represent.

The miners have a strong tendency to distrust men not brought up in the mine when the interests of their class are concerned. We find therefore that however various the duties now required of them, the whole of the officials of the miners' unions have risen from underground employment. The man in the street does not ordinarily realise that to run a big Trade Union requires as great business and administrative skill as the management of a large commercial undertaking ;   and it is correct to say that the ordinary Trade Union official is, in his department, as highly skilled as the man who controls big industrial enterprises.

It must not be forgotten also, that nearly all the men responsible for the management of mines, except mining engineers, are recruited mainly from amongst the men employed at the coal face ;   and that the success of numerous mining undertakings is entirely due to the knowledge and experience of men who, though lacking the polish and culture acquired by college-trained officials, are highly efficient in their particular departments of work.

It is not only in politics and industry that miners have distinguished themselves. Although but few colliers change their vocation, when they do so it is generally to good purpose.   Numbers of them, by their own saving, and that of their parents, have obtained some college training and have become schoolmasters, or have entered the Nonconformist ministry.   There are miners on the staffs of Universities and Colleges, and a few in positions in the Civil Service.   Miners have also come to the front as authors and on the stage, whilst the amateur draughts-championship of England was recently won by a Nottinghamshire miner.   It is interesting to note that the miners also believe strongly in higher education, particularly with a view to equipping for their life's work men who will be employed as trade union officials, or otherwise, in the interests of labour. The miners have for many years maintained scholarships at Ruskin College, Oxford, and since 1909 at the Central Labour College which was recently removed to London. The recent decision of the South Wales Miners' Federation and the National Union of Railway-men to make themselves responsible for the future of this College, each sending a number of students, may well prove a landmark in the history of the trade union movement. It will certainly provide the opportunity for a number of young miners of intellectual ability to rise to positions of influence in the worlds of labour and politics.

CHAPTER XXII

The Housing of Miners

Development and the Housing Problem

THE rapid development which has taken place in the mining industry during recent years has led to a considerable influx of population to the new colliery districts, and the inability of ordinary building agencies to provide for the housing needs of the people has given rise to a congestion which has not only resulted in high sickness and death-rates, but has also tended in some districts to inflate rents abnormally. In areas where growth has been slow and steady the shortage of dwellings has been least apparent, as local private enterprise has usually been able to meet the demand. In districts where development has been exceedingly rapid, however, as in South Wales and in South Yorkshire, building enterprise has hopelessly failed, and a serious house famine has resulted. The increase in population has in some instances been nearly 200 per cent., whereas the increase in the number of dwellings has been considerably less.

The Census figures show that during the last intercensal period the most rapid growth took place in South Wales, South Yorkshire, and in

Fifeshire and Dumbartonshire. From the following tables, which are based on the recently published Census returns for 1911, the extent of the housing problem in the various mining counties may be ascertained, and a rough comparison established between them. In reading these tables, however, it must be clearly understood that no regard is paid to the character of the existing accommodation. In some counties, for example, a large proportion of the dwellings are small and old and unfit for habitation. The proportion of agricultural land to the whole area also varies in different counties so that the figures given cannot be taken as an exact estimate of the position. The conditions, however, are sufficiently similar to give the comparison a substantial value.

Census Housing Revelations

The following general conclusions may be derived from an examination of these tables :—

(1) In all the coal mining counties except Fife, Warwick, and the four South Wales counties, the increase of accommodation during the intercensal period was in excess of the increase of population. In some of the counties—Edinburgh, for instance—building enterprise was so active that more than twice as many dwellings as were necessary to meet the needs of the increased population were provided. It is assumed that demand and supply balance if one new dwelling is provided for each increase of five in

Coalfield.	County.	Per cent. of Increase of Population. 1901–1911	Per cent. of Increase of Accommodation. 1901–1911	Increase of Population for each additional dwelling.
Scotch	Ayr	5·45	10·90	2· 8
	Lanark	8·04	16·11	2·39
	Dumbarton	22·80	31·15	3·47
	Stirling	13·14	14·64	4·27
	Linlithgow	..	..	..
	Edinburgh	3·86	11·13	1·58
	Fife	22·34	21·06	4·45
Northern	Northumberland	15·54	19·34	4·94
	Durham	15·36	21·09	4·13
Yorkshire	York (W.R.)	10·28	12·68	3·48
	Nottingham	17·42	21·84	3·47
	Derby	13·96	16·18	3·96
Lancashire	Lancashire	8·89	9·62	4·26
	Cheshire	13·80	14·50	4·10
Midland	Leicester	8·90	9·84	3·87
	Stafford	9·00	11·57	3·67
	Warwick	10·57	10·47	4·47
	Shropshire	2·90	4·90	2·60
South Wales	Glamorgan	30·34	19·72	6·34
	Monmouth	32·75	27·22	6·07
	Carmarthen	18·53	14·15	5·70
	Brecon	9·35	5·90	6·80

population, which is almost precisely the average number of persons per dwelling in the United Kingdom. In particular cases, like Edinburgh, for example, there may be an emptying of old large houses which have contained two or three families, in order to fill new cottages.

Coalfield: County.	No. of occupants per house.		Percentage of sub-letting.		Percentage of uninhabited dwellings.	
	1901	1911	1901	1911	1901	1911
Scotch—						
Ayr	4·93	4·79	1·57	0·99	6·70	8·59
Lanark .	5·02	4·93	3·81	2·37	4·25	9·29
Dumbarton	5·06	4·95	2·50	1·54	6·35	10·35
Stirling .	5·06	4·97	2·85	2·10	5·85	5·40
Linlithgow						
Edinburgh	4·84	4·65	6·80	5·07	5·67	8·16
Fife	4·53	4·55	3·54	2·10	7·42	6·82
Northern—						
Northumberland	6·53	6·28	26·24	24·53	5·84	5·26
Durham	5·93	5·69	16·15	14·75	4·20	4·89
Yorkshire—						
York (W.R.) .	4·57	4·43	0·73	0·98	5·96	5·14
Nottingham	4·61	4·48	1·34	1·03	5·55	6·32
Derby .	4·85	4·70	1·26	1·28	5·26	4·08
Lancashire—						
Lancashire	4·97	4·82	3·38	3·54	7·04	4·70
Cheshire	4·80	4·63	3·55	2·57	7·68	4·89
Midland—						
Leicester	4·59	4·46	1·09	0·81	6·99	4·96
Stafford .	4·98	4·86	1·74	1·44	5·19	5·05
Warwick	4·74	4·69	1·88	2·00	6·65	5·62
Shropshire	4·65	4·54	1·74	1·34	6·11	4·54
South Wales—						
Glamorgan	5·46	5·52	7·93	9·49	5·94	2·58
Monmouth	5·35	5·43	6·58	8·05	9·44	2·93
Carmarthen	4·65	4·73	2·06	2·78	6·13	4·29
Brecon .	4·68	4·74	2·31	2·25	7·22	5·34

PERCENTAGE OF OVERCROWDED DISTRICT COUNCIL
AREAS IN ENGLISH COALFIELDS

Coalfield : County.	Total No. of Authorities.	No. of Authorities with over 5 persons per dwelling	Percentage Column.	No. of Authorities where average has increased during past 10 years, 1901-1911.	Percentage Column.
Northern—					
Northumberland	34	20	58·8	7	20·5
Durham . .	48	33	68·7	12	27·0
Yorkshire—					
York (W.R.) .	165	24	14·5	23	13·9
Nottingham .	27	1	3·7	5	18·5
Derby . .	44	4	9·0	4	9·0
Lancashire—					
Lancashire .	25	..	..	3	12·0
Stafford . .	57	12	21·0	10	17·5
Warwick . .	28	1	3·5	8	28·5
Shropshire .	31	1	3·2	4	12·9
South Wales—					
Glamorgan .	30	23	76·6	18	60·0
Monmouth .	27	17	70·3	10	37·0
Carmarthen .	15	3	20·0	6	40·0
Brecon . .	13	1	7·6	7	53·8

(2) Assuming an average of five persons per
dwelling as the standard of safety from the public
health standpoint, it would appear from the table on
p. 640 that in four counties only—Northumberland,
Durham, Glamorgan, and Monmouth—is there an
excess of population over dwellings. It should be
stated, however, that in the other counties, although
the averages may not be high, there may be, and

indeed there usually is, considerable overcrowding in particular areas. The comparatively high figures for Northumberland and Durham are due to the existence of a large number of flats or tenement buildings.

(3) The number of occupants per dwelling was less in 1911 than in 1901 in all the mining counties except Glamorgan, Monmouth, Carmarthen and Brecknock. From this it would appear that whereas the problem is becoming less acute in the other coalfields, in South Wales the congestion is becoming more serious.

(4) The percentages of dwellings sublet are highest in Northumberland and Durham and in South Wales. In the northern counties, however, the flat system prevails very largely, whereas in South Wales the number of flats is exceedingly small, and two families crowd into a cottage. Thus in Northumberland no fewer than 176,852 out of a population of 696,893, or over 25 per cent., are flat dwellers. In Durham the proportion is 200,387 out of 1,369,860. In order to compare the two coalfields mentioned, which are undoubtedly the worst-housed of all the coal-fields in respect to sub-letting, the following table based on the Census totals for "ordinary dwellings" only are given. In this table the figures for flats, shops, hotels, public institutions, etc., are omitted.

Counties.	Population.	Families.	No. per dwelling.	Percentage of sub-letting.
Northumberland and Durham .	1,569,425	330,597	5·32	10·82
Glamorgan and Monmouth .	1,383,478	282,537	5·38	9·06

This table shows that although the percentage of houses containing two or more families is higher in the Northern coalfield, the congestion is really slightly greater in South Wales. The larger number of occupants per house in Glamorgan and Monmouthshire is partly due to the great number of men resident there as lodgers, one or two with each family. It should be stated that, as a rule, the South Wales cottages are distinctly larger and better built than those of the Northern coalfield.

(5) A fairly accurate index of the extent of a house-famine is the percentage of vacancies. When dwellings are scarce vacancies are few ; and houses, which under other circumstances would be regarded as unsuitable for working-class occupation, are pressed into service. It will be noted that the highest percentages of vacancies prevail in Scotland, where building activity seems to have been in excess of the increased demand, and the lowest percentages in Glamorgan and Monmouth. These figures seem to indicate that the problem is least acute in Scotland and most acute in South Wales.

Evil Effects of Bad Housing Conditions

The bad effects of inadequate housing upon public health and morals are well known. It is not generally realised, however, that the existence of a house famine also acts as a serious handicap on industry. In some of the coalfields mining development is much hindered by the lack of a sufficient supply of labour ; and the supply is short because no house accommodation is available. In such areas the provision of a large number of new dwellings would considerably quicken trade activity, and advance the prosperity of the coalfields.

The difficulty arising from a shortage of suitable dwellings near a new colliery has been severely felt in many localities ; and in many cases where ordinary building agencies have not proved adequate, the colliery companies themselves have undertaken building. It is thought, however, that in some cases it was not solely in order to increase the supply of labour that colliery companies have adopted this course. Mining companies, naturally anxious to keep down the cost of production to the minimum, have sought every possible means of combatting the tendency of their employees to combine and strike for higher wages ; and they have not been blind to the fact that workmen resident in " company " houses, in areas where the dwelling supply is restricted, would hesitate to strike for fear of getting notice to quit their homes. The weapon of eviction during disputes has in a few cases been freely used in order to compel the men to come to terms. Several instances of such evictions are known in Yorkshire and in Scotland. In the latter country—where colliery companies own the bulk of miners' cottages—the eviction question has received considerable attention ; and the M.F.G.B., on the motion of Scottish delegates, have passed numerous resolutions in favour of legislation to prevent the eviction of miners from cottages during trade disputes. Public opinion has recently been growing very strong against this unfair method of bringing rebellious miners to heel ; and colliery companies are learning that investment in house property for strike-breaking purposes is diminishing in value. For this reason partly, therefore, numbers of companies are disposing of their cottages.

There is also a tendency to diminish the investment, in the building of dwellings, of capital raised by colliery companies. Most mining companies need all their capital for developing their mining undertakings ; and they naturally prefer not to tie any of their share capital up in buildings, which will as a rule yield a much smaller return than coal. The rise of the rate of interest has also made it difficult to raise money on debentures at a rate which can be paid out of the rentals received at the customary level of the district—one prominent South Wales colliery company having, for instance, issued 5 per cent. debentures for building houses on which it only gets a return of 4 per cent. In localities where existing building activity is inadequate, however, mining companies are still compelled, in the interests of their collieries, to embark on housing enterprises. During the past few years this has been made easier, as it has been made possible to obtain State loans at a low rate of interest for the building of workmen's cottages ; and new methods of house ownership have also been devised in which colliery companies can exercise influence more in proportion than the extent of their financial interests. To some of these methods reference will be made later.

Housing Agencies in South Wales

In most of the coalfields the housing needs of the people have been catered for by colliery companies, private building speculators, and building clubs. As the Census figures already given show, there does not seem to have been any considerable failure on the part of speculators in most of the mining districts to keep up with the demand. In South Wales, however, this agency has been quite inadequate, and as colliery companies here have not to any considerable extent embarked in housing enterprise, the miner has been obliged to provide accommodation as best he could for himself. The method selected has been through the formation of building clubs, a system imported from Yorkshire. A number of miners club together, and with the assistance of a secretary, who is usually an accountant, arrange for a large number of houses to be built in one contract. Each member pays from £10 to £20 down, and thereafter monthly instalments of from 10s. to 25s. for each " share," that is, house. When about one-fourth of the cost of each house has been paid in, the club " divides," and each member takes over his house, which is allotted him by ballot, subject to a mortgage which he can pay off gradually like an advance from a building society. The houses are built all the same, so that the allocation by ballot may be fair. Some wealthy gentlemen of the district are appointed trustees, and the building would usually not be financed without their guarantee at the bank. Owing to the widespread depreciation of property in recent years fewer guarantors are now available, which may be one cause of the decreased building by this method in South Wales.

To meet the financial claims of these clubs men have had to save large sums from their wages to pay for the cost of their houses over a series of from 15 to 25 years, the usual rate of contribution being at the rate of from 15s. to 24s. per lunar month. In parts of South Wales where the club system flourishes hundreds of cottages have been built through the medium of building clubs, and the proportion of owner-occupiers is often very large indeed. In many cases workmen own three or four houses. The fact that in South Wales capital for the provision of dwellings has had to come out of wages rather than —as in the Northern and Scottish coalfields—out of the company's funds, is probably one reason why the rates of wages paid in South Wales are higher than in the other coalfields. The club system, however, useful though it has been, is now, for various reasons, diminishing in popularity, and it seems probable that during the next few years co-operative house-owning societies will become the chief method of catering for the supply. In addition a considerable increase in municipal housing activity is anticipated.

Housing in the Scottish Coalfields

Whilst the Scotch coalfields, except in a few places, do not suffer from a house-famine like South Wales, the quality of the accommodation leaves much to be desired. The standard cottage in England and Wales—at any rate south of the Tyne—is a four or five-roomed cottage with a small or large scullery, and offices in addition. There is the living-room, where all the cooking is done, usually a small parlour, and two or three bedrooms upstairs, one being over the scullery, or back-kitchen and offices in a back extension of the cottage—an objectionable feature which often shuts out light and air from the living-room.

In Scotland, however, the prevailing type is a cottage of one storey only, and consists of but two or three rooms, one very large, in which the family live and cook by day, and two or three persons sleep at night. Dr. J. C. McVail has thus described the typical miner's cottage in Dumbartonshire :—

" A collier's house consists typically of two apartments, a ' room and kitchen,' with a connecting door between. The entrance from the roadway is not usually quite direct, but by a small square lobby not much more than enough to allow the door to open. The wall facing the open door is the gable end of a set-in bedplace. There are two such bedplaces in the kitchen, along one wall, and on the opposite side is the fireplace, perhaps with a good, though small, cooking range. The kitchen has a window looking to the front, and the room another looking to the back, the open door between the apartments permitting some degree of through ventilation. The windows have an upper and a lower sash, which may or may not be hung on cords with pulleys. The lower sash opens upwards ; the upper sash may or may not open downwards. In addition to the two apartments, there may be a small porch built out in front of the kitchen, and used as a store or a scullery, with, or more commonly without, an indoor water supply. Sometimes there are a water tap and sink in the kitchen window place. The press or storage accommodation is limited, but is supplemented by utilising the space under the beds. In older rows this space may be the only coal store. The bedplaces are separated by a brick partition reaching to the ceiling, and are structurally open from floor to ceiling and from side to side, but are partially closed in by curtains. The ' room ' has a fireplace on one side and a single bed-place on the other, so that in the two apartments there are usually three beds.

" Cottages are now built occasionally in Scotland with one room upstairs ; and in some cases flats have been built, two of the common single-floor type of cottages being superposed, with an outside stone staircase. It cannot be said that the practice of sleeping in the living-room, usually two in each bed, is a healthy practice ; but the Scotch miners in common with the labourers of all trades in Scotland cling tenaciously to the custom. Probably the coldness of the climate, and the expense of thick bedding material needed in an unwarmed bedroom is chiefly responsible for this preference."

The following extracts from the reports of the Medical Officers of Health for Ayrshire and for Kinross-shire (1911) are of interest, the former as giving a general idea of the conditions, the latter as giving some description of the sanitary defects of bad types of cottages still inhabited :—

Extracts from the Report by the County Medical Officer on the Housing of Ayrshire Miners

" The entire population, including men, women, and children, representing the mining industry in Ayrshire, may be estimated at fully 40,000. . . . It might be roughly estimated that of the whole mining community 30,000 live in the mining rows or villages belonging to the mining companies, while the remaining 10,000 reside in the ordinary villages, towns or burghs. The proportion of miners and their families living in the latter is small, except in the Burgh of Galston, where close on half of the population consists of the mining class. . . .

" With the exception of two-storey dwellings at

Dreghorn, Barrmill, and Dalmellington, the houses which have been erected for miners by the coal companies are all of one storey, while the accommodation provided consists as a rule of two apartments, namely, a room and a kitchen. There are, however, several rows, such as at Annbank and other collieries, with only one-apartment houses, while in a few isolated instances there are miners' dwellings of more than two apartments each.

" The rental of the miners' houses varies from £2 10s. to a comparatively few at £9 10s. per annum, the former figure applies to the older houses of a more or less inferior type, while the latter refers to a somewhat superior class of houses erected at Townend. Taking miners' dwellings all over the rental is between £4 and £5 per annum. The rates on these rents are as a rule paid by the occupiers. In towns and villages other than mining rows and villages the rental is higher. The average might be put at £6 per annum. It may be noted that in the Burgh of Galston, where over 400 houses are occupied by miners, the miners in a number of cases are themselves proprietors of their houses. In the latter instance the yearly valuation of the house is about £8. . . .

" Mining villages are frequently arranged in several parallel rows all facing the same direction ; others may face each other on opposite sides of the road ; or they may be found in single rows or squares. The sites of the miners' rows have evidently been considered simply from the standpoint of convenience, and not as to aspect, nature of soil, etc. A large number of them are therefore built on the somewhat damp, impervious, cold clay subsoil which generally overlies the coal measures throughout the county ; and as the majority of these buildings—practically the whole of the older ones—have no damp-proof course in the walls, the latter being generally solid without strapping and lathing, they tend to be more or less damp. . . .

" Excessive overcrowding as a rule only occurs temporarily, such as when additional pits are opened before sufficient house accommodation is provided for the workers. In miners' rows with one-apartment houses, such as at the Annbank Colliery, there is a greater tendency to overcrowding than in the two-room dwellings."

Extract from the Report on the Housing of Miners in Kinross-shire (1911)

" Some of the buildings were old. These had been solidly built in their day, and the outer walls, though their joints were open, still stood firm ; but the internal fittings in many cases had fallen into disrepair.

" Kitchens had tile floors ; rooms not kitchens had floors of wood. A few wooden floors were under-ventilated, and dry rot affected several. Floor joists were frequently laid on earth. In some houses joists or bridges were absent ; the boards lay on the earth, and were kept in position by a groove and tongue.

" Some floors were incomplete. Flooring ceased in line with the beds ; beyond there was bare earth."

Many more descriptions could be quoted, showing the deplorable character of the housing accommodation provided by some of the companies for their miners in Scotland.

Co-operative Garden Villages

A very pleasing feature in some of the mining-areas is the awakening of public opinion in favour of more attractive houses and pleasanter surroundings. The Garden City ideal seems to have permeated the community, and a revolt is taking place against building ugly and monotonous cottages in crowded rows. The co-operative or co-partnership method of ownership, combined with Garden City methods of development, have yielded such excellent results at Hampstead, Ealing, Harborne, and elsewhere, that its gradual extension to the coalfields is inevitable. In South Wales, indeed, the method has already attained considerable popularity. Several organisations for the promotion of co-partnership housing societies have been established,[1] and some eight or ten societies are already actually engaged in building.

[1] The most important of these is the Welsh Town-Planning and Housing Trust, Ltd., a semi-philanthropic company limiting its dividends to a maximum of 5 per cent. This organisation was established by a philanthropic coal owner, Mr. David Davies, M.P., of the Ocean Colliery Co. It seeks to acquire and develop on Garden City lines large estates, and to promote housing schemes on co-operative lines.

These schemes differ considerably from building
clubs. The object of the latter is to enable individuals to become house-owners ; in the case of the
co-operative schemes, however, the houses are
always jointly owned, and the tenants pay rents to
the society, and receive interest on the shares they
hold in it. Co-operative housing societies give the
miners all the advantages of individual ownership
—security of tenure, fixity of rent, monetary profit
—without the disadvantage of being tied to a locality
and having a house on one's hands. There are also
several special advantages, not the least important
of which is that societies registered under the
Industrial and Provident Societies Act, and limiting
their dividends to a maximum of five per cent., are
able to borrow two-thirds of their capital from the
Government for periods of 30 or 40 years at $3\frac{1}{2}$
or $3\frac{3}{4}$ per cent. respectively.[1] Another great
advantage is that co-partnership societies dealing
with fairly large areas of land are able to apply with
excellent effect Garden City methods of development. In connection with the co-partnership Garden
Villages now being developed in the South Yorkshire and South Wales colliery districts, for example,
provision is made for ample garden spaces, recreation grounds, tree-lined streets, and social institutes,
every care being taken to foster the co-operative
spirit and to make the villages pleasant communities.

---

[1] The rates of interest were raised 10s. per cent. shortly after the
outbreak of the European War.

Mining Companies and Model Housing Schemes

As has already been mentioned several colliery companies have realised the importance of Garden City principles, and during quite recent years have attempted praiseworthy experiments in housing their miners under conditions which are more conducive to health and refinement than those which have prevailed in the past. In some cases the collieries have provided the whole of the necessary capital, in others they have obtained part by means of Government loans. A number of colliery companies also have entered into arrangements to guarantee the rents for 25, 30 or 40 years to syndicates of private investors, while others have invested in the funds of local co-partnership societies. Whichever method has been adopted, however, recent building schemes have shown a vast improvement on those which were formerly carried out ; and, whilst most of them do not comply with strict Garden City principles, the style of building and the character of the surroundings provided have been far superior to the usual mining town.

The most promising, and the best known of all the new model colliery villages, is that established at Woodlands, near Doncaster, by the Brodsworth Main Colliery Co., of which Sir Arthur Markham, M.P., is managing director. This company acquired a large area of land near the collieries, laid it out on Garden City lines with pleasant streets, lined with trees and grass plots, and with open spaces and recreation grounds. The cottages, of which there are now over 630, are all provided with three bedrooms, and a bath with hot and cold water service. They are built in pairs, and in small groups of four or five ; and the average number per acre is less than six, which is to be compared with 25 and 30 to the acre by the present-day speculative builder. All the cottages have good-sized gardens, back and front ; and they let at rents which vary from 5s. to 6s. 3d. inclusive of rates. Sir Arthur Markham has also been instrumental in establishing a model village at Oakdale, near Blackwood, in Monmouthshire. Another interesting colliery garden village is being built at Kirkconnell (Dumfries-shire) for the Sanquhar and Kirkconnell Collieries Co. In this village an excellent site is being developed, with miners' cottages built 15 to the acre, with games-field and village hall. In addition a hostel for the accommodation of 40 unmarried miners has been provided. In the Kent coalfield there have been two or three garden villages started by the Kent Coal Concessions, Ltd., over 120 substantial cottages having been built to date. It is much to be hoped that capital will be forthcoming to continue building in this excellent manner, as every one will wish that Kent may be spared from the hideous rows of slate-roofed cottages which have defiled the beautiful valleys of the South Wales and Northern coalfields.

The Housing of Lodgers

The lack of adequate accommodation for lodgers constitutes a real and serious difficulty in some of the coalfields, notably in South Wales, where the very rapid development attracts to the mining centres large numbers of young men from the agricultural areas and from other coalfields. Lodging-houses for casual workers and the lowest grade of labourer are to be found in most of the towns ; but no decent provision has been made for the accommodation of young unmarried men in regular employment. As a rule these find lodgings in the houses of married miners. This arrangement is not always desirable ; as a rule both householders and lodgers are subject to inconvenience and discomfort, and family life is often much disorganised. In many areas, for example the Rhondda Valley, the percentage of lodgers is particularly high, and some of the five and six-roomed cottages are overcrowded to such an extent as to militate against the general health of the community.

In some localities where house accommodation is scarce, lodging accommodation is quite impossible to obtain, and men wishing to work there must either give up the idea, or they must walk or come by train several miles. This scarcity of lodging accommodation often handicaps the activities of the colliery management to a very great degree ; and instances are known where tenants of " company " dwellings have been compelled to take in

2 T

lodgers nominated by the employers. A possible solution of the problem consists in the erection of large hostels providing separate sleeping accommodation, but common sitting and recreation rooms, etc. Numerous institutions of this kind have been provided in the coal mining areas of Germany, and have proved very satisfactory.

Town-Planning of Mining Areas

Another matter which needs emphasis is the importance of the proper planning of new mining areas. In several of the newer coalfields—for example, around Doncaster and in South Wales and in Kent—new districts are being opened up for the first time ; and land formerly of no value except for agricultural purposes is now being utilised for building purposes. The ordinary method of development employed is to crowd as many dwellings as possible on the smallest area of land, the streets being laid out just as comes convenient at the time, without regard to any plan of the town or village as a whole. No adequate provision is made for access between the various parts of the village or town which are brought into existence, or for through communication with adjoining townships. The effect of ill-considered development in the past has been highly disadvantageous to public health and convenience ; but unfortunately the same lack of method still prevails in many localities. Powers to prevent the worst evils of individual development were granted to local authorities by the Housing

and Town Planning Act, 1909 ; but the importance
of their application does not appear to have been
realised in most of the new areas. The Medical
Officer of Health for Doncaster Rural District in a
recent report urged the need for town-planning
schemes in that area. He writes : " It is in such
places as Rossington with a model village as its
nucleus that a town plan might be considered. If a
town plan had been adopted at Carcroft embracing
an area lying within a circle, say a radius of a mile
and a half from the pit shaft, the development
which is taking place in the neighbourhood would
have been on orderly lines instead of the present
unsystematic method . . . It is particularly desirable when a colliery company has made provision
for employees by the establishment of a model
village that private speculators should not be in a
position to build property anywhere or anyhow
without due regard to the amenities of the district."

CHAPTER XXIII

Uses of Anthracite

As was explained in the chapter dealing with the various kinds of coal, anthracite is the hardest and "driest" coal possible, being practically pure carbon, and properly containing only a slight trace of the hydrocarbons which form the bituminous element of ordinary coal. For this reason it burns slowly and with difficulty, so that it is necessary to start an anthracite fire with a small amount of bituminous coal or a good quantity of wood. In a closed stove, or in a large open grate well lined with fire-brick, it makes a fine glowing hot fire with no flame.

For household purposes anthracite has the following special advantages. It is clean to handle, as it does not soil the fingers like ordinary coal; and it produces little dust and no smoke. It burns very slowly and never requires poking or attention, except to replenish the stove. A number of different stoves for domestic and office use are on the market, of French, German, and English make, generally arranged to burn nuts or cobbles; but there is one type of stove designed to burn anthracite "peas", which cost less per ton than nuts. It is

660

claimed by the makers that such an anthracite stove is just as convenient and clean as a gas stove, but heats a room at less than one-third the cost of using gas.

During the past ten years there has been a considerable increase in the demand for anthracite, broken, washed and sized for domestic and other purposes. Some London merchants even go so far as to sell anthracite nuts packed in small bags, each one of which just contains a refill for a stove, so that the householder can handle it with the maximum of convenience and cleanliness. The use of anthracite for domestic purposes is not yet nearly so extensive in England as on the Continent, and far more of the "nuts" size is sent to France and Germany than is used in this country.

Anthracite is sold almost entirely in special sizes for each particular purpose, and the following is a brief list of the chief market sizes of anthracite coal with their dimensions :—

Name.	Purpose.	Sort.	Size.
Cobbles	(Malting)	Machine-made	$2\frac{1}{4}$ to $3\frac{1}{2}$ or 4 inches.
		Screened	Cubes which pass over longitudinal bars $1\frac{1}{4}$ inches apart and through bars 3 inches apart.
Nuts	(Domestic)	German	$\frac{3}{4}$ inches to $1\frac{3}{4}$ inches, or 1 inch to 2 inches.
		Paris	$\frac{3}{4}$ inches to $2\frac{1}{4}$ inches, or 1 inch to $2\frac{1}{4}$ inches.
		French	$1\frac{3}{4}$ inches to $2\frac{1}{4}$ inches, or $1\frac{1}{4}$ inches to $2\frac{1}{4}$ inches.

Name.	Purpose.	Sort.	Size.
Pea-nuts	(Plantations)		$\frac{1}{2}$ inch to 1 inch.
Beans	(Producer Gas)		$\frac{1}{2}$ inch to 1 inch.
Peas	ditto		$\frac{1}{4}$ inch to $\frac{5}{8}$ inch.
Grains	ditto		$\frac{1}{8}$ inch to $\frac{1}{4}$ inch.

Rubbly Culm or *Small* includes all that passes through longi-
tudinal bars placed 1¼ inches apart, and is mainly a waste
product of the collieries which do not screen and wash the
smaller of the above sizes. Some of it is sold to patent-fuel
works, which can make use of a certain proportion.

Anthracite is valuable for a number of other
purposes besides domestic heating. It is used for
heating greenhouses, for curing hops in Kent, and
it is the best fuel for making " producer " gas, which
consists of carbon monoxide, made by burning it in
a slow current of air, and of " water-gas," which
consists of carbon monoxide and hydrogen, and is
obtained by passing steam over red hot anthracite
or coke. " Producer gas " is used for driving
internal combustion engines, for which purpose it is
much cheaper than coal-gas or any kind of oil ;
and now that the " producer " for making the gas
can be obtained in compact form at moderate cost
and requires comparatively little attention, this
source of power is rapidly becoming more popular
than the steam engine, except where great power
is required. Anthracite beans are the sort commonly
used for power purposes ; and as this fuel is convenient to handle, does not deteriorate, and is
moderate in price, its use is likely to extend.

Anthracite is also much used in tropical plantations, particularly by the rubber companies in the

Malay States. Anthracite possesses properties which are advantageous for so many purposes that it is assured an ever-increasing popularity as a fuel.

Where Anthracite is Found

Anthracite is not nearly so widespread in its occurrence as other kinds of coal. In Great Britain, although small quantities are mined in various parts of the country, the mineral is practically a monopoly of South Wales. Here the anthracite district forms the north-western and western border of the coalfield ; and covers an area of about 137,000 acres in West Glamorgan and Breconshire, Carmarthenshire and Pembrokeshire. As pointed out previously,[1] there is no sharp line of demarcation between the anthracite region and the steam coal area, for the composition of each seam changes gradually towards the west and north-west from the standard steam coal in the centre of the field, to a drier and harder steam coal, and finally to anthracite at the border. In the easterly direction the extension of the anthracite seams is cut off by a great fault running from north-east to south-west in the Vale of Neath. From this line of demarcation at Glyn Neath, anthracite coal seams extend westward for a distance of about 30 miles to Kidwelly. The mean width of the field is considered to be from eight to ten miles. There are a number of seams, of which the chief ones worked are known as the Red Vein or Stanllyd Vein, and the Brass Vein. There

[1] Chapter V, p. 97.

are some splendid seams, such as the Pumpquart,[1] beneath these and so far almost untouched. The thickness of the seams is very variable, running from about nine inches to 13 feet. The average thickness of the majority of the seams is from 2 feet 6 inches to 3 feet. The extreme thickness is usually $4\frac{1}{2}$ feet, but in some collieries certain of the seams worked are 12, 18, 28, and even 30 feet thick. Such local thickening is usually due to earth movements, the seam being doubled by being pushed over itself.

Anthracite occurs in small quantities in Derbyshire and Scotland, appearing, however, there to be simply a local variation of the ordinary coal, due to the coal seam having been highly heated, owing to the intrusion in its neighbourhood of a white-hot molten volcanic rock.

The principal Anthracite coalfield of the world is that of Pennsylvania (U.S.A.) ; but there are also smaller and much less important anthracite areas in Colorado and New Mexico, Spain and Portugal, Italy, Germany, France, and in Shan Si (China). In Germany and France, and probably in most of the occurrences other than Pennsylvania and South Wales, the quality is poor.

The official statistics of the production and export of anthracite are set out in the accompanying table, p. 665, which shows the production of the United Kingdom, and of South Wales, the total exports, and the percentages which these two bear

---

[1] Welsh for five-fourths of a yard, referring to the thickness.

TOTAL OUTPUT OF ANTHRACITE

Year.	Total Output United Kingdom.	Total Output South Wales.		Total Exports United Kingdom.	
	Tons.	Tons.	Per cent. of U.K.	Tons.	Per cent. of output.
1894	1,795,939	1,580,154	86·3	..	..
1895	2,072,210	1,761,186	85·0	..	..
1896	2,077,578	1,784,963	85·9	..	..
1897	2,129,423	1,830,600	85·9	..	..
1898	2,112,736	1,805,,490	85·4	..	..
1899	2,418,507	2,113,720	87·4	..	..
1900	2,523,150	2,263,468	87·3	..	..
1901	2,565,462	2,254,066	87·9	..	..
1902	2,922,651	2,596,665	88·7	..	..
1903	2,901,006	2,572,800	88·7	1,254,445	43·2
1904	2,962,252	2,626,851	88·7	1,315,735	44·4
1905	3,112,054	2,789,178	89·6	1,478,576	47·5
1906	3,377,523	3,042,216	90·1	1,852,025	54·8
1907	3,850,437	3,498,258	90·8	2,127,903	55·3
1908	4,080,460	3,731,074	91·4	2,275,492	55·5
1909	4,258,980	3,914,400	91·9	2,535,903	59·5
1910	4,379,490	4,032,212	92·1	2,425,932	55·4
1911	4,350,479	3,992,763	91·8	2,454,523	56·4
1912	4,696,691	4,353,010	92·7	2,547,712	54·2
1913	5,194,620	4,833,159	93·0	2,976,050	57·3

to the total production. It will be seen that the
production in South Wales is more than 90 per
cent. of the total anthracite mined in this country ;
and the importance of the Welsh anthracite is
further emphasised when it is remembered that the

quality is in general superior to that mined in other parts of the United Kingdom. Of the total quantity of anthracite mined, it will be seen that an average of about 55 per cent. has been pretty regularly exported in recent years. This is a distinctly higher proportion than the average percentage of export of the total of all kinds of coal in the United Kingdom, which fluctuates around 33 per cent. It would seem that foreign nations have learnt to appreciate the good qualities of anthracite more readily than the English. Probably this is due to the long-established custom of heating rooms with stoves. If a stove is used, there is no doubt as to the convenience and cleanliness of anthracite as a fuel. Hence it is popular in Paris and larger towns of Germany at a price of about £3 per ton.

The figures show that the annual anthracite production of South Wales has more than doubled during the past twelve years, whilst in the same period the exports of anthracite have very nearly doubled.

The chief ports of shipment of anthracite are Swansea and Llanelly, though a few little cargoes are loaded from the smaller towns further west. Over 90 per cent. of the total shipment goes from Swansea. The following table gives particulars of the principal markets and the amounts exported to each during the eight years 1904–1911. The relative proportion of the figures alters little from year to year.

ANTHRACITE EXPORTS FOR EIGHT YEARS, 1904–11

Countries of Destination.	From Swansea.	From Llanelly.	Total.
Russia . .	90,861	3,488	94,349
Sweden . .	680,649	206,771	887,420
Norway . .	126,147	8,709	134,856
Denmark .	81,106	24,442	105,548
Germany . .	1,434,320	293,248	1,727,568
Holland . .	660,734	118,070	778,804
France . .	5,589,354	443,365	6,032,719
Spain & Portugal	506,562	19,349	525,911
Italy . .	3,008,473	2,781	3,011,254
Channel Isles .	251,121	111,005	362,126
Other Countries .	927,833	2,496	930,329
	13,357,160	1,233,724	14,590,844

It will be seen that France is our best customer, taking over 6,000,000 tons, whilst Italy is the next, and Germany comes third, with Sweden and Holland fourth and fifth. Germany mines a certain amount of anthracite in her own coal fields.

The distribution of the anthracite production in the United Kingdom is shown by the following table:

PRODUCTION AND EXPORT OF ANTHRACITE, 1913

	Output (Tons).	Exports (Tons).
South Wales . .	4,833,159	2,804,544
Scotland . . .	291,245	163,286
Ireland . . .	70,216	8,210
England . . .	..	..
	5,194,620	2,976,040

Although many countries produce anthracite, there is only one of which the total production exceeds that of South Wales, namely, the Pennsylvania anthracite coalfield, which is far the most important in the world, probably not excepting that of China, though the latter is as yet very little developed. The total production of anthracite in the United States is about eighty million tons per annum, as compared with the 4½ million tons in South Wales.

Anthracite Collieries

Considered as a class, anthracite collieries are relatively small in size and in the scale of working and management. The following table gives a comparison between the anthracite collieries and the collieries of the South Wales coalfield as a whole.[1]

COMPARISON OF ANTHRACITE COLLIERIES WITH ALL COLLIERIES IN SOUTH WALES

	All Collieries.	Anthracite Collieries.
Number of collieries . .	478[2]	76
Number of men employed .	222,823	19,506
Average number of men employed . . .	466	263

[1] A list of the principal Anthracite Colliery Companies is given in Appendix III.

[2] This figure does not include collieries employing fewer than 10 men.

TABLE OF ABOVE COLLIERIES GRADED BY SIZES

No. of employees.	All Collieries.	Anthracite Collieries.
Below 200 . . .	197	28
Between 200 and 300 .	56	19
,, 300 and 400 .	42	13
,, 400 and 500 .	26	7
Above 500 . . .	157	9

The latest statistics indicate that although the size of the collieries has been growing rapidly in the last few years, there are still only two or three mines which approach employing 1000 workmen, and there are only two companies owning collieries which jointly employ over 1000 men.

Another characteristic difference between the anthracite mines and the other South Wales mines consists in the fact that anthracite is usually worked by levels or slants, driven from the outcrop of the seam and following its dip, whilst in all the steam coal areas the coals are mined almost entirely by vertical shafts.   Two or three pits are now in process of sinking two or three miles south of the outcrop, and it is probable that a good deal of anthracite will be raised in the future through the usual vertical pits.   At present, however, these pits for working on a large scale and some little distance south of the outcrop are regarded purely as an experiment and as rather speculative business, because it is found that the ground becomes very broken by faults as the seams are followed southwards, and there is, also some question as to the quality of the anthracite being maintained.

The reason for the existence of small collieries and the general use of levels and inclined planes in the anthracite district is probably the fact that the greater demand for the Welsh steam coals has in the past attracted into the areas where these are mined most of the capital available for mining enterprises. As a rule, the amounts of capital invested in individual anthracite collieries are not large, and for this reason the cheaper method of development by level or slant is preferred, especially as it yields a speedier return on the capital expenditure than the sinking of pits. It seems safe to assume, however, that in the future the working of anthracite on a small scale will become less general, as experience shows that a much greater expenditure is now necessary on new takings before a satisfactory yield can be obtained. A much greater amount of capital is becoming available for investment in anthracite mining enterprise than formerly owing to the expansion of the trade in this country and the marked success of one or two of the larger companies. There is to some extent, therefore, a movement for buying up and combining the smaller collieries with the neighbouring larger ones whereby they can be worked much more economically. No great combinations of anthracite colliery undertakings have been effected as yet, however, although, as is shown in the chapter on Amalgamations, an attempt was made in 1903 to establish a trust or combine to be known as Anthracite Ltd.

Anthracite mining in South Wales is a particularly difficult and risky business for colliery proprietors, as the strata are so much disturbed that the seams are constantly disappearing through small local faults, and much expenditure on cutting through dead-rock is necessary to reach the coal again. Frequently the main roads have almost the character of a switchback, having had to be driven up-hill and down-hill to follow the jumps of the seam. This makes haulage very expensive ; and it is questionable whether it would not be far more advantageous to adopt the Continental system of mining by horizontal main roads at different levels, with subsidiary roads working back up the seams wherever they are struck. Another source of difficulty in working the anthracite mines has been the small size and extraordinary shape of many of the takings. They are held under lease often from different owners, and conform with the boundaries of the farms on the surface. Combined schemes for pumping and ventilating might do much to reduce the expense ; but there is much to be said for establishing by new legislation some kind of judicial court whereby the boundaries of takings and the terms of leases to prospective colliery owners could be compulsorily altered to meet the reasonable needs of the mining industry.

The anthracite mining trade must be regarded as being decidedly speculative owing to the conditions of the strata and the takings ; but with the formation of larger companies much of the risk would be removed. Putting it more exactly, I mean that the sums of money liable to be lost through meeting difficulties say, from £1000 to £10,000—whilst enough seriously to affect a small company, would be of little moment to a company of, say, £250,000 capital. By carrying the operations on over a large enough area, the risks can be over-ridden or averaged out, the lucky discoveries of small areas which can be very cheaply worked being set off against those areas which are very costly.

The general result is that under present conditions the mining of anthracite is a much more costly business than that of mining steam or house coals, and it must always remain somewhat more costly. Besides the large amount of deadwork, there are other causes of the high cost of anthracite. It is, generally speaking, rather harder to work than most kinds of coal ; also the mines are situated in remote valleys, and the hewers must be highly skilled work-men, so that the tonnage rate paid is liberal, and the earnings are large. It is by no means unusual for a man to make 70s. to 90s. per week when the colliery is working full time. Unlike the steam-coal trade, too, anthracite mining is to some extent a seasonal trade, orders being scarce in the spring and early summer, and particularly so after a mild winter, when Continental merchants have stocks left on hand. Many of the collieries are then obliged to work for weeks or months only three or four days a week.[1] The French and German merchants largely control the anthracite market. There are no merchants in Wales who carry such large stocks as they do, mainly because convenient storage accommodation has never been provided at the Welsh ports of shipment. The Continental merchants who have the storage room buy largely at the low prices of June, July and August for the following winter.

A question which has considerably exercised the minds of anthracite colliery owners during past years has been that of reducing the wastage from small coal. On account of the small amount of volatile constituents, anthracite duff possesses little commercial value, and is not capable of being used to any considerable extent for the manufacture of patent fuel briquettes. Up to a few years ago from 10 to 30 per cent. of the total yield was small, and was practically given away. On the average probably about 15 per cent. of the total quantity of coal mined was left underground in the gob. The obvious loss of potential wealth resulting from this practice caused considerable attention to be paid to the question, and the mine owners took steps to follow the German practice of washing and sizing the small coal for various purposes. In many of the larger collieries elaborate machinery has been installed for crushing, washing, and screening all the coal brought to bank, and much of the coal

[1] The outbreak of the great European war has just caused a period of extreme depression in the anthracite field, with serious distress, whilst the steam coal areas of South Wales were exceptionally busy.

2 U

which under previous conditions would have been packed into gobs or deposited in surface rubbish heaps now fetches a good price in the market. Probably 75 per cent. of the anthracite output is now treated by such modern equipment, and it has been found advantageous by many collieries even to crush up all the large coal and grade the products into various sizes of nuts, beans and peas. On account of the inadequacy of the water-supply in some localities, the washing process is not possible at a number of collieries, and some of these, therefore, consign their output to the ports for treatment at washeries adjacent to the docks.

CHAPTER XXIV

FOREIGN TRADE IN COAL

THE outstanding feature of the trade of the great manufacturing countries of Germany, the United Kingdom and the United States during the past quarter of a century has been the remarkable growth in coal production. In Britain the rate of increase has not been so great as in America and Germany, but the development of the export trade has been far greater. In our foreign commerce coal now occupies an exceedingly important place, and the value of coal exported in 1913 was no less than $8\frac{1}{2}$ per cent. of the total value of all goods produced in the United Kingdom which were exported. The following table illustrates the growth of our coal export trade over a period of nearly sixty years.

From this table it will be seen that this important branch of our foreign trade is of comparatively recent growth, and is constantly absorbing a larger and larger part of our total production of coal. It will be observed that from 1855 to the present date the value of our exports of coal has risen from $2\frac{1}{2}$ to 10 per cent. of the total value of the produce of the United Kingdom exported, whilst it has increased nearly twenty-two times in actual magni-

Table Illustrating the Growth of the Export Trade in Coal from the United Kingdom[1]

Year.	Total Value of Exports of Produce of United K. 000's[2] omitted.	Value of Coal Exported. 000's omitted.	Percentage Value of Coal to Total Exports	Total Production of Coal in U.K. 000's omitted.	Quantity of Coal Exported[3] 000's omitted (including bunker).	Percentage Quantity of Coal Export'd to Total Raised.
	£	£		Tons.	Tons.	
1855 .	95,688	2,446	2·55	64,307	4,977	7·74
1856–60 .	124,161	3,145	2·46	69,690	6,695	9·60
1861–65 .	144,396	3,933	2·72	88,660	8,482	9·46
1866–70 .	187,820	5,330	2·84	105,325	10,313	9·79
1871–75 .	239,502	10,304	4·30	125,885	16,644	13·22
1876–80 .	201,395	7,932	3·93	136,321	20,790	15·25
1881–85 .	232,286	10,096	4·34	158,906	28,451	17·90
1886–90 .	236,328	13,031	5·51	169,621	34,490	20·33
1891–95 .	226,969	16,577	7·30	181,906	40,517	22·27
1896–1900	239,125	22,332	9·34	208,964	51,039	24·42
1901–05 .	291,117	27,620	9·49	229,007	62,994	27·51
1906–10 .	388,693	38,036	9·78	261,727	83,115	31·75
1900 .	282,604	38,620	13·66	225,181	58,405	25·93
1901 .	270,873	30,335	11·20	219,047	57,783	26·38
1902 .	277,552	27,581	9·94	227,095	60,400	26·59
1903 .	286,517	27,263	9·51	230,334	63,805	27·70
1904 .	296,256	26,862	9·06	232,428	65,822	28·32
1905 .	324,387	26,061	8·03	263,129	67,161	28·44
1906 .	366,931	31,504	8·59	251,068	76,788	30·28
1907 .	416,016	42,119	10·12	267,831	84,682	31·62
1908 .	366,653	41,616	11·35	261,529	84,655	32·37
1909 .	372,253	37,130	9·97	263,774	85,408	32·37
1910 .	421,615	37,813	8·97	264,433	84,046	31·78
1911 .	448,456	38,448	8·55	271,892	86,536	31·83
1912 .	480,196	42,584	8·86	260,416[4]	85,327	32·76
1913 .	514,430	53,659	10·43	287,412	97,719	34·00

[1] From Coal Tables and Statistical Abstract for United Kingdom.

[2] Ships and boats were not included in the official returns till 1899, and have therefore been omitted from the latter totals here. The export of ships in 1913, £11,031,236, was the highest on record; but this only reduces the percentage of the value of coal exported from 10.43 to 10.21.

[3] Coal shipped for bunkers of foreign-going vessels is included in these figures from 1873 onwards. It amounted to about 3,000,000 tons then, and now reaches nearly 20,000,000 tons.

[4] In this and succeeding years the amount of dirt in the coal has been deducted, whereas in previous years no estimate was returned. In 1911 it was 2,268,000 tons.

tude. At the same time the quantity of coal
exported has also increased nearly twenty-fold, and
has grown from about $7\frac{3}{4}$ per cent. to 34 per cent. of
the total quantity raised in the country.

Coal Exporting Countries

At the present time the United Kingdom stands
second (the United States being first) among the
nations of the world in respect of the amount of
coal produced ; but our own country is easily first
in respect of the amount of its export.   The latter
fact is shown in the following table, which gives
the total production of coal in each of the chief
coal-producing countries, also the quantity exported
from each, including coke reckoned in equivalent
coal, and the percentage of the exports to the total
production.

Much confusion is introduced into the statistics
of the exports of coal from different countries by the
differing practice with regard to coal shipped in the
bunkers of ships for their own use on the voyage.
The general practice is to treat bunker coal shipped
by steamers going to places abroad as exports, whilst
coal used in the coasting trade is not so considered.
Some countries, like Germany, treat as exports
only the coal shipped in their bunkers by vessels
sailing for foreign destinations and under foreign
flags.   This largely reduces the amount included
among exports.   Although the principle adopted
by Germany of taking the nationality of the shipowner as the distinguishing feature would seem to

PRODUCTION AND EXPORT OF THE CHIEF COAL-FIELDS OF THE WORLD IN 1912

Country.	Coal Production.	Exports.[4]	Percentage of Exports to Production.
	Tons.	Tons.	
United States .	477,202,000	26,600,000	5·57
United Kingdom .	260,416,000	85,843,000[5]	32·96
Germany . .	172,065,000	42,671,000[6]	24·79
France . . .	39,745,000	2,407,000	6·05
Belgium . .	22,603,000	7,194,000	31·83
Russia . . .	25,998,000	202,000	0·78
Austria-Hungary[1] .	16,813,000	1,351,000	6·81
Japan[2] . . .	17,349,000	5,757,000	33·18
India . . .	14,606,000	881,000	6·03
Canada . . .	12,958,000	1,836,000	14·16
Australia . .	11,730,000	3,823,000	32·58
South Africa . .	7,248,000	1,423,000	17·63
Spain[3] . . .	3,605,000	7,000	1·94
New Zealand . .	2,178,000	233,000	10·69
Other Countries .	18,000,000	..	..
	1,102,516,000	180,228,000	

[1] Hungary : Output for 1911, 1,269,000 tons, is included, the figures for 1912 not being available.

[2] Mainly 1911.

[3] Production in 1911.

[4] The figures for Exports include " bunker coal " in case of the following countries : United Kingdom, Australia, New Zealand, South Africa, Germany, France, Belgium, Italy, and Japan. According to the Introduction to *Coal Tables* (1912), " No statement with regard to bunker coal is made in the export returns of British India, the Dominion of Canada, Russia, Sweden, Spain, or Austria-Hungary, but as the exports of coal from the three first-mentioned European countries were small and the quantities shipped by sea from Austria-Hungary were also small, it appears probable that the exports from these four countries are exclusive of bunker coal."

[5] This figure differs from that given in Table I, owing to the fact that the quantities of coke and patent fuel exported are here converted into the equivalents in coal, but were not in Table I.

[6] Include coke and patent fuel exports converted into their equivalents in coal.

be the soundest, it is not general. The particulars usually published are not sufficient for us to disentangle the bunker coal and put the figures for all countries upon a uniform basis. In the foregoing table the bunker coal is included with the exports, but in some countries, like Germany, the amount is small because of the nationality test. This does not make much difference for our purpose, however, as the quantities involved are comparatively small and will alter little the relative importance of the figures. The table shows that our own country exports more coal than the four next largest exporting countries taken together, and is not far from claiming one-half of the whole of the world's foreign trade in coal.

Overseas Exports

To the Englishman accustomed to regard exports as necessarily oversea, the above figures may in some instances be rather misleading. The great bulk of the coal exported from Germany for instance, goes by rail, canal, or river, overland to contiguous countries—to France, Austria, Switzerland, Holland, etc. In like manner much of the French coal goes overland, whilst the greater part of that exported by the United States goes overland to Canada. This feature must not be overlooked when comparing the oversea exports of different countries, as it may happen that coal transported by land from one country to another may again be exported by sea and reckoned as part of the exports of the country from which it was shipped. It is probable, for example, that a large part of the coal cleared from Germany into Holland is destined to be re-exported from Holland.

The total exports of each country are easily learnt from the last table, but they do not state what proportion of the trade is carried oversea, and thus comes into undoubted competition with British coal. We must, therefore, consider the trade of each country separately. The total exports of Japan, New Zealand and Australia must be entirely oversea ; the only figures which we have for British India are for exports by sea ; and probably nearly the whole export from South Africa goes by sea. In the case of countries which do not export wholly by land or by sea, special study is necessary of the countries of destination of their exports ; but a difficulty arises when exports go to the same country both by land and sea, as from the United States to Mexico, and Germany to France.

For the United States I have assumed the exports to Canada to be wholly by land, and half of those to Mexico to be by land ; whilst exports to Cuba and other countries must be wholly by water ; and I have added the coal equivalent of coke exported. German exports to Denmark, Great Britain, Norway, Sweden, Finland, Spain, etc., must be transported almost entirely over sea, together with probably about 25 per cent. of the exports to France. Probably 80 per cent. of the exports to Holland are re-exported by sea, and, though appearing officially as exports by sea from Holland in Dutch statistics, are really German exports and may, for our present purpose, be so considered.   Allowing $1\frac{1}{4}$ million tons for the exports of other countries, we may sum up the sea-borne coal trade of the world thus :—

WORLD'S SEA-BORNE FOREIGN COAL TRADE

	1906. Tons.	1912. Tons.
Germany . . .	4,000,000	10,360,000
United States . . .	2,060,000	4,140,000
Australia . . .	2,062,000	3,823,000
Japan	2,402,000	3,442,000
Belgium	688,000	1,280,000
South Africa [1] . .	730,000	1,423,000
France (about) . .	900,000	1,000,000
India	940,000	881,000
New Zealand . . .	142,000	230,000
Other countries . .	1,000,000	1,250,000
Total Non-British . .	14,924,000	27,829,000
United Kingdom . .	58,197,000	67,552,000
	73,121,000	95,381,000

Percentage of United Kingdom to the whole, *70.8*.

It is immediately evident that the British overseas trade in coal is more than twice that of the whole of the rest of the world put together.   Our most formidable competitor is Germany, whilst the United States probably comes next, as she is making

[1] Natal only.

some shipments to the Mediterranean and South America, which compete with the British trade. On comparing the figures for 1912 with those of 1906, it is evident that Australia and Japan are both increasing their trade. Japan supplies the China Seas, the Pacific Islands and Australia ; and besides these markets, sends large cargoes to the Pacific Coast of South America, and even to ports on the Indian Ocean. With the development of distant coalfields, and the growth of shipping on all ocean routes, we are naturally losing the position of almost complete monopoly of the overseas trade in coal which we held in the nineteenth century. In 1900 our overseas trade was still about 85 per cent. of the world's total. By 1906 it had fallen to 80 per cent., and in 1912 was 71 per cent. The year 1912 was, however, an unfortunate one for this country, owing to the miners' national strike, which reduced our own exports and deflected orders to our competitors. If I had the figures for 1913 complete, they would show a slight recovery ; but the general tendency will continue to be a fall in our proportion of the whole world's sea-borne trade, at the same time that we continue to increase the total amount of our exports.

The World's Markets for British Coal

The quantities of British coal exported to various destinations can be conveniently ascertained from the Return of Coal Exports annually prepared by the Board of Trade, and I give below a table

QUANTITIES OF COAL EXPORTED TO VARIOUS MARKETS FROM THE PRINCIPAL DISTRICTS OF U.K. IN 1887 AND 1912.[1]

(000's omitted.)

Ports.	Baltic and North Sea.		France and Mediterranean.		Africa and India.		Ceylon and Far East.		A. America (Atlantic Ports), W. Indies, and Gulf of Mexico.		Brazil, Argentine, Uruguay, and Paraguay.		N. and S. America, Pacific Coasts.	
	1887.	1912.	1887.	1912.	1887.	1912.	1887.	1912.	1887.	1912.	1887.	1912.	1887.	1912.
Bristol Channel	233	1,621	7,492	17,717	1,033	884	563	377	325	74	897	4,972	83	480
North-Western	9	87	172	415	154	25	19	3	66	9	100	130	114	5
North-Eastern	3,460	10,891	3,108	9,629	100	44	7	4	31	12	94	186	86	85
Humber Ports	1,391	5,298	419	1,191	35	23	..	..	18	6	15	249	4	3
East Scotland	1,942	6,647	241	978	10	2	..	1	60	5	39	163	2	..
West Scotland	185	422	380	1,186	46	22	6	4	149	74	56	197	9	2
All other Ports	76	305	2	16	7	..	1	..	3	..	2	..	1	..
Total	7,296	25,271	11,814	31,132	1,385	1,000	606	389	652	180	1,203	5,897	300	575

[1] There was a shrinkage of exports in 1912 as compared with the previous years due to the National Coal Strike.

extracted from this return, showing the quantity
of coal exported for each group of ports to the
principal markets of the world in 1912, with
comparative figures for 1887, a quarter of a century
earlier.

Inspection of this table shows that the European
and Mediterranean market takes over 87 per cent.,
or nearly the whole of our exports of coal, ports
in Belgium and to the north of it taking some 26
million tons, and the ports of France and the Mediterranean (including North Africa) requiring 30
million tons. The only other market of any real
importance is that of South America, which takes
well over 4 million tons. The distant markets, as a
whole, are, indeed, comparatively unimportant,
except to one coalfield, South Wales. The high
freights required for long voyages are responsible
for this fact ; and it is only the high quality of the
Welsh steam coal which gives it its exceptional
position. Our other coals, burdened with high
freights, cannot meet the competition of native
coals in distant parts ; but a certain quantity of
Welsh coal is required almost everywhere for coaling
the navy and fast liners.

The question of freight rates is of the greatest
importance to the export trade in coal, as the relative
distances of potential markets, and consequently
the freights charged from the coal exporting ports,
largely determine the extent and nature of the trade
which may be opened with such markets. The
average freight rates to ports in various parts of the

world given in the following table will be of interest in this connection :—

AVERAGE FREIGHT RATES FROM CARDIFF TO GROUPS OF FOREIGN PORTS FOR THE THREE YEARS 1909—1911 [1]

Port.	Distance (Nautical miles).	Average Freight Rate per ton.		Rate per ton per mile.
N. Europe—		s.	d.	d.
Copenhagen . .	1,110	4	7	·049
Stockholm . .	1,500	5	1	·041
Hamburg [2] . .	822	4	0	·058
France—				
Dieppe . . .	415	4	5	·128
St. Nazaire . .	404	4	3	·126
Bordeaux . .	540	4	$8\frac{1}{2}$	·087
Marseilles . .	1,846	6	$8\frac{1}{2}$	·043
Mediterranean—				
Gibraltar . . .	1,155	5	4	·055
Algiers . . .	1,560	6	$2\frac{1}{2}$	·048
Barcelona . . .	1,665	7	5	·053
Genoa . . .	2,022	7	1	·042
Malta . . .	2,135	5	5	·030
Trieste . . .	2,808	7	0	·030
Port Said . .	3,075	6	5	·025
Constantinople . .	2,929	6	7	·027
Danube . . .	3,275	6	10	·025
Smyrna . . .	2,777	7	1	·031
Indian Ocean (via Suez)—				
Aden . . .	4,490	8	8	·023
Bombay . . .	6,155	9	1	·018
Colombo . . .	6,608	9	6	·017

[1] *South Wales Coal Annual.*   [2] 1908–10.

TABLE (*continued*)

Port.	Distance (Nautical miles.)	Average Freight Rate per ton.		Rate per ton per mile.
		s.	d.	d.
Far East (via Suez)—				
Singapore	8,188	10	3	·015
Hong Kong	9,718	12	6	·015
Shanghai .	10,470	15	3	·017
Africa—				
Madeira .	1,300	7	2	·066
Cape Town [1]	6,000	11	9	·024
Bona	1,800	7	2½	·048
Tunis	1,943	8	0	·049
Oran	1,385	6	5	·056
West Indies—				
Havana .	4,025	8	5	·025
S. America—				
Pernambuco	3,945	13	9	·042
Rio de Janeiro .	5,030	14	0	·033
Monte Video	6,140	13	11	·027
Buenos Ayres .	6,250	14	9	·028
America (Pacific)—				
San Francisco .	13,000	23	0	·020
Valparaiso [2]	8,870	19	9	·027

The above table shows that the freight-rate to the nearer European ports is about 4s. 6d. per ton on the average, and to the Mediterranean from about 5s. 6d. to 7s. per ton. As the distance increases so does the freight, but not in proportion.

A freight-rate which varies round 8s. 6d. per ton to the West Indies gives American coal a great

[1] 1905–7.   [2] 1907.

advantage, whilst 9s. is enough to prevent much of
our coal being sold in the Indian Ocean, now that so
much Indian coal is mined. In time, no doubt,
Indian competition will beat us off the Indian Ocean
altogether, except, perhaps, for a little Welsh coal.
Similarly in the Far East, and along the Pacific
Coast of North and South America, it is practically
impossible for our ordinary qualities to compete
with Japanese and Australian coal. Practically
the only British coal which is exported to the Pacific
Ocean is the best Welsh steam coal.

Freight-rates fluctuate very considerably from
year to year according to the state of foreign trade.
A sudden boom of trade occasioned by abundant
harvests in the great grain-producing regions of the
world, means a great demand for tonnage, not only
for carrying the grain but also for sending out the
manufactured goods and the coal purchased with
the proceeds of its sale. Good harvests in the
Argentine need much more coal than usual to haul
the additional grain to the coast ; and the railways
immediately purchase more from South Wales
than in normal years. Freights rise rapidly to a
very profitable figure with increased demand,
and as it takes twelve to eighteen months to build
a number of new ships, the boom in freights is safe
to last this period. With the new tonnage in use,
there comes a slump in freights. The oscillations
in recent years are well illustrated by the graphs
in the accompanying diagram showing the freight
rates for coal from the Tyne Ports to Hamburg

and Rouen, and Cardiff to Bordeaux and Buenos Ayres for the past twenty years.[1]

The great increase of our exports during recent years has been occasioned almost entirely by the growth of the demand in Europe and South America. In other markets local supplies have been developed to meet the expanding wants, and our coal has lost ground, or remained practically stationary. A comparison of the figures of coal exported in 1887 with those of 1912 (see table on p. 683) shows that our trade has fallen off with North America, the West Indies, India, and in the Middle and Far East, whilst it has remained stationary in East Africa and the Pacific coast of the Americas. In regard to the last market the following figures giving decennial averages from 1858 to 1907 and the average for the six years 1908–1912 are interesting :—

Period.	Tons (Average).	Rate of Increase per cent.
1858–67 . . .	123,100	..
1868–77 . . .	317,400	157·8
1878–87 . . .	375,400	18·3
1888–97 . . .	453,200	20·7
1898–1907 . .	493,500	8·9
1908–1912 . .	710,700	44·0

The rate of growth, which had become very slow at the beginning of the present century, has shown some recovery, which is due partly to the remarkably rapid economic development of the Pacific slopes

[1] Other useful diagrams showing fluctuations of freight rates are given in Appendix XVI of Prof. Kirkaldy's *British Shipping*.

of America, and partly to the slump in ocean freight-rates from this country in 1909 and 1910. When the demand for outward tonnage is great, as in 1912, the shorter voyages are so profitable to ship-owners that freights to distant markets become prohibitive.

Those who would follow further the question of the relative growth of the markets for British coal cannot do better than consult a paper by Mr. D. A. Thomas, read ten years ago before the Royal Statistical Society and the useful tables in his appendix.[1] The figures can be carried up to date by reference to the tables of Coal Shipments issued annually as a Parliamentary Paper.

The third column in the table on p. 685 gives the average freight-rate per ton per nautical mile to the various foreign ports. There appear to be three principal factors influencing the relative magnitude of the rates per mile : (1) The length of the voyage ; (2) the profitableness of return cargoes ; (3) the cost of labour obtainable to work the vessels. As regards the first, we notice that the rates to the nearer European ports are from two to four times as high as to the majority of ports distant more than 4,000 miles. This is because much of the expense of running the boat is incurred in entering and leaving ports, and in idle time spent there. Touching the second factor, we observe that it is the outward and return voyages together which the owner seeks to make profitable, and he is therefore able to offer

[1] Journal Stat. Soc., 1903, pp. 486–522.

2 x

a low rate when going to a port where he is sure of a valuable return cargo, or to one near a third port offering a good return freight. The extraordinary lowness of the rates per mile to India and the Far East is probably due to the fact that vessels trading in these directions are manned chiefly by Chinese, Japanese, Lascars, and Malaysians, at low wages.

Causes of the Growth of Our Coal Exports

In seeking the causes of the astonishing growth of British exports of coal during the last half century, we find two pre-eminent factors : (1) The growth of a foreign demand ; (2) the improvement of means of transportation. The increase of demand abroad is largely due to the improved political situation, which gradually became established throughout the Continent during the last century, and especially since 1870. When peace and liberty were assured there were opportunities of developing industry and building railways. As the peoples of Europe have gradually become more wealthy, so have they been able to purchase more coal for domestic as well as industrial purposes. Of course the energies of British merchants had to be directed towards bringing the merits of this country's coal to the notice of foreign buyers ; so that the growth of our exports may also be said to be due to the enterprise of men like John Nixon, who taught the French the value of Welsh coal.

The cheapening of the cost of transporting the coal to the foreign markets has meant an enormous increase in our foreign coal trade.   Not only have we now cheaper carriage to the dock and improved means of loading, but ocean freights have been very remarkably reduced during the last forty years, as shown by the table on p. 692, which gives the freight rates from Cardiff to some of the principal foreign markets.   It will be seen that the recent and present figures are little more than one-third of what they were in 1863–5, and in the case of distant markets distinctly less than one-third.

This wonderful fall of freight-rates far exceeds the reduction in the cost of railway transport, or in the price of any of the staple commodities of trade. It has resulted from the manifold improvements in steam navigation accomplished by British engineers.   Vessels have been built of steel, and much larger, and with engines more economical in fuel consumption ; so that both the initial cost and the cost of running per ton of carrying capacity have been greatly reduced.   The economy of large ships arises partly from the fact that their carrying capacity increases as the cube of their dimensions (length or width), whilst the resistance to the water increases approximately only as the square of such dimension, which means much saving in fuel ; but also largely from the fact that the expense of the navigating staff increases much less than proportionally with the size of the ship, for there is little or no addition to the number of officers required, and only a small increase in their salaries.

OUTWARD FREIGHT RATES FROM CARDIFF TO

	Cronstadt.		Bordeaux.	Lisbon.	
Distance in Nautical Miles.	1778		540	880	
Year.	s.	d.	Fcs.	s.	d.
1863–5.	11	8	14·50	15	3
1873–5.	9	6	11·00	9	6
1883–5.	6	6	7·25	7	0
1886	5	9	6·00	6	4½
1887	6	6	5·50	6	0
1888	5	10½	7·20	7	8
1889	6	3	7·00	8	6
1890	6	9	6·50	7	0
1888–1890	6	4½	6·90	7	8½
1891	6	3	6·25	6	6
1892	6	10½	5·37½	6	3
1893	5	5	5·75	5	4½
1894	5	3	5·10	5	3
1895	5	6	5·10	5	1
1893–1895	5	4½	5·32	5	3
1896	4	8	5·10	5	2
1897	5	5	4·65	5	4
1898	7	0	6·45	6	10½
1899	7	8	5·68	7	2½
1900	8	10	7·50	7	11
1898–1900	7	10	6·54	7	8
1901	5	6	5·60	5	2
1902	5	0	5·00	4	6
1903	4	7	4·78	4	5
1904	4	9	4·60	4	6
1905	5	4½	4·85	4	10
1903–1905	4	11	4·74	4	7
1906	5	6	5·20	5	2
1907	5	10	5·80	5	6½
1908	4	4½	4·87	4	10
1909	4	6	4·96	4	9
1910	4	6½	5·30	5	0½
1911	5	2	6·60	6	4
1912	7	5½	8·08	7	3½
1913	6	4½	6·94	7	0½

Representative Foreign Ports, 1863–1913.

Genoa.		Port Said.		Bombay.		Singapore.		Buenos Ayres	
2,022.		3,075.		By Suez 6,135. By Cape 10,445.		By Suez 8,188. By Cape 11,520.		6,250.	
s.	d.	s.	d.	s.	d.	s.	d.	s.	d.
22	0	24	6	27	6	37	6	35	6
16	0	18	6	24	6	34	6	33	6
10	3	12	0	16	6	26	6	27	6
9	6	10	3	14	9	20	0	22	6
9	10½	10	0	14	9	17	0	21	6
10	0	10	0	21	8	20	6	26	6
10	6	11	6	19	6	19	6	23	0
8	9	9	0	15	0	17	3	21	6
9	9	10	2	18	9	19	1	23	8
7	0	6	7½	7	3	16	0	19	3
7	6	8	0	10	9	11	3	14	0
7	0	5	7½	9	7½	10	9	10	3
5	8	5	4	7	11	10	5	9	6
5	10	5	5	9	5	10	10	9	6
6	2	5	5½	8	0	10	8	9	9
6	9	6	9	16	1	14	0	14	3
7	8	7	7	16	6	16	11	16	5
8	9	9	3	13	9	18	8	16	1
8	10½	10	0½	12	0	14	5	11	3
10	4	12	1	20	10	18	11	16	0
9	4	10	9½	15	10	17	4	14	5
6	9	7	4	12	6	14	7	13	1
5	5	5	4	10	0	12	0	11	0
5	9	5	7½	9	3	10	9	8	2
5	9	5	3½	8	4	9	8	7	2
6	6	5	10½	9	3	9	3	8	6½
6	0	5	7	8	11½	9	11	7	11½
6	10½	6	1½	10	3	11	3	13	10
7	5	6	5	9	8	10	9	14	1½
6	1½	6	3	11	4½	10	4½	9	10
6	6	5	9	8	0	8	7½	10	10
6	8½	6	0	9	1	9	4½	15	1½
8	1	7	5½	10	3	12	9	18	4
11	8	11	2	13	1	15	7½	20	5½
8	8	6	2	12	9½	12	9	12	10½

CHAPTER XXV

OIL FUEL

Coal Commission's Views

ONE of the questions submitted to the Royal Commission on Coal Supplies (1903–05) for investigation was that of the economies which might be effected by the utilising for various purposes other forms of fuel, and amongst the possible substitutes for coal which received careful consideration was oil. The Commission obtained much valuable evidence with regard to the use of liquid fuels, but decided that although the demand for coal might be reduced by the adoption of other sources of power no real substitute for coal existed. Since the report of the Coal Commission was published some important developments in the use of liquid fuel have taken place, and the question of the possible injury which the widespread use of oil may do to the coal trade has given rise to some amount of alarm, which has been accentuated by the scare writings of people of vivid imagination who see in recent engineering progress " the dawn of the oil age."

The Royal Commission, with all the facts before them, were agreed that for certain purposes oil was far superior to coal, but that on account of its strictly limited supply it could never become a serious competitor for general purposes. It is probable that the Commission took a less serious view of the matter than recent facts warrant, but they are probably quite correct in their main conclusion. In this chapter it is proposed to review the chief features of the question and to inquire into the possible results to the coal industry which will arise from the increasing use of oil fuel.

Use of Oil Fuel

There are two ways in which oil can be used as a substitute for coal :—

1. *As a steam raiser.*—In this case the oil is burnt in furnaces instead of coal for the purpose of making steam. The oil is injected into the fire by means of steam or compressed air through jets which break it up into a very fine spray which is almost gaseous in form. It is in this way that oil is used in the navy and mercantile service, usually as an auxiliary to coal ; but in torpedo boats and destroyers and other small craft independently of coal. The advantages claimed for oil over coal as a steam generator are numerous. It takes up considerably less space for storage, thus giving greater cargo room and adding to the earning power of mercantile vessels. It renders unnecessary the dirty and inconvenient process of coaling, and may save much of the time often spent on a long deviation of the voyage to call at a coaling station. The oil bunkers of vessels can be easily replenished at sea from tank steamers by means of a syphon or pump, which is of enormous

importance in the Navy. A great saving of labour is also effected in every ship equipped for oil fuel, probably not less than 75 per cent., through stoking and trimming being unnecessary ; and the work in the stokehole is rendered less disagreeable and exhausting. Much of the advantage of oil over coal, however, consists in its higher calorific power ; for one ton is estimated to be equivalent to $1\frac{1}{2}$ tons of coal.

2. *As a direct power-raiser.*—The method of obtaining power through the medium of steam is a wasteful one, and for years past engineers have been working out and perfecting methods of burning oil directly in a cylinder to supply the driving force of an engine. The principle of such an internal combustion oil engine is practically the same as that of the gas engine,[1] except that it is necessary to vapourise the oil, either before it is mixed with air to form the explosive mixture, or at the same time. For this purpose a variety of forms of vapourisers or carburettors have been invented, whilst the ignition is obtained either by a heated tube or by an electric spark.

Oil engines were first used, like gas engines, as small stationary power-producers, and in the closing years of the nineteenth century their utility for farmers and for a variety of purposes in country districts where coal-gas was not obtainable was coming to be widely recognised. For such engines a medium oil was used, a little heavier than the ordinary paraffin or petroleum illuminating oil. Then arose

[1] See Chapter III, Uses of Coal, p. 51.

the idea that with a lighter oil like petrol a more powerful explosive mixture could be obtained, so that engines could be constructed of much lighter weight and smaller size for the same horse-power, and thus be applied to mechanical transport on roads. The motor-car engine was evolved ; and finally, by further special reduction of the weight of the engine in relation to its power, the aeroplane engine. For power production upon a large scale by internal combustion engines, whether for stationary or marine purposes, oil did not make much progress for some years both because of certain technical difficulties and because the price of the refined medium-weight oil required was too high. For furnace use in generating steam, heavy or crude oils costing much less could be used.

To some extent the difficulty of cost was overcome by Dr. Rudolph Diesel, whose internal combustion engine contained several new and ingenious principles which greatly increased the efficiency of burning the oil and so reduced the cost. Briefly described, his process is to inject oil by an automatic pump through a jet, which breaks it up into a fine spray into the cylinder. In the latter the air is so highly compressed by the returning piston as to acquire sufficient heat to fire the oil vapour. It should be noted that the oil vapour burns as it is squirted into the hot air, and technically there is not an explosion as in other types of oil engine.

The efficiency of the Diesel engine is remarkable. It is from three and a half to four times as efficient as the steam engine in its use of the energy contained in the fuel; and it can be shown by calculations that with the price of oil five times as high per ton as that of coal it would still be more economical to operate a ship with oil in Diesel engines than with coal and the latest reciprocating steam engines. I show later in this chapter the special advantages of oil for the purpose of the mercantile marine. Including those advantages which would be common to other engines, it is estimated that a ship fitted with Diesel engines will carry 30 per cent. more load, cost distinctly less to operate owing to reduction of the wages bill, and owing to a saving in the cost of fuel as compared with the ordinary merchant ship fitted with good steam engines of the present day. The Diesel engine is of importance not only when we are considering the competition of petroleum with coal; for if approaching exhaustion of the world's oil-fields makes the mineral fuel too costly, ships fitted with Diesel engines might be driven with oil distilled from coal, a development which the late Dr. Diesel himself anticipated in the following words:—

" Since tar oil can be employed three to five times more efficiently in the Diesel motor than coal in the steam engine, it follows that coal can be much more economically utilised when it is not burnt barbarously under boilers or grates but converted into coke and tar by distillation." [1]

[1] Dr. Diesel's Introduction to *Diesel Engines for Land and Engine Work,* by A. P. Chalkley, London, 1913.

Oil for Steamers and Railway Locomotives

There has been a remarkably steady growth in the use of oil as a fuel for motive power during the last few decades. It was only about 1860 that paraffin, which was derived from the distillation of oil shale in Scotland, and petroleum from American deposits, began to be used for purposes of illumination. The art of distillation had to be discovered as a necessary preliminary to the separation of the lighter oil from heavy oils and waxes, which are all mixed in the crude oil. The mechanical difficulties of drilling the oil-wells were rapidly overcome, once the commercial importance of the mineral oil was realised. Hence the exploitation of the oil-fields on the Caspian Sea began only a few years after well-sinking in America, and by 1870 there were already steamships on the Caspian Sea whose furnaces were fired exclusively by oil, the cost of coal being very high in this district. Since then the use of oil in firing boilers of lake and river steamers has grown rapidly in many parts of the world, largely owing to the cost of transport of oil to remote regions being less than that of coal. It is now used by coasting and river vessels on the Californian coast, on the Danube, on South American, Russian, Persian, and some Chinese and African rivers. Many ocean-going vessels are now fitted for burning oil, including over forty steamers of the Shell Transport Company, many boats of each of the commercial lines, as well as some of the steam packet services plying shorter distances. The use of oil for firing railway locomotives followed soon after its use in ships, both in Southern Russia and in the United States, and for the past twenty years it has been extensively used for this purpose in many parts of the world, Austrian, Mexican and South American Railways being conspicuous in addition to those mentioned. In our own country the Great Eastern Railway has for several years fired many of its locomotives with oil, and the London and North Western Railway has adopted it to a small extent.

The economy of the use of oil for firing the boilers of steamships is particularly marked with modern appliances.[1] With proper burners oil gives from 30 to 50 per cent. greater heat convertible to power in the furnace than the same weight of best steam coal, and it also occupies less space, because there are no voids filled with air as in a heap of coal. It

---

[1] The following interesting comparison of the cost of running the same steamer with coal and, after conversion, with oil, is given by Prof. A. W. Kirkaldy in *British Shipping* in the National Industries series, of which the present volume forms a part. The comparison refers to the Canadian Pacific Company's steamer " Princess Charlotte." The trial took place early in 1913 :

Cost of one day's running with :

(i) *Coal.*		(ii) *Oil.*	
100 tons coal at $4.50 =	$450.00	344.17 barrels oil at $90 =	$314.25
13 Firemen at $55 =	23.80	6 Firemen at =	11.10
10 Trimmers at $45 =	15.00	Food for 6 men =	2.52
Food for 23 men =	9.56		
	$498.36		$327.87

There was a saving of $170.47 on one day's steaming.

has been estimated that if we compare the space
occupied by oil which would develop the same
amount of power as a quantity of coal, 36 cubic
feet of oil are equivalent to 60 cubic feet of coal.
This means a very great saving in the bunker space
of ships ; or, what is equally important, that with
the same bunker accommodation, a longer voyage
can be made without going to a port for coaling.
The foregoing figures, however, by no means measure
the saving in bunker space in the use of oil, for oil,
being a liquid, can be stored in reservoirs of all
shapes and sizes in various parts of the ship's hold
or shell, just as the ship's architect may find convenient, the various reservoirs being connected
by pipes and drained by pumps feeding the burners.
Coal bunkers, on the other hand, must be of suitable
shapes and situation to form storage heaps on which
men can work with their shovels ; and thus they
occupy valuable space in the centre of the ship.
It will be seen, therefore, that a ship planned from
the commencement for the burning of oil as its sole
fuel can effect economies of space which will equal
the addition of five or six hundred tons to the cargo
accommodation, the ship costing very slightly more
to build and less to run. The ship built specially
for oil firing has a very great advantage over one
built for coal and converted. For this very reason,
however, the extension of the use of oil is not likely
to be rapid ; for shipowners fear to commit themselves to the sole use of a fuel the price of which
may rise considerably.

Besides the saving in storage space, oil is much easier to handle and to transport than coal. The requisite oil can be pumped into a ship in less than half the time taken to coal it, and there is no dirt or dust which it costs money to clean away afterwards. In burning oil with properly adjusted burners there is practically no smoke. Oil burners can be lighted instantly so that steam can be got up to full pressure more rapidly than with coal. There are no ashes or clinkers to be removed, and no large force of stokers is required. So enormous are the advantages of the use of oil that there can be little doubt that its use will extend even though the oil has to be made by the coking of coal. At present the development has been mainly in the direction of using oil for the raising of steam ; but it is probable that with the technical improvement of the Diesel or other internal combustion engines they will be more extensively used. The *Selandia*, an ocean-going ship of 10,000 tons register, was the first fitted with Diesel engines to make long voyages some five years ago. She was so successful that several ships have since been fitted with Diesel engines ; and more would have been so fitted except for a nervousness as to the price of oil. The Diesel engines give to a ship all the advantages above stated as arising from the use of oil as fuel, and also go further by saving the great space occupied by the boilers, and by the fact that much less heat arises from the engine room, whilst it is also said that starting, or altering of the speed of the ship can be more quickly effected.

The British Admiralty was quick to recognise the many advantages of oil. It saw that the available storage space would allow the carrying of fuel for longer voyages, and thus give greater radius of action. Furthermore, oil can be taken on board at sea from tank steamers by a pipe line, whereas coaling in the open sea is generally impossible, requiring a dead calm. The rapidity with which full steam can be raised at short notice was also an important consideration, whilst the oil would be as smokeless as the best Welsh coal. It is for naval purposes that oil has been most widely used as a motive power in the United Kingdom. In 1909 was established the first flotilla of ocean-going torpedo boats and destroyers fitted solely with oil furnaces. A further hundred are now in course of construction. Several battleships have also been fitted with oil bunkers as an auxiliary to coal; and it is now the avowed policy of the Admiralty to adopt oil for practically all new ships.

The only cause of hesitation in adopting this policy was the question of an adequate supply of oil, especially in time of war. The shale-oil industry of Scotland produces nothing like sufficient oil for the whole of the British Navy; and we could not afford to be dependent upon foreign supplies which in time of peace might be controlled by a gigantic commercial monopoly, and in time of war might be

cut off by the activities of the enemy's fleet. The
Admiralty therefore decided upon the twofold policy
of acquiring control of their own oilfields and of
creating immense storage reservoirs at all the great
naval ports of these islands, and our foreign stations.
In pursuance of these ends enormous storage tanks
have been built which in the aggregate contain a
supply sufficient to last the ships now fitted, or
intended for burning oil, for a period of three years.
The object of storing upon such a large scale is not
only to safeguard our position in time of war, but
also to tide over any market fluctuations when the
price of oil may rise much above normal. To obtain
its own oilfields the British Government has entered
into a long term contract with the Anglo-Persian Oil
Company, and has agreed to subscribe for shares to
the amount of £2,000,000, which will give it absolute
control of the company and its operations. The oil
derived from these Persian fields will not for some
years, if ever, be sufficient to meet the requirements
of the navy ; but it is anticipated that this supply,
together with the immense storage at our ports, will
enable the Government to meet any combination of
producers to rig the market, by holding off for as
long as may be necessary. Supplies for replenish-
ing the store would be purchased only at lowest
prices. The saving thereby effected will probably
be equivalent to a substantial rate of interest upon
the large amount of capital lying idle in the
stored up oil ; so that the Government is following quite a sound commercial policy as well as

one most necessary in the interests of national defence.

It is an interesting question how far the adoption of oil as the principal fuel of the British Navy would diminish the demand for the best South Wales steam coal. There are thirty-one South Wales collieries on the Admiralty list, and the amount of coal purchased each year for naval purposes is between two and two and a half million tons. This is about $3\frac{1}{2}$ per cent. of the total output of the coalfield, but nearly 10 per cent. of the output of the steam coal area. The total consumption of oil on British naval vessels is approximately 400,000 tons per annum, or the equivalent of 600,000 tons of coal. We may say, therefore, that oil has displaced coal to the extent of less than 25 per cent. of the navy's total fuel requirements, and the amount of coal displaced by oil is less than 3 per cent. of the output of best steam coal in South Wales. The transition from coal to oil must necessarily be a gradual one, and, to quote Mr. Winston Churchill's words, "Coal will continue to be the main basis of motive power in the line of battle for the present." No stagnation of the South Wales mining industry need be anticipated, therefore, from the adoption by the Admiralty of the new policy, as the general demand for steam coal now exceeds the supply, and the loss of naval demand will simply mean the diversion of supplies to other directions at only a small reduction of price.

2 Y

The World's Oil Supplies

Far more serious for the coal industry than the loss of the British Navy as a consumer would be an extensive general adoption of mineral oil as a power producer for all purposes. This, however, is a very unlikely contingency, for, so far as our present knowledge extends, the supply of oil is strictly limited. Nearly all of the known oilfields are already being exploited, though some of them not fully. It is quite probable that new oilfields will be discovered in South America, Africa and parts of Asia. I think it can be shown conclusively, however, that even if the world's production of petroleum should be increased to ten times the present annual output, the effect upon the world's coal trade would only be a comparatively slight restriction of the growth of the total output of coal.

The present output of oil in the various producing countries of the world is shown in the following table, which contains the latest figures available for all countries.

WORLD'S PETROLEUM PRODUCTION [1]
(000's omitted)

Country.		1911. Imperial Gallons.	1912. Imperial Gallons.
Russian Empire .	.	2,273,603	2,336,658
United States .	.	7,712,642	7,770,853
Other Countries—			
Germany .	.	34,124	33,612
Roumania .	.	424,875	473,679
Austria . .	.	368,661	284,867
Japan . .	.	60,725	55,768
Mexico . .	.	491,611	584,432
Peru . .	.	47,871	61,265
Canada . .	.	10,188	8,517
British India .	.	225,792	249,084
Trinidad .	.	7,685[2]	4,296[2]
Dutch East Indies .		395,113	349,578
Total .	.	12,052,890	12,212,609

It will be seen that the United States is responsible for more than one-half of the world's total supplies, and that Russia comes next with less than one-third of the production of the United States. The production in many of the new fields of other countries is increasing fast, and that of the United States has probably nearly reached its zenith, so that these proportions will not hold good for long. The following table shows clearly the increase of

[1] From *The Coal Tables*, 1912.
[2] Domestic exports, particulars of production not being available.

output in recent years in Russia, the United States
and other countries, the numbers denoting thousands
of imperial gallons :—

GROWTH OF PRODUCTION OF PETROLEUM [1]

(Thousands Imperial Gallons)

Year.	Russia.	United States.	Other Countries.[2]	Total.
1900	2,612,876	2,261,808	..	..
1901	2,898,030	2,427,650	396,858	5,722,538
1902	2,797,251	1,105,599	400,096	6,302,946
1903	2,954,614	3,514,740	570,001	7,039,355
1904	3,040,372	4,095,251	715,125	7,850,748
1905	1,923,600	4,715,115	596,880	7,235,595
1906	2,061,395	4,427,290	945,900	7,434,585
1907	2,164,785	5,811,612	1,171,972	9,148,369
1908	2,140,356	6,245,958	1,498,014	9,884,328
1909	2,342,436	6,408,416	1,839,732	10,590,584
1910	2,423,547	7,331,528	1,586,732	11,341,807
1911	2,273,603	7,712,642	2,066,645	12,052,890
1912	2,336,658	7,770,853	2,173,872	12,281,383

The table shows that the yield of oil has more than
doubled in the past twelve years, but whilst that of
the United States has more than trebled, the Russian
output shows a net decrease. The sudden fall of
output in Russia from 3,040 million gallons in 1904
to 1,923 million gallons in 1905 was due to the
revolution of 1905, when there was much rioting

[1] From *The Coal Tables.*

[2] The figures contained in this column do not necessarily indicate
the *actual* production of *all* other countries, but only the production of
countries for which records are available.

and destruction of the oil companies' property in the
Baku oilfields. From this destruction of the whole
of their surface plant many of the oil-producing
companies never recovered. The increase of the
world's production has so far been predominantly
due to the influence of the United States supplies ;
and the larger part of the increase recorded is due
not so much to the opening up of new fields as to
an increase in the number of wells in existing oil-
bearing districts. It may, therefore, be confidently
anticipated that with the opening up of oilfields
in all parts of the world and the multiplication of
wells thereon, the world's supply will rapidly and
greatly increase.

For how long such increase of the world's aggregate output can go on is, however, an open matter.
It has to be borne in mind that an oilfield tapped
by numerous wells can show signs of exhaustion in
a few years. Every oilfield known is found to be
capable of extension, however, and it is probable that what is commercially termed an oilfield,
i.e., a district where oil is proved by wells, is generally only a small part of the whole field geologically
considered. Thus whilst areas which are studded
with wells close together may be exhausted in from
five to ten years, it will generally be found to be
possible to keep on opening up new almost conterminous " fields " which are simply other parts of
the geological field. The latter may, therefore,
last for several decades ; but under modern competitive conditions it does not seem likely that any

fields would last more than 60 or 70 years. It is
possible that the Baku, Grosny, Maikop and other
oil " fields " on the Caspian are only parts of a great
field of oil-bearing strata running along the north
of the Caucasus, and even possibly conterminous
with the other great area south-east of the Car-
pathians, which includes the fields of Roumania and
Galicia. In the same way oil-bearing strata may
be discovered eventually over very wide areas in
other continents. With the rising price of petrol,
however, and the strenuous search for new fields,
it is hardly conceivable that any oil-fields of any
importance will remain undiscovered and undeveloped after the next 30 or 40 years. Hence
practically complete exhaustion of the world's
petroleum reserves may be looked for within 100
years. Dr. Enger estimates the world's reserves
at only 5,000 million tons, which is less than 100
times the world's present and growing output ;
but this is probably under the truth, for oil keeps
turning up in unexpected places.

The falling off of the production of certain of the
American oilfields through approaching exhaustion
is well shown by the following figures :—

PRODUCTION OF CRUDE PETROLEUM

	Colorado. In barrels of 42 gallons.	Canada. In Imperial gallons.
1906 .	600,000	19,941,000
1907 .	400,000	27,611,000
1908 .	411,800	18,480,000
1909 .	310,770	14,726,000
1910 .	239,900	11,056,000
1911 .	226,900	10,188,000
1912 .	206,050	8,517,000

The life of the supplies of natural gas found in Western Canada and the Western States of North America is still more fleeting.

Taking all the facts into consideration, it seems to me unlikely that the annual oil production of the world could be increased at any future date to more than ten times its present volume, that is to say, to more than 120,000 million gallons. This is approximately equivalent to 480 million tons of oil per annum, and would represent about 720 million tons of coal, or about half the present world's output of coal and lignite.

It must not be thought, however, that this great output of oil, if ever reached, could be maintained for more than a few years. Nor must it be supposed that even a large part of all this oil would be applied in the displacement of coal; for the uses to which oil is put are legion, and it is only a small fraction of the total output which goes as fuel for ships, locomotives and stationary engines which would

otherwise burn coal.[1] The following figures as to
the output of oil produced in the shale oil works
of Scotland are of distinct interest.

PRODUCE OF SHALE OIL WORKS, SCOTLAND, 1907.[2]

Output of crude oil : 39,423,000 gallons.

Manufactured products :	Quantity. galls.	Value. £
Lamp oils . . .	16,977,000	376,000
Heavy or lubricating oils	6,463,000	117,000
Gas oils . . .	12,259,000	160,000
Spirit (petrol) . .	4,496,000	140,000
	tons.	£
Paraffin wax . .	25,000	601,000
Sulphate of ammonia .	52,000	590,000
Coke	5,000	12,000
Lubricating greases and other products . .	..	18,000

It may be noticed that the total volume of products is greater than the amount of the crude oil

[1] The following interesting analysis of the oil imports of the
United Kingdom during 1913 is given in *Business Prospects*, 1915 :—

Oil Imports of the United Kingdom, 1913.

Variety.	Quantity in Gallons.	Value. £
Crude	1,109,800	14,780
Lamp Oil	157,141,250	2,679,500
Motor Spirit	100,982,750	3,806,450
Lubricating Oil . . .	67,974,550	2,472,850
Gas Oil	65,949,700	735,400
Fuel Oil	95,062,150	1,149,800
Other Varieties . . .	24,200	1,400
	488,244,400	10,860,180

[2] Census of Production for 1907, Final Report, 1912, p. 40.

distilled from the shale ; and this must be accounted
for by the fact that it is found useful to use a certain
amount of petroleum imported from abroad in
making some of the various kinds of oils sold by the
works.

Heavy oils used for power purposes are a comparatively small fraction of the whole output.
These, however, are the only oils which come
seriously into competition with coal for power
purposes. I shall not be far wrong in saying that
at least four-fifths of the petrol, and of the intermediate oils used for power purposes in stationary
engines or motor boats, are in no way displacing
coal, for much the greater part of motor transport
upon roads, the smaller kinds of motor launches,
and, of course, all aeroplanes, would have been
impossible without the use of petrol. This light oil
or spirit has, in fact, created a demand of its own ;
a demand, indeed, which is fast outstripping the
supply.

The result of the great and rapidly growing demand
for the lighter oils or motor spirit is that many
experiments are being made in processes of " cracking," as it is technically called, the heavier oils
into lighter oils of the petrol character, which is
found to be possible by subjecting heavy oil to
various high temperatures together with steam
during the process of distillation. Here we come
to a very interesting economic question of the
balance of supply and demand. Assuming, as is
practically certain now, the success upon a large

commercial scale of this operation of cracking the heavy oils, the demand for the light petrol oils will tend to affect the price of the heavier oils. The extent to which the latter are used for power purposes in displacement of coal will therefore depend upon a balance between the price of coal on the one hand, and the price of light oils upon the other hand. My own view is that as we have such vast supplies of coal, we shall find the tendency of commerce in general is to utilise oil for those purposes which coal cannot directly meet, and that the oil production of the world will, for the larger part, be divided into the heavy lubricating oils, greases and waxes, and into the light spirits and illuminating oil on the other hand, not more than about, say, 10 per cent. of the total production coming into competition with coal as a power producer.

It often happens that after a great increase of demand for a commodity the supply may over-reach itself, so that for a time the price is much reduced, until the increase of demand thereby induced gradually takes effect through new users installing the necessary plant. It may be that the rapid increase of demand now being experienced will lead to new oilfields being opened up everywhere, and to the construction of the necessary tank steamers for carrying the oil to the consuming markets.[1] There may be thus produced a comparative over-supply

---

[1] High freights through shortage of tonnage is one of the causes of the present rise of the price of oil in the English market. There is actually a temporary fall of price of the heavier oils in some of

and a lowering of the price of oil which would last for a few years. So long as it proved possible to open up new fields without greatly increased capital expenditure or working costs, the supply would go on increasing and the world's output might remain very large for 50 or 60 years. I do not say that this increase of output will ever reach ten times the present volume. That is the figure I have suggested above as being the outside limit to which the world's annual production is likely to increase. If it ever does increase to 120,000 million gallons of crude oil per annum, it could not be maintained for more than a very few years at that figure, and would probably soon begin to decline. For the purpose of showing the insignificance of the possible output of fuel oil in comparison with the world's output of coal, I may take the figure named as the probable maximum output of oil.

Of this annual output of crude oil it is very unlikely that more than 25 per cent. would be utilised for purposes under which it would be displacing coal, even if we assume an extensive adoption of the Diesel type of engine and that the demand for petrol is not so great as to lead to much cracking of the heavier oils. Assuming, then, this proportion of one-fourth, the corresponding maximum of coal which would at any time be displaced

the great producing countries ; for just now it pays much better to ship the petrol, which is worth more per unit volume because it is less easily substituted, and stands the freight better for that reason.

by oil would be one-fourth of 720 million tons. But it would be thirty years hence at least before the world's oil supply could be multiplied to ten times its present yearly volume ; and by that time the world's total output of coal for uses quite apart from those which could be served by oil, is likely to have doubled at least, for the world's total output of coal has actually doubled itself in the last sixteen years. In figures this means that by the time the oil used for fuel could become equivalent to 180 million tons of coal per annum, the coal output would have grown to about 2,600 million tons, equivalent to 2,780 million tons including the oil output. It appears, therefore, to be very unlikely that the total production of petroleum could ever reduce the world's output of coal by more than about $6\frac{1}{2}$ per cent. Coalfields which happen to be situated near great oilfields are likely, of course, to suffer more severely ; but that, fortunately, is not the case with Great Britain, where the coal lies near great markets and oil must be brought very long distances.

I do not wish the foregoing to be taken in any way as a prophecy on my part of the future of oil production. It is rather an attempt to give definiteness in figures to arguments which show conclusively the relative magnitude of the figures affecting coal and oil. This view as to the future of oil for fuel purposes is held by a number of authorities. I may refer readers to an inexpensive little book by Mr. Vivian Lewes, which deals in a popular manner with

the whole subject of the use of oil as fuel.[1]   He
expresses the view that the world's petroleum
supplies will certainly give out at no very distant
date ; and he proceeds to consider what substitutes
are possible, of which a more permanent supply
may be anticipated.   Though excellent motor
spirit may be made by the cracking of benzol, which
is derived from the distillation of coal, the purging
such spirits from sulphur, to free them from obnoxious
smell when burning, may continue, as now, to be
an expensive process.   The ideal motor spirit, he
contends, is alcohol, which would be obtained from
the distillation of wood refuse, as is being done
so widely in France, and in this country in the
Forest of Dean.   Such alcohol would be denatured
by the addition of 10 per cent. of benzol : and
thus there would be available an excellent motor
spirit of never-failing supply.   Mr. Lewes believes
that the heavier coal tar oils will be extensively
used and will serve admirably for fuel purposes in
ships, and wherever convenient, whether used in
engines of the Diesel type or in furnaces.   But here
we are thrown back upon coal again !   It still
remains our only great ultimate source of power.

[1] *Oil Fuel*, Arnold & Co., 1913.

CHAPTER XXVI

THE COAL QUESTION

NEARLY half a century has passed since the late Professor Jevons published his well-known book on the Coal Question. Writing of the national welfare in relation to the supply of coal, he stated : " We cannot long maintain our present rate of increase of consumption," and that " the check to our progress must become perceptible considerably within a century from the present time." His contention was much misunderstood then, however ; and it is so even at the present day. He did not mean that our coal seams would be worked out by 1965—in fact, he did not dispute the conclusion of the Royal Commission on Coal Supplies that our deposits of coal could very well last for more than 300 years.

The crux of the question, as this author pointed out, is the price of coal, which is mainly dependent upon the cost of mining it ; and the fact that there is a natural tendency to mine first the coal which is cheapest to work. The cost of mining increases with the depth of coal from the surface, and some seams are for various reasons more expensive to work than others. Hence pits have been sunk to

718

work just those seams which were thought cheapest to work, where they were nearest to the surface, or otherwise easiest to get at. Many collieries have failed because the coal when reached proved to be too expensive to work to be profitable. Fifteen or twenty years later the price of coal has risen sufficiently to make it worth while in several cases to re-open the pit. In Staffordshire and in Durham and Northumberland the best seams are already practically worked out.

A further cause of increased cost of working lies in the natural progression of the working face away from the shaft. When the mining of any seam was commenced the coal was got near the bottom of the shaft; but the cost of mining that seam now has been increased by the expense of hauling the coal more than a mile underground from the face to the shaft bottom; and this is a cause operating continually in all collieries.

The real coal question turns, therefore, upon the working out of the easily getable coal; because the increase of the price of coal in this country relatively to the price in other countries which would result must seriously prejudice British industries. Until late in the eighteenth century England was not a great industrial and commercial nation. She bought from, and learned of, foreigners. It was the discovery of how to use coal, more especially the success at length attained in smelting iron with coal instead of charcoal, which launched England on her career of greatness. With cheap iron, machines and engines could be cheaply built, and here was the coal alongside to run them.   There followed a prodigious growth in the production of iron, and later of steel ; and of every kind of goods which could be made by machinery driven by steam power—at first chiefly textiles.   Our great industrial and commercial expansion in the nineteenth century was largely the result of the abundant supplies of cheaply obtainable coal with which Nature had endowed us.

The building of iron ships enormously increased the mercantile and naval supremacy of England. A great trade in the export of coal also developed ; and it is important to note that this was much assisted by the low freights ruling from England, owing to the convenience of coal as ballast to make up weight with the light finished goods exported. England has for long imported bulky raw materials for her manufactures, and exported finished textiles and other materials occupying much less of the ship's space than the raw material composing them. Coal has been for the past 150 years chiefly carried from our shores from this cause, as shipowners offered low freight-rates for cargo to make additional weight.

For three-quarters of a century the natural advantages of our country, coupled with the political troubles of our neighbours on the Continent, kept us free from effective competition, so that our wealth became the envy of all the world.   In 1865, however, the late Professor Jevons, knowing the marvellous natural resources of the United States and the enterprising character of its people, prophesied that before many years had passed we should find that country a serious competitor in every branch of the iron and steel industry, and later in many other industries. As everybody knows, this prophecy has been fully justified by events. In Germany, and particularly in the United States, in Canada, and in China, there are vast deposits of coal lying so near the surface that they can be mined at very low cost. The inevitable result is that industries requiring much coal, such as the smelting of iron and manufacture of steel, can be carried on there more cheaply than here. In 1862 we turned out five times as much pig-iron as the United States ; but now their output is nearly three times as great as ours, and it may be only a question of time for us to lose practically the whole of the heavy steel trade. The inference is that with the increasing cost of coal we shall gradually lose all trades closely dependent upon cheap fuel ; and we shall be obliged to specialise on the lighter and more intricate forms of manu- facture, in which, however, we have no special advantage.

The author of *The Coal Question* showed very clearly that none of the supposed substitutes for coal, such as water-power, the tides, air, or direct use of sunbeams, could be made a commercial substitute for coal ; and that, if they could be, we as a nation should have no natural advantages, but rather relative disadvantages, as compared with

2 z

other nations in regard to their supply.   He demon-
strated clearly also that economy in the use of coal,
such as has been since achieved, whereby one ton will
do the work done by two tons fifty years ago, is no
means of delaying the exhaustion of our coal, for
the possibility of cheaper power, due to the more
efficient use of coal, leads to the constant application
of power to new uses, and an evergrowing demand
for coal, which is added to the increased demand due
to growth of population.

The Several Coal Questions

The foregoing is the thesis of the late Professor
Jevons' book *The Coal Question*.   It will be observed
that the question which he raised is one of
national economics—it is a question of the industry
and commerce of this country in its relation to the
industry and trade of other countries.   He was not
concerned as to the date or effects of the ultimate
exhaustion of our coalfields ; because he knew that
that was a very remote event, whilst the loss of our
relative advantage of cheap coal, which England
was bound to lose, was an event to be expected in a
few generations.

In reality there are several coal questions, depending upon the point of view from which one is considering the exhaustion of reserves.   The coal
question, as it presents itself to the public is : How
long will our British stock of coal last ?   This,
indeed, is what the Royal Commission of 1871,
appointed as the result of Professor Jevons' book,

devoted itself to. It shirked the issue he raised ;
and studied that which appealed to the popular
mind. We have then two quite different coal
questions in this country.

There are, however, two much greater coal
questions. Firstly, as to what will be the result
of the partial exhaustion of coal supplies of the
whole world, and secondly, as to when their practically complete exhaustion will occur, and what
will be its effect upon the human race. These are
academic questions compared with the question of
partial exhaustion as it affects British national
economy ; but yet these world-wide questions have
a speculative interest, and it is worth while to make
some effort to forecast the progress of events, if only
to show how closely the world's trade of the future
must link the nations together. In the next chapter
I shall deal with the world's coal resources, and the
probable rate of their exhaustion.

It has long been recognised that the quantity of
coal available to be mined in this country is limited ;
and that in mining annually many millions of tons
we are drawing on a reserve of natural wealth which
it is impossible for us to replace—in other words
that we are in a national sense living upon our
capital. So long ago as 1789, John Williams, a
mineral surveyor, included in his *Natural History
of the Mineral Kingdom,* a chapter on " The Limited
Quantity of Coal in Britain " ; and from time to
time in the first half of the nineteenth century
various authors have alluded to the possible exhaus-

tion of our coal, and have hazarded estimates of the duration of the Durham coalfield. It was not, however, until the publication of *The Coal Question*, in 1865, that a measure of public interest and concern in the matter was aroused. The result was the first systematic examination of our coalfields with a view to estimating the quantity of coal remaining unworked in the strata of this country which had ever been made, besides an extensive inquiry into the probable future consumption and production of coal and the possible methods of economy in its use. The geological survey of the United Kingdom had been at work mapping out the whole country, and particularly the coalfields, since 1832; but beyond defining the coalfields, indicating where coals cropped out, and the dip of the strata, and collecting some information from collieries, they had not done anything towards making an estimate of quantity. The result of the labours of the Commission was to show that up to 1869 there had then probably been mined in this country about 4,300 million tons of coal, and that reserves remaining were probably not less than 146,480 million tons. They also estimated that the consumption of coal would greatly increase; but their estimate of the future export trade of coal was ridiculously below the mark.

In the early years of the twentieth century it was felt that so much exploration of our coalfields had been accomplished in the past thirty years, such great economies had been effected in coal consumption, and the export trade in coal was growing so rapidly, that there was a reason for the appointment of a new Commission to make a similar inquiry into all the new evidence. This Commission was appointed in 1902 under the chairmanship of Lord Allerton ; and it issued a number of interim reports and a final report with several volumes of evidence in 1905. These contained not only an elaborate and detailed estimate of the reserves of coal still existing in this country, but also a great mass of evidence on many matters connected with the British Coal Trade. This Commission's report is still the standard authority on matters not subsequently dealt with by special authors.

The estimate of the actual duration of our coal supplies was not attempted by the Allerton Commission, because the estimates of future coal consumption by the late Professor Jevons, and of future exports by the Royal Commission of 1871, had proved quite unreliable. The Allerton Commission indicated, however, the factors which must be taken account of to make any estimate of duration ; and their report will be frequently referred to in this chapter. A very useful revision of the estimates of coal reserves made by the latter Commission was carried out two years ago by Mr. Aubrey Strahan, for a general report on the World's Coal Resources.[1]

See *post*, pp. 735 and 772.

Reserves of Coal

We may now proceed to an examination of the estimates which have been made of the reserves of coal likely to be available for commercial purposes in these islands.   We shall then turn to the question of the growth of home consumption and exports with a view to making some numerical forecast of that growth which will indicate in a general way when the maximum annual output is likely to be attained and when the output is likely to begin falling off, and the price of coal to rise considerably, through exhaustion of many important mines.

The first detailed estimate of the coal reserves of this country was made by the Royal Commission of 1871, and the results of their investigation were published in three large volumes.   They found that there was more coal available to be worked than had been generally anticipated, although they only took into account the concealed coalfields so far as there was good evidence for the existence of coal.

The Commission did not merely estimate how much coal was then remaining in the ground.   Their estimate had to be that of the " available " coal, by which is meant all such coal as can be brought out of the pit at less than prohibitive cost, and is not commercially useless through containing too much dirt.   In most coalfields the " available " coal is probably only about half of all the coal that exists in the strata.   Much of it may be in very thin seams ; and in seams of workable thickness, much coal may be lost for a variety of reasons, such as faults and other natural causes, besides the coal which has to be left underground as barriers between mines, or for the support of the pit shaft and surface works, and under towns or important buildings.

There were three important factors to be ascertained before the amount of coal available in each coalfield could be estimated, namely :—

(1) The maximum depth at which it might be expected that it would be possible to work coal without the cost being prohibitive.

(2) The least thickness of seam which could be profitably worked.

(3) The percentage of deduction to be made for the various causes of waste in working, and other losses.

The depth of working increases the cost for several reasons.   Not only is the shaft more expensive to start with, but to wind the same output per day far more powerful machinery is required.   The pressure of the strata at great depths is so enormous that the roof requires more support, and the floors of the roads tend to squeeze up, costing much more in repairs.   The coal is also crushed, which decreases its value owing to the larger percentage of small ; but there is a compensation in the fact that the coal is worked much more easily.   A further serious difficulty is the high temperature of the rocks, which warms the air to 90° or 100°, and renders manual labour most exhausting to Europeans. The rate of increase of temperature in a boring or mine as we descend towards the interior of the earth is different in different places. At 50 feet below the surface the effect of the seasons is lost and the temperature remains constant at 50° Fahr. in this country. The average rate of increase for the first 3,000 feet is 1° Fahr. for every 60 feet of depth below 50 feet, so that at 3,050 feet the temperature of the rock would be about 100° Fahr. The rate of increase falls off slightly, however, with increasing depth, so that at 4,000 feet the temperature might not be more than about 115° Fahr.

After going into the matter with great care the Commission decided to regard 4,000 feet below the surface as the limit of practicable working. The Royal Commission of 1904 re-examined the question and adopted the same figure. Some very interesting evidence was given as to conditions of working in deep mines. At Pendleton, in Lancashire, is a colliery where the workings slope downwards from the bottom of the shaft and have been carried to a depth of 3,500 feet. In some of the working places the temperature is 94°, but fortunately it is a dry heat. As the witness, Mr. H. Bramall, said: "The men work in that drier atmosphere with less fatigue at a temperature of 94° than they do in the other mine when the temperature never gets above 85°." A Belgian mine, the " Produits " Colliery at Flénu, near Mons, Belgium, has a shaft 3,773 feet deep.

The rock temperature at the bottom is 113° Fahr. and the highest air temperature found in the workings whilst ventilated is 104°.[1] Some of the mining engineers who gave evidence before the 1904 Commission were of the opinion that the high temperature at great depths could best be overcome by ventilating with brisk currents of air carried straight to the working places, instead of by rather circuitous routes as is the usual and cheapest method. Experts are unanimously of opinion that artificial cooling of the enormous volume of air required to ventilate a colliery (like a London tube) would be so costly that it need not be considered, unless the value of the coal be quite extraordinary. Belgian engineers propose to extend their workings experimentally to a depth of 4,500 feet, and think even 5,000 feet practicable ; but I believe it remains to be shown that this can be done at a profit without a great rise in the price of coal. The practicability of very deep workings is amply proved by experience in metalliferous mines. In Michigan (U.S.A.), near Lake Superior, the copper mines have been carried to a great depth, as, for example, the Tamaroc mine, which is 6,000 feet deep, whilst the Hecla mine at Kalumet is nearly 5,400 feet deep. Artificial cooling is used in some cases ; or else the men work short shifts. It is entirely a question of cost, and for all practical purposes we shall be suffering from the effects of the exhaustion of our coal before

---

[1] Royal Commission on the Coal Supplies. Final Report. Part X (Cd. 2362), 1905, p. 354.

it becomes valuable enough to work at such great depths. At the same time it is quite probable that it may ultimately prove profitable, where collieries are working inclined seams, to continue the workings downwards from the shaft bottom, although the depth of 4,000 feet may be thereby exceeded. Such instances are not likely to be important, however, in the amount of coal so rendered accessible at reasonable cost ; and for the purpose of estimating the available coal reserves, the limit of 4,000 feet adopted by the Royal Commission of 1871 may still be regarded as holding good.

In regard to the least thickness of seam which it is profitable to work, it is to be observed : (1) that where, in the absence of disturbed strata, coal-cutting machines can be used, the cost of working thin seams is greatly reduced ; (2) the thickness of the seams is only one of many factors determining the cost of working, and it is the value of the coal produced relatively to the total cost of working it which determines whether it will be worked ; (3) thin seams occurring close together can often be worked in conjunction with one another, or thin coal occurring with a bed of ironstone or fireclay can be profitably worked with it. Although it is only in exceptional cases that it is profitable at the present time to work coal in seams less than 18 inches in thickness, there is no doubt that a permanent rise of 4s. or 5s. per ton in the price of coal relatively to the cost of labour and materials would render it profitable to work a number of thinner seams down to 12 inches in thickness. One foot was in fact taken as the limit of workable thinness by both the Royal Commissions, and they excluded altogether all thinner seams from their estimate of available coal.

The percentage of deduction to be made for loss in working and on account of barriers, pillars, and other causes for which good coal must be left underground, was variously estimated for the different coalfields, or parts of coalfields, according to the local conditions of the coal measures and of the surface, and according to local mining customs. The Commission of 1871 appears to have made deductions for all causes varying from 10 per cent. to as much as 40 per cent. of the total coal in seams over 12 inches thick, the average deduction being apparently about 30 per cent. The 1904 Commission, finding that considerable economies had been effected in recent years, both in actual working and as regards the practice of leaving barriers, and in unwatering drowned sections of mines, felt justified in allowing its district commissioners to use lower percentage deductions, ranging from about 15 to 30 per cent. in the different coalfields, and averaging 22·7 per cent of the coal in fields for which details are given.

With this introduction, the figures given by the two Commissions as their estimate of the reserves in the United Kingdom will be intelligible. The first Commission divided the coalfields of the country into the *proved* and the *unproved* coalfields and portions of fields.   In the former were included all areas in which workable coal was definitely known to occur as proved by actual outcrop, or by mining operations, or by boring.   In the unproved coal-fields were included all areas where the presence of commercially workable measures beneath newer formations might be inferred with reasonable probability from geological evidence.   The estimates were separately made for these two classes of fields with the following result :—

ESTIMATE OF ROYAL COMMISSION, 1871

Depth.	Proved Coalfields.	Unproved Coalfields.	Total.
	Millions of tons.	*Millions of tons.*	*Millions of tons.*
Not exceeding 4,000 ft. .	90,207	56,273	146,480
From 4,000 ft. to 6,000 ft. .	5,922	29,341	35,263
From 6,000 ft. to 10,000 ft. .	1,397	15,303	16,700
Totals . .	97,526	100,917	198,443

The important figures are those contained in the first line, practically all of this 146,480,000,000 tons of coal being assumed to be existent (there being good reasons for this assumption), and being within reachable depth.   The Royal Commission of 1871 proceeded to estimate the probable future growth of the annual output of coal from the mines of this country by a rather crude method, which under-estimated it, and showed that on these assumptions this store of 146,480 million tons of available coal would last 360 years. Further on I shall show that this was an over-estimate of the duration of this quantity ; but first we must see that later estimates have shown that the available British reserves of coal are larger than could be discovered in 1871.

The Royal Commission of 1904 was able, owing to the progress of new sinkings and further boring, to denote as *proved* many of the coalfields or parts of fields which were necessarily classed as *unproved* in 1871 ; and certain new or extended areas were included amongst the unproved coalfields, with the following results :—

ESTIMATE OF ROYAL COMMISSION, 1904

Depth.	Proved Coalfields.	Unproved Coalfields.	Total.
	Millions of tons.	*Millions of tons.*	*Millions of tons.*
Not exceeding 4,000 ft. .	100,914	40,721	141,635
From 4,000 to 10,000 ft. .	5,239[1]	..	5,239
Totals . .	106,153	40,721	146,874

[1] Not including a very large amount of coal which is assumed to be under the Cheshire Basin (see Final Report of Royal Commission, Part I, p. 26).

The amount of coal raised in the United Kingdom from January 1, 1870, to the end of 1903 was 5,695 million tons. If we add this to the 141,635 million tons estimated to be the reserve in 1904, we have 147,330 million tons as the reserve which existed in 1870, according to the estimate of the later Commission. It agrees remarkably closely with the figure 146,480 million tons found by the 1871 Commission.

This close agreement is due to the well-known law that a number of unbiassed errors tend to cancel one another; for the estimates of the two Commissions for the different coalfields differ widely in a number of cases. In several cases the second Commission found the reserves somewhat smaller than had been estimated, after allowing for coal raised in the meantime, but in a few coalfields (notably Scotland, Cumberland and the East Midlands) they were able considerably to increase the estimate.

The 1904 Commission were particularly anxious to publish only figures for which they had good evidence; and this has led to their omitting altogether much coal in unproved fields, and much more coal lying at depths below 4,000 feet. In the first place they omitted altogether the Kent coalfield, now known to be very rich, as well as probable concealed extensions of the Gloucester and Somerset coalfields. A very large amount of coal which may be assumed to underlie the Cheshire basin at depths exceeding 4,000 feet was also omitted.

Another authoritative estimate of the coal reserves of these islands has been made subsequently by Mr. Aubrey Strahan, Assistant Director of the Geological Survey of England and Wales, for a very remarkable Report in three volumes on the coal resources of the world, compiled by the Geological Survey of Canada for the Twelfth International Geological Congress, held at Toronto in 1913. Mr. Strahan based his estimate (made apparently early in 1912) on the report of the Royal Commission of 1904, modifying their estimates in accordance with a considerable accumulation of fresh data obtained in the following eight years, during which the exploration of concealed coalfields by boring had been rapidly proceeding. Mr. Strahan, for example, included 2,000,000,000 tons in the Kent coalfield, and estimated a much larger quantity present in the concealed coalfields of North and South Staffordshire, and of Cumberland and Westmoreland, than did the 1904 Commission. The classification adopted for this International Report differed somewhat from that of our Royal Commissions, the reporters of each country being requested to divide their estimates amongst *Actual*, *Probable* and *Possible* reserves. The actual reserve coincides closely with the definition of *proved* coalfields ; and the probable and possible reserves together may be taken to include the Royal Commission's class of *unproved* coalfields ; though it is to be noted that amongst *possible* reserves are included some coal measures whose existence or coal-bearing value is doubtful, and which were omitted altogether by the second Royal Commission. On the other hand, Mr. Strahan appears to have been conservative in his estimates, and has, in some instances, reduced the figures adopted by the 1904 Commission. His figures may be summarised as follows :—

ESTIMATE OF A. STRAHAN FROM " WORLD'S COAL RESOURCES," 1912

Depth.	Actual Reserves.	Probable and Possible Reserves.	Total.
	Millions of tons.	*Millions of tons.*	*Millions of tons.*
Not exceeding 4,000 feet .	133,117	45,610	178,727
From 4,000 to 6,000 feet .	6,108	1,659	7,767
Totals . .	139,225	47,269	186,494

In comparing these figures with those of the second Commission it will be seen that the resources in proved coalfields (actual reserves) largely exceed those of the 1904 Commission, which is due to transference of several areas from the unproved to the proved, and to giving some of the latter a greater thickness of workable coal. Particularly important in this respect is the increased total for the East Midland field,[1] which Mr. Strahan takes at nearly

[1] Yorkshire, Derbyshire, Nottinghamshire, and part of Lincolnshire.

40,000 million tons in the proved coalfield, as against
26,500 million tons estimated by Mr. Currer Briggs
for the 1904 Commission, with, I think, much too
liberal deductions and a smaller area. Mr. Strahan
has allowed 2,000 million tons for Kent in the
probable reserves.

Even Mr. Strahan's estimate does not serve to
give a complete picture of the coal which will probably ultimately be found to be available in this
country, as he is probably too conservative in his
estimates as regards Kent and some other concealed coalfields, and he attempts no estimate
of the possible large reserves of coal at from 4,000
to 6,000 feet depth in South Staffordshire, Warwickshire, the Cheshire basin and Solway Firth. The
only estimates I have found of these possible large
reserves are those made by the 1871 Commission, and
included in the table on page 732 of this chapter,
and which I adopt for addition to Mr. Strahan's
figures. In a previous chapter I have shown that
the reserve of the Kent coalfield may be put at
7,600 million tons more than Mr. Strahan takes it at.
Study of a recent memoir of the Geological Survey[1]
dealing with the concealed coalfield of Yorkshire
and Nottinghamshire has convinced me that Mr.
Strahan has unnecessarily contracted the boundary
of this unproved field, which he takes in every direction well within the limit assumed by the 1901
Commission. There appears to be good reason for

---

[1] Memoirs of the Geol. Survey : *The Concealed Coalfield of
Yorkshire and Nottinghamshire*, by Walcot Gibson, D.Sc. 1913.

3 A

contracting the eastern boundary ; but on the other hand there appears to be equally good reason for extending the southern boundary nearly to the limits preferred by Professor Kendall. Upon this ground I add 11,000 million tons to Mr. Strahan's figure. There are still several possible concealed extensions of coalfields not included in Mr. Strahan's figures, particularly in the South-West and North of England, and in Scotland, to say nothing of the possibility of coal being found under Oxfordshire and Buckinghamshire. Here we depart from the scope of numerical estimates to pure speculation, or judgment of probabilities. Seeing that the possible total deposits of coal in these unexplored areas would be at least 40,000 million tons, I do not think I can be going far wrong in assuming that at least 15,000 million tons does exist in these or other undiscovered or unproved fields not taken account of by Mr. Strahan. Upon this basis of supposition I have therefore added to Mr. Strahan's estimate the above-mentioned figures so as to get a kind of picture of what the total reserves of coal in these islands in seams of one foot and over in thickness, and at depths not exceeding 6,000 feet, probably are. The Royal Commissions and Mr. Strahan have tried to be well on the safe side. I am now trying to be as near erring in excess of the truth as in falling short of it ; and it is with this explanation that I give the following figures of the total coal reserves which may be available for mining in these islands :—

PRESENT AUTHOR'S ESTIMATE OF PROBABLE TOTAL
    RESERVES IN UNITED KINGDOM AT LESS
    THAN 6,000 FEET DEPTH

Depth.	Actual Reserves.	Probable and Possible Reserves.	Total.
	Millions of tons.	*Millions of tons.*	*Millions of tons.*
Not exceeding **4,000** feet .	**136,000**	**61,000**	**197,000**
From 4,000 to 6,000 ft. .	9,000	14,000	23,000
Probable undiscovered coal .	..	15,000	15,000
Totals . .	145,000	90,000	235,000

The foregoing survey clearly proves that there is
still a vast store of coal to be mined in this country,
and it is interesting to note that from the beginning
of mining operations in this country to the present
date we have consumed but a small fraction of our
original store. The Royal Commission of 1871
made a careful estimate of the probable amount of
coal mined in this country prior to 1854 when the
official record of mining statistics commenced.
Using their figures,[1] and carrying the figures of total

[1] I find no definite figure given for the years 1851-3, and have
estimated the output of these years from the data given by the 1871
Commission.

output up to the end of 1913 we have the following sum :—

				Total Output. Tons.
Prior to 1800	.	.	.	850,000,000
1800 to 1850	.	.	.	2,000,000,000
1851 to 1853	.	.	.	153,000,000
1854 to 1913[1]	.	.	.	9,574,133,000
Grand total	.	.	.	12,577,133,000

It will be seen that during the last sixty years the total amount of coal raised far exceeds all that was ever mined before in this country. Adding the grand total of coal mined to the present date to the 197,000 million tons which I have estimated as available at less than 4,000 feet depth we get the estimate of the total quantity of coal originally deposited in the strata of these islands, and lying at less than 4,000 feet depth, namely, 209,577 million tons. The grand total mined to the present date is indeed only just 6 per cent. of 209,577 million tons.

It is evident that in centuries of mining activity we have done little more than nibble at the surface of our great stores of coal ; and it is perhaps a comforting reflection that we have grown a wealthy nation by using so little of the whole. At the same time it must be remembered that we have been

[1] In 1912 for the first time the Home Office issued the output statistics as coal free of dirt, instead of stating the weight of coal including dirt as was always done previously. The percentage of dirt in 1912 was 0·871 per cent., and this amount I have added for 1912 and 1913.

using the cream of our stock.   The warning which my father sought to convey in his book, *The Coal Question*, was that the nation would suffer in its industry and commerce by the exhaustion of this "cream," namely, the most valuable and cheaply workable coal.   Before we follow out the results of partial exhaustion, however, it will be best to try to make some forecast of the future growth of our consumption and output of coal, which will give us an idea of the ultimate duration of our supplies.

CHAPTER XXVII

Growth of Home Consumption

THE duration of our reserves of coal will depend upon the rate at which we draw upon them—that is to say, upon the volume of annual production in future years. To form even the most approximate estimate of duration, it is necessary to make a forecast of how long and how rapidly the annual production will continue to increase, before becoming stationary and finally declining.

The total output of coal is disposed of in two principal markets : the home country and foreign countries respectively, including in the latter coal supplied to bunkers of foreign-going ships. Hence an estimate of future output must involve a forecast of the probable growth of demand both for home consumption and for export.

In estimating the probable future home consumption, there are two factors to determine, namely, the growth of population and the amount of coal likely to be consumed per head of the population. Let us examine the question of population first.

A simple method of estimating the probable

future population is to take the percentage of increase in each decade over a long period as shown by the census figures, and to assume that this percentage rate of increase will be maintained. Going back to the early part of the nineteenth century we find, however, that a continuance at the same percentage rate of increase which prevailed during the first half of that century would have produced a very much greater population than we now have. There is a continued, though not steady, falling off in the percentage rate of increase, and this it is convenient to call a *decrement* of the rate of increase.

The probable future growth of population in Great Britain was estimated by Mr. Price-Williams, an expert on the subject, for the Royal Commission of 1871 ; and he was invited again to make another estimate for the 1904 Commission in the light of the actual growth of population as revealed by the census up to 1901.[1] In the following table I quote Mr. Price-Williams' estimates for both Commissions, together with the actual census figures.

It will be observed that the estimate of the Royal Commission of 1871 has proved too low ; and Mr. Price-Williams explained this by the discovery of an omission in the preliminary figures of the 1871 Census supplied to the Commission, which reduced the apparent below the then actual rate of increase of the population.

[1] Royal Commission on Coal Supplies (1871), General Report, Vol. I, pages 14 *et seq.* Final Report of Royal Commission on Coal Supplies, 1904. Appendix V.

ESTIMATES OF FUTURE POPULATION

Year.	Census.	Royal Commission 1871.	Price-Williams for 1904 Commission.	Present Author (1914).
1871 .	26,25	26,06	..	..
1881 .	29,71	28,94	..	..
1891 .	33,03	31,96	..	..
1901 .	37,00	35,09	..	..
1911 .	40,83	38,33	41,18	..
1921 .	..	41,56	45,56	44,77
1931 .	..	44,86	50,12	48,76
1941 .	..	48,32	54,85	52,73
1951 .	..	51,82	59,71	56,62
2001 .	..	69,62	85,33	72,96
2051 .	..	86,58	111,32	81,26
2101 .	..	102,01	135,59	83,34
2151 .	..	115,25	156,94	84,09
2201 .	..	126,20	174,98	84,00

The estimate for the second Commission was based upon the average decrement of the previous eight decades, which he found to be 5·917 per cent., and he applied this to the percentage of increase of the population in the period 1891–1901, which was 12·02 per cent. It is distinctly apparent already that this basis is an over-estimate. Both emigration and a falling birth-rate are reducing the rate of increase of our population; and Mr. Williams' estimate is no doubt very wide of the mark after 100 years hence.

At the same time there is almost certain to be a continuous and considerable increase of the population of Great Britain during the next 100 years. The difficulty is to find a satisfactory numerical basis for representing what may be considered probable. Guided by the experience of Mr. Williams' estimate, I have attempted my own forecast based upon the experience of the past thirty years, but with the addition of an arbitrary gradual increase of the decrement to represent my estimate of the probable results of all the tendencies reacting on the growth of population. My estimate is given in the fifth column of the above table, and in greater detail in the table on page 752.[1]

Perhaps some of my readers will be alarmed at the probable population (70 millions) of these islands in 2001, only eighty-seven years hence, and may think the estimate extravagant. Doubtless there are many factors, such as the growth of education and of a higher standard of living, as well as emigration, tending to reduce the rate of increase ; but I think full allowance has been made for them. On the other hand, the reduction of infant mortality, and better housing, besides the progress of medical practice, all tend to lengthen life.

---

[1] The estimate has been made in the following manner. The average decrement of the rate of increase during the past 30 years (7·1 per cent.) has been taken as 7 per cent., and it has been increased each decade by 10 per cent. This is a purely arbitrary figure, chosen because it seems to represent with a considerable degree of probability the resultant of many forces now bearing on the growth of population.

It is to be observed that the country will not be unduly crowded in the year 2111, or even much later. In 1911 Lancashire had 4,767,832 inhabitants, being 3·92 to the acre. But the population of Great Britain, which in 1911 averaged 0·71 to the acre, will, in two hundred years later, by my estimate, be only 1·48 per acre. The increase probably will be mainly in our coalfields, at sea-ports, and in the two metropolitan areas of London and Manchester ; and the towns will probably spread out in a series of garden cities and garden suburbs wherever transit facilities are provided.

Turning now to consider the probable future consumption of coal within this country per head of population (which for shortness is termed " consumption per head "), it is instructive to see how continuously it has grown during the past sixty years as shown by the following table.[1]

The increase during 1912–13 is to some extent misleading, as these were years of good trade, when the consumption always rises. The safest basis is to take only the decennial averages. As to the future change of the consumption per head we may note that the economies of the use of coal which have been effected for steam boilers in stationary plants, in railway locomotives and in gas and steel making

---

[1] I am here using Mr. Price-Williams' basis, which is not the true consumption per head of the United Kingdom or of Great Britain, but is the total home consumption of the United Kingdom divided by the population of Great Britain. Mr. Williams doubtless found it impossible to estimate the future population of Ireland, or to get statistics of the home consumption of Great Britain only.

AVERAGE HOME CONSUMPTION OF COAL

Decade.	Average Home Consumption.	Average Coal Consumption per head.
	1000's of tons.	*tons.*
1855–60 . . .	62,385	2·775
1861–70 . . .	88,095	3·579
1871–80 . . .	112,386	4·047
1881–90 . . .	132,793	4·260
1891–1900 . . .	149,657	4·303
1901–10 . . .	172,108	4·423
1911	184,852	4·527

have been paralleled by the economies of slow combustion grates, gas stoves, etc., in household use. There is still great room for these economical appliances to be more widely used; but, on the other hand, with the increasing spending power of the working classes there will probably be an increase of their consumption. Hence one cannot assume any important decrease of consumption per head of population; and there may even be an increase. The change either way will probably be slight; hence the simplest course is to assume that the consumption per head will continue increasing slowly for forty years, till it reaches 4·8 tons per head per annum, and then for a long while remain constant, so that the total home consumption will afterwards increase proportionally with the growth of population.

Mr. Price-Williams has, in one of his statements before the 1904 Commission, assumed that the consumption per head will increase until the year 2001, and thereafter remain fixed at 4·82 tons per head per annum : but in a second statement of his before the same Commission the consumption per head increases to 5·06 tons per head in the decade 1941–50, and thereafter falls off continuously and rather rapidly. This is explained by the fact that he has made on an independent basis a very conservative estimate of the future total annual home consumption ; and he has proceeded to divide it by his estimate of future population, which is doubtless greatly exaggerated after the next fifty years.

Mr. Williams' independent estimate of the future growth of home consumption is based upon the average decrement in the recorded rate of increase of home consumption during the period 1871–1900. Personally, I think that the falling off in the rate of increase of production during the nineties of last century was abnormal and due to the exceptionally depressed conditions of trade with very low prices ; and that Mr. Williams has therefore taken somewhat too large a rate of average decrement. This would not make much difference at present, but brings his figures probably much too low fifty to sixty years hence and thereafter.

My own estimate of the probable future home consumption is made by multiplying the consumption per head (rising to 4·8) into the estimated population, the resulting figures being as shown in the table below (page 752).

Probable Growth of Export Trade

The future growth of our exports of coal is more difficult to predict than the growth of home consumption, for it depends on the growth of foreign populations, and their use of coal for power and heat, upon the cost of ocean transport, and upon the development of competing foreign coalfields. My own view is that our exports are likely to go on increasing in the future at nearly the same percentage rate of increase as they have grown at during the past few decades. Let me state my reasons.

On the whole, I think it is likely that ocean freight-rates will continue to decrease relatively to railway rates, which will tend to increase the migration, so manifest in recent years, of the heavier manufacturing industries to the port towns. The trade across the seas between the ports of different nations will tend to grow in volume relatively to inland trade ; and this growth will be enormously accentuated by the gradual adoption of free-trade by all the great nations of the world, which is likely to come about, I believe, within the next thirty or forty years.

Once universal free-trade is established our seaports will be situated, as it were, on the shores of a great lake. So considerable is the advantage of water carriage for long distances that the world's merchandise will be carried from continent to continent across this great lake of the world's oceans. The advantage for manufacturing will then lie with ports at the mouths of great rivers, and other coastal points well situated to serve an existing dense population. There are many such points around the Mediterranean and North European coasts, and some such points on the Asiatic, African, and North and South American Atlantic coasts. To these strategic points the raw material will come in ships from distant lands, and coal also will be cheaply brought ; and to a very great extent it will be British coal. It might be thought that it would be cheaper for the raw material, once on board ship, to be brought to British ports, there to meet the coal ; and in industries requiring much coal for heat as well as power, and only unskilled labour, this will be the case. In a whole series of manufacturing industries, however, I believe that the cost of skilled or semi-skilled labour will be the decisive factor, together with proximity to big local markets near these foreign ports.

The conditions I imagine likely to exist a century hence will practically unite all nations into one in an economic sense ; and our coalfields of East Scotland, Northumberland and Durham, Yorkshire and Kent, will belong just as much to the coastal regions of the Continent as to us. Welsh coal will continue to have an advantage for far-distant ports where quality per bulk (*i.e.*, freight rate) becomes an important factor.

Giving due weight to all these probabilities I cannot

but think that our children's children will live to
see a vast export trade in coal—much more than
double the present figure, which is already approaching 100,000,000 tons per annum.

Our exports have already exceeded Mr. Williams'
estimate. I have calculated the figures for future
exports by the same method as he adopted, but
making due allowance for the recent more rapid
growth. This estimate of exports, together with
my estimate of future population and home consumption, are set out in some detail in the following
table.

In the last column is shown the total output, which
is the sum of the home consumption and the exports,
as the whole of the coal raised must be disposed of
either at home or abroad.

It will be seen, on examining the figures in this
table, that the growth of population is taken to be
very slow after about 150 years hence, so that the
home consumption becomes practically stationary ;
a bit later on it begins to fall off slightly through
the increase of prices necessitating considerable
economies. The export trade also is assumed to
cease growing about the same time ; although I
must admit having doubts whether it would not
probably continue growing whilst the home consumption was decreasing slightly, unless artificial
measures, such as an export tax, were taken to check
the trade, which is sure always to have continually
growing markets in some accessible parts of the
world.

FUTURE POPULATION, HOME CONSUMPTION, EXPORT AND TOTAL OUTPUT.

Year.	Population of Great Britain.	Consumption per head.	Annual Home Consumption.	Exports (including bunkers).	Total Annual Output.	Total Output to Date.
1911	40,83	4·43	180,9	87,1	268,0	900,[1]
21	44,77	4·52	202,5	125,0	327,5	4,175,[2]
31	48,76	4·6	224,0	172,2	396,2	8,137,
41	52,73	4·65	245,1	227,1	472,2	12,859,
1951	56,62	4·7	266,1	272,0	538,1	18,340,
61	60,38	4·75	286,8	314,0	600,8	24,248,
71	63,94	4·8	306,9	347,0	653,9	30,787,
81	67,25	,,	322,7	375,2	697,9	37,766,
91	70,27	,,	337,3	394,8	732,1	45,087,
2001	72,96	,,	347,0	411,0	758,0	52,667,
11	75,30	,,	361,4	423,0	784,4	60,511,
21	77,29	,,	371,0	435,0	806,0	68,571,
31	78,93	,,	378,7	443,5	822,2	76,793,
41	80,24	,,	385,0	456,0	841,0	85,203,
2051	81,26	,,	390,0	468,0	858,0	93,783,
61	82,03	,,	393,7	479,5	873,2	102,515,
71	82,58	,,	396,4	488,0	884,4	111,359,
81	82,95	,,	398,2	497,0	895,2	120,311,
91	83,19	4·76	395,0	505,0	901,0	129,321,
2101	83,34	4.73	394,0	508,0	902,0	138,341,
11	83,49	4·70	392,4	501,5	893,9	146,380,
21	83,64	4·68	391,3	488·0	879·3	155,173,
31	83,79	4·66	390,5	457,0	847,5	163,648,
41	83,94	4·64	389,1	413,0	802,1	171,669,
2151	84,09	4·61	387,7	355,0	742,7	179,096,
61	84,24	4·58	385,7	294,0	697,7	185,893,
71	84,39	4·55	384,0	208,0	592,0	191,813,
81	84,50	4·50	380,2	132,0	512,2	196,935,
91	84,50	4·45	376,0	72,0	448,0	201,415,
2201	84,00	4·40	371,0	38,0	409,0	205,505,

[1] Estimates for 1913 to 1915 inclusivé
[2] Estimate for 1916 to 1925 inclusive, and so on.

Duration of the Reserves of Coal

Having made an estimate of the future output of coal it only remains to sum up the total quantity likely to be mined each decade and century in order to see by what time the exhaustion of our reserves will be approaching. It would be, of course, absurd to imagine that a big output would be continued until suddenly the whole is exhausted. I cannot do better than quote part of the concluding paragraph of the Report of the Royal Commission on Coal in 1871 on this point.

" The absolute exhaustion of coal is a stage which will probably never be reached. In the natural order of events the best and most accessible coal is that which is first to be worked, and nearly all the coal which has hitherto been raised in this country has been taken from the most valuable seams, many of which have in consequence suffered great diminution. Vast deposits of excellent and easily available coal still remain, but a preference will continue to be given to the best and cheapest beds, and as we approach exhaustion the country will by slow degrees lose the advantageous position it now enjoys in regard to its coal supply. Much of the coal included in the returns could never be worked except under conditions of scarcity and high prices. A time must even be anticipated when it will be more economical to import part of our coal than to raise the whole of it from our residual coal beds ; and before complete exhaustion is reached, the importation of coal will become the rule, and not the exception, of our practice."

We may perhaps take it that the effects of exhaustion will be seriously felt, and the price of coal will have risen sufficiently to induce a considerable import trade from the United States and elsewhere, by the time that three-fourths of our whole store of coal has been mined. There will then remain probably but little coal in seams of more than 2 feet thick at a less depth than 3,000 feet, and many of the thicker seams will have been worked out to a depth of 4,500 feet or more. Hence the remaining coal must be mined only at a cost which is likely to reduce our home consumption per head and considerably reduce our exports except to the Continental coast, where Belgian, French, and German mines will probably have reached much the same condition.

If we take our total store of coal to be 235,000 million tons, as I have indicated is probable, we may ask ourselves when three-fourths of this, that is about 175,000 million tons, will have been used. In Table III, column 6 gives the total annual production in each decade from the present time onwards, according to my estimate, and in the last column is the total up to each date. It will be seen that 175,000 million tons will not have been consumed until the year 2150, over 235 years hence.

This date, when we shall have exhausted three-fourths of our reserve of coal, is very remote, and it may be doubted whether, if all the areas which are now believed to be concealed coalfields are proved to yield good coal in workable seams, their exhaustion may not proceed more rapidly than I have estimated. In other words, I cannot but regard this as the probable maximum of duration. It will be seen that I have estimated the total output of the country to reach in the year 2111 a total of over 900 millions per annum. This is not much more than three times the present output, which is nearly 300 millions, whereas the output of Great Britain has more than trebled itself in the last 50 years. It may be expected that if methods of boring to prove coal are perfected and cheapened, very wealthy companies will begin extensive exploration over possible coalfields and lease large areas. The modern tendency is to sink large pits with very big winding capacity, so as to work the coal in any taking as rapidly as possible. In another 50 years probably all the older pits, with small winding capacity, will have been worked out or rendered unprofitable, and the large new pits will be winding at full speed over a much greater area of coalfield than is now developed. The rapid growth of production from the new pits may much more than counterbalance the falling off from those worked out, so that the total annual production of the country 60 or 70 years hence may greatly exceed my estimated figure. If that is regarded as probable it must be inquired where all the coal raised is to go, as I cannot believe that the home consumption will increase much more than I have estimated. The reply is that if the supply is so enormously increased by the advent of large scale mining over new areas with huge aggregations of capital, the surplus output will find its way abroad as new markets will be opening up more rapidly there than here, and surplus coal will probably be sold abroad at a lower price than in the home market. If this development be regarded as probable my estimate of future exports, based on present experience, is altogether inadequate ; and it is manifest that by this means the time which will elapse before three-fourths of our store of coal is exhausted may be shortened by 50 or 60 years. It is quite possible, indeed, that we shall be suffering from this degree of exhaustion of our mines in less than 200 years.

Effects of Exhausting the Best Seams

As mentioned early in this chapter the question of when the almost complete exhaustion of our coal mines will take place is really subordinate to the far more important national question of what will be the result of exhausting the seams which are most easily workable, and those containing the best quality of coal. The grave warning uttered by my father when writing his book on *The Coal Question*, referred to this, and the handicap which our national industries would be placed under in competition with America and other countries which would still have large stores of very easily accessible coal.

It is now just 50 years since my father was

writing his book on *The Coal Question;* and the progress of this period has greatly added to our knowledge and altered many factors which have to be taken into account. Perhaps the most outstanding effect is the discovery of far greater deposits of coal than could then have been expected. In fact, as explorations are continued throughout the concealed coalfields, every new estimate increases the proved resources, so that during the last 50 years we have been discovering new coal deposits faster than we have been raising coal from all the coalfields of the country. Whilst it is true that in some parts of the concealed coalfields the coal is only reached at all at depths over 2,000 feet, it is also true that over large areas of the newly explored fields there are thick and valuable seams at depths which compare very favourably with large areas of the long-known, visible fields. There is also to be taken into account the very great economies which can and are being effected by large-scale working of mines. In manufacturing industries the so-called law of increasing returns from the growth of internal and external economies is well known, the most important factor in such economies being the greatly increased average size of the factories, and the combination or amalgamation of firms to secure economies in management. All these economies and all the most advanced business methods are now beginning to be increasingly applied in coal mining, and it is interesting to note that in the competitive struggle it is not the deep mines which are on the margin of profit-

ableness. When the slump of coal prices occurs
after a boom, it is the small, old collieries, usually
working pretty good seams, which are stopped,
often to be permanently closed. They are handicapped by their comparatively old-fashioned and
inefficient mechanical equipment, by their shafts
being too narrow, or by an original, uneconomical
lay-out of the main haulage roads. They even get
inferior labour, because they are more dangerous.
Most of the pits sunk 40 or 50 years ago were not
planned and equipped to deal with an output of
more than 500 tons per day ; and it is pretty obvious
that the overhead costs of all kinds, when reckoned
per ton of coal, will be much higher than for a modern
colliery equipped to wind 5,000 tons a day. It is
probable that pits will be sunk in the near future
large enough to raise 8,000 tons per day, or 15,000
tons per day if working double shift at the face of
coal. The necessary capital expenditure is, of
course, greatly increased, but not in so large a proportion as the output.

I want very clearly to show that increased depths
of working, and the working of poorer quality and
thinner seams, is only one of a number of factors
affecting the cost of mining and marketing the
coal ; because in his great work on " The Coal
Question " my father did not, I think, give sufficient
weight to these other factors. Besides the growing
practice of working upon a much larger scale, there
is the possibility I have just alluded to of working
double shift, or even three shifts. Working double

shift upon the face of coal has long been a practice in Durham and Northumberland, and since the Eight Hours Act came into force the three-shift system has been extensively introduced in Durham. In other coalfields, generally speaking, only a small proportion of the collieries work double shift ; and I am not aware that the three-shift system is in operation anywhere else than in Durham. In South Wales just a few collieries work double shift on the coal ; but it is quite a small proportion because the extension of this practice is most strenuously resisted by the Miners' Federation, partly because they fear the lowering of the price of coal, and partly because of the extreme disorganisation of home and social life which it causes in the mining towns. At present colliery proprietors are making no special effort to extend the double shift system, because there is not the labour available for working the coal in a second shift. To make the double shift universal throughout South Wales, the population would have to be increased by nearly 50 per cent. Should there be an influx of labour from other coalfields which are worked out, or from older collieries closed, such surplus of labour would probably be at once absorbed by working double shift. It may be that, during a generation following upon the complete taking up of the measures and development of mines throughout the South Wales coalfield, a sufficient population will develop to allow of a fairly general introduction of the double shift.

I must not be taken as urging that it is desirable
to introduce the double shift system. I merely
wish to point out the conditions which render it
possible, and that it will mean a very considerable
reduction of the cost of putting coal upon the market
to have the overhead charges of deep and expensive
collieries spread over the output of two shifts instead
of one. If the nation felt it desirable to reduce the
cost of coal by encouraging the double shift system,
the first step it should take should be to provide
sufficient new housing accommodation of a high
standard in order to attract thither a greater population. At the present time it is impossible to increase
the labour supply in many of our coalfields owing to
the impossibility of housing the larger number of
workers. There are many other ways in which an
intelligent Government could improve the amenities
of life in the mining towns, whereby living could be
so arranged as to minimise the disadvantages of the
double shift system.

The foregoing observations will make it plain how
much the Coal Question is in reality a labour question, for the price of coal depends more upon the
cost of labour than upon anything else. The miners
are now becoming so well organised that they may
very well decide that wages shall not fall below a
certain minimum, and they may, and probably will,
take measures to restrict the output in order to
protect the minimum which they consider essential
to maintain their standard of life. The price of coal
in the future is likely to be affected in greater degree

by the attitude of labour, and by the cost of meeting
more stringent legislative requirements, than by the
mere cost of deeper working and poorer seams.
If the Miners' Federation of Great Britain so decided,
the annual output of the country could very soon be
considerably increased, and it could be sold at a
reduced price.   It will depend very much upon the
policy of labour whether the country will suffer
seriously or not when an advanced stage of the
exhaustion of our coal resources has been reached.
I am not overlooking the fact that there will be a
continued increase in the use of labour-saving
machinery, both in winding the coal and in conveying
it from the face to the pit bottom ; for machines
must be controlled by workmen, and we should
simply be substituting a less numerous and more
efficient class of workmen for the more numerous
body of handworkers and pony drivers.

Writing in *The Coal Question* in 1865, my father
stated " that the cost of fuel must rise, perhaps
within a lifetime, to a rate injurious to our commercial and manufacturing supremacy " ; and " that
the check to our progress must become perceptible
within a century from the present time." [1]   Half of
this time has already elapsed, and it is interesting
to note to what extent the rise in the price of coal
which he anticipated has already come about.
Making use of Mr. Sauerbeck's series of index
numbers,[2] I find that the average price of exported

[1] *The Coal Question*, third edition, 1906, Chapter XII, p. 274.
[2] *Statistical Society's Journal*, Vol. 77, 1914, pp. 556 and 568.

coal is stated, and also the price of " Wallsend Hetton " Coal in London, and of Newcastle Steam Coal. Comparing the price during the last decade, 1904–13, with the eleven years, 1867–77, it appears that there has been in each case a decrease of price by the following percentage :—

Average export price.	Wallsend Hetton in London.	Newcastle steam coal.
4·7%	8·3%	5·0%

No conclusion can be drawn, however, from the rise or fall of the price of one commodity without comparing it with the price of the other principal commodities of commerce which are all subject to a general rise or fall on account of conditions of currency or credit. Mr. Sauerbeck's general index number for 1913 is 85, the number 100 applying to the base period 1867–77. Taking the average export price of coal we find that its average index number for the decade 1904–13 is 95·3 ; so that there is evidently a relative rise in the price of coal to the extent of 12 per cent. In other words, if the general average of prices had remained constant from 1867 to the present at 100, the average export price of coal for the decade 1904–13 would have stood at 112. This increase of the average export price is doubtless partly due to increased cost of mining, but it is partly also accounted for by the increase of the export trade from South Wales. The growth of foreign trade in South Wales has been very great during recent years, and as the coal of this area is of higher value than that of other coalfields the average export price for the whole country has thereby been increased.

The net result is that there certainly has been some relative rise in the cost of coal since the period 1867–77, but not a great increase. A very great part of this increase must be attributed to the increased cost of labour, although it is difficult to say exactly to what extent wages have risen.

So far as it is possible to form an opinion I am led to believe that we shall continue to see a gradual increase of the price of coal relatively to prices in general for the next 40 or 50 years; and that after that period there will be a more marked relative rise in price. It is to be remembered that the price of any article of commerce is always determined by the cost of production on the margin of profitable supply—that is to say, by the cost of obtaining any increase of the supply. In enlarging the total output of the country we shall be constantly opening up new coalfields and new coal seams. I anticipate that after 40 or 50 years the whole of the best seams of coal in the country will be in course of being mined even in the, at present, unproved coalfields; and that for addition to the supply it will be necessary to open up much deeper or thinner seams. It is then that the effect upon prices will begin to be felt, and such collieries as may be still working upon large untouched reserves of the better seams will become correspondingly valuable as investments, because there will be a large margin between current prices and the cost of working the better seams. No such effect will come into play until collieries have reached the limit of effective increase in size, and I think this may be expected to have occurred within 50 years.

In a paper on the Coal Question read by him before the British Association in 1898, the well-known mining engineer, Mr. T. Foster Brown, gave it as his opinion that we should have in 1900 only 15,000 million tons of best coal remaining within a depth below the surface of 2,000 feet, and he calculated that by the year 1950 we should have exhausted eleven-fifteenths of this reserve of best coal. He further states that the "increased cost of our coal may be gradual and comparatively imperceptible for 50 years, or thereabouts, owing to the continual development of the unopened portions of our coalfields and the resources of good thick seams still unworked in existing collieries, but after 50 years or so this progress will be much more rapid."

My general conclusion is that there has been during the past 50 years a distinct relative rise in the price of coal, though it is probably not mainly due to exhaustion of the more cheaply worked coal, there being other more important factors tending both to decrease and increase the price. During the next 50 years I think we shall see a continued slow relative increase of the price of coal due only partly to the necessity of mining poorer coal and at greater depth. After about 50 years hence I anticipate that there will be a more marked rise of the relative price of coal, this being mainly due to the exhaustion of the better seams and to the margin of supply being pushed to greater depths, and thinner or lower quality seams.

In *The Coal Question* my father indicated that grave consequences would follow from the rise of the price of coal in this country relatively to the price in other countries which would be competitors with us in the world's manufacturing industries and trade. He anticipated that the United States of America, with their enormous untouched resources of coal, would be our keenest competitor, and in that he has proved to be perfectly right. Whether we shall in reality be much handicapped by the rise in the price of coal in this country, relatively to the cost in other countries, is to my mind a matter of some doubt, because we have to remember that all the mining activities of this country, from the early centuries to the present date, have done little more than scratch the surface of our great natural stores of coal. It is only at the present time, now that huge mines working at depths down to 2,000 or 2,500 feet are becoming the standard type, that we can really be said to be developing a great mining industry in this country. In this respect none of the European countries have really any advantage over us. France, Belgium and Germany have all very largely exhausted whatever supplies were cheaply available close to the surface, and they, too, must depend upon deep and large scale mining operations ; and if they produce coal more cheaply (which they hardly do at the present time), it will only be because of labour rates being lower. The United States, on the other hand, is in a different position, having still large untouched areas where coal crops out at the surface and can be mined cheaply from levels. The coal-mining industry is growing very rapidly, however, in the States, the output having gone up by leaps and bounds during the past few years, until it is now 80 per cent. greater than that of this country. The output of the United States has, since 1876, been doubling itself in every 12 or 13 years. Hence, by 1926, the annual output is likely to be no less than 900 million tons, and by 1950 at least 1,800 million tons. In other words, the United States is going through the same period of rapid development of its coalfields which we did 50 years ago ; and with so great an industry as is rapidly developing in the United States, they cannot long have the advantage of very cheaply workable coal. Much the same observations apply with some postponement to Canada, and to other countries now opening up great coal resources. In all such countries where there is already a considerable population, they are likely in 50 or 60 years to have worked out all their most easily accessible coal ; and there will not, therefore, be any great differential advantage in their favour as compared with us in the accessibility of coal. We shall still be opening up new concealed coalfields, where we have to commence mining at 2,000 or 2,500 feet deep, at the same time that the

Americans are beginning deep working of 1,500 to
2,000 feet or more in the coalfields where they will
then have exhausted the shallower seams. The
greatest asset of Britain will remain in the future,
as in the past, in having coalfields with high quality
coal actually intersected by the sea, with numerous
convenient ports of shipment.

I do not underrate the probability of serious
competition from America, both in the manufacture
of crude and finished steel, and in many manufactures, and also in the export of coal ; but such
serious competition will be due, not mainly to the
possession of a cheaper coal supply, but rather to
their vastly more efficient business organisation and
mechanical appliances. If they are handicapped by
long distances from their coalfields to the seaports,
they overcome the obstacle by carrying coal at one-
fourth or even one-fifth the rate per ton per mile
which prevails in this country. Their manufacture of
steel is conducted on a vastly greater scale and more
economically than the same industry in this country,
and this is due almost entirely to the better designed,
larger, and more costly plant which they employ.
I am, therefore, of opinion that under modern conditions of industry the question of whether or not
we achieve the economies possible by improved
organisation and by new inventions, due to the
progress of science, is of far greater importance
than the question of national disadvantage due to
the exhaustion of our more cheaply workable coal
supplies. It is enterprise, science and organisation

which will win the commercial race of the future. Even countries possessing no coal can yet enter the arena of industrial competition, for they can develop at their ports those manufacturing industries which produce goods of much value in proportion to the cost of the coal which they have to import to drive the machinery. The only question is whether their population has the enterprise, the educated labour, and the capital required, to launch such industries.

Conservation of National Capital

Although I have no great fear that exhausting our more cheaply workable coal will handicap us seriously in the commerce of nations, it is certainly right that some consideration should be given to the welfare of our descendants, even in the distant future of our country. Some 50 years hence the price of coal will be rising distinctly, relatively to prices of commodities in general. One hundred years hence there will have been a further relative rise of the price of coal throughout the world, as well as in this country ; and in proportion to material wealth and the well-being of the people, less coal will be consumed than at present. There will also be less coal consumed in each particular use ; though the use of coal for a variety of new purposes will probably maintain the consumption per head of population.

Two hundred years hence, when the annual output has reached, or very possibly, passed its maximum, and is steadily though slowly declining, coal will be becoming a valuable, though still very necessary commodity. The expense of every operation of transport and manufacture will be increased by the high price of coal ; and this cannot but adversely affect the wealth-producing powers of the whole community. Emigration will be stimulated to tropical regions and to countries still possessing vast stores of unworked coal. We shall have used up the bulk of our natural resources—our national capital—and shall be gradually falling behind in the standard of civilisation as compared with the great nations of the world.

What course can we as a nation pursue in the meantime in order to mitigate, so far as may be, the burden which must gradually be thrust upon us ? There are two main directions in which I think national enterprise may be at once directed ; and they have the advantage that both can be of immense utility in the present, besides building up a great store of wealth for the future. In the first place, great attention should be paid to scientific and technical education, and to the encouragement by Government of all kinds of scientific and technical research. So many inventions of the greatest utility to civilisation are the direct result of scientific experiment and investigation, that our Government could make no more useful expenditure of public money than in fostering scientific research. Nor, in my opinion, should Government grants be confined purely to research in the natural or applied sciences. Inventors are constantly conferring great benefits upon mankind, and often for little reward. At best their work is highly speculative : and lack of adequate financial backing leads to the great majority of promising inventions never securing an adequate trial. A body of Inventions Commissioners might well be established having power to advise and finance inventors giving promise of success, recouping the funds of the commission by a tax upon the profits of commercially successful inventions assisted by them. Thus could many trains of thought initiated by professors of pure science in university laboratories be matured through experiments in technical institutes, and by the efforts of practical inventors, to be finally launched through the channels of trade in the service of man.

The second direction in which national enterprise ought to be directed is in the acquisition and aggregation of capital in national ownership of the means of transport and production, so that in the long run the services rendered by that capital may be cheapened. During the next hundred years we shall be probably enjoying unexampled wealth, and shall be using up all the coal available to be mined without a decided increase of cost. Whilst the service of coal must become more and more expensive, cannot we do something to make other services, now expensive, cheaper for future generations ? In no direction is economy more desirable in this country, and at the same time more practicable, than in the transport of merchandise by railway and canal. In nearly every class of traffic our rates are higher than in the principal countries of Europe and America. Nearly one-half of the gross revenue of English railways goes to pay interest and dividends on capital. If the railways were purchased by the State, economies in working would probably allow some immediate reduction of rates, whilst the economy in interest charge would provide a sinking fund by which the whole of the railways would be owned free of cost to the nation 80 or 100 years hence. Railway rates might then be reduced, relatively to general prices, to half their present level.

By national ownership of land, and possibly by national ownership of coal mines, and of large industrial plants, other enormous economies can gradually be effected : with the result that the prices of many commodities would be cheapened, and the cost of public services reduced, to an extent far outweighing the increased costs arising from the progressive exhaustion of our coal. Whilst we are wasting our wealth by burning our coal we must build up a store of national capital to replace it. The wealth given to us in the bowels of the earth should not perish with the people who consume it, but rather should it be converted through the results of men's handiwork into a permanent possession for the good of all the people.

CHAPTER XXVIII

THE WORLD'S COAL RESOURCES

ALTHOUGH any detailed description or consideration of the coal resources of the world would be out of place in this book dealing with the British trade, there is every reason to make a brief survey of the resources of other countries and the probable course of their development. The export trade in coal, and the character of much of our foreign trade in other goods, will depend upon the accessibility and upon the development of foreign coalfields.

The task of making a survey of the coalfields of the world has been considerably lightened by the authoritative publication referred to in the last chapter, entitled " The Coal Resources of the World." This topic was made the chief subject of discussion at the Twelfth International Geological Congress, held in Canada in 1913 ; and this monograph in three volumes was prepared especially for the Congress. It was edited by gentlemen connected with the Geological Survey of Canada ; and they had the assistance of the geological surveys of different countries, and numerous mining engineers and geologists.

It will be convenient to describe the coal resources

of the various countries according to continents, giving a summary table for each continent. In the Report the coal reserves of the different countries are classified according to the quality of the coal, and also according to whether they are considered as actual reserves, or probable and possible reserves, and also according to depth. Thus we have first of all the broad distinction between Group I, consisting of coal seams of not less than 1 foot of merchantable coal occurring not more than 4,000 feet below the surface, and including workable submarine areas; and Group II, which includes seams of not less than 2 feet thickness and occurring at a depth of between 4,000 and 6,000 feet. The separation of the reserves into actual, probable and possible, is determined by including in the actual reserves all coal of which the calculation of the amount is based on a knowledge of the actual thickness and extent of the seams. In the probable reserves are placed cases in which only an approximate estimate can be arrived at, and in the possible reserves, cases in which not even an approximate estimate in figures can be given. Thus the possible reserves are simply distinguished by the words " large," " moderate," " small," etc. The coal estimated for the actual and probable reserves is classified according to its kind in four classes denoted A, B, C and D, which are very carefully defined. Class A corresponds generally with anthracite and hard and dry steam coals. Class B contains bituminous coals, subdivided into coking and non-coking coals. Class C

COAL RESERVES OF EUROPE

(IN MILLIONS OF METRIC TONS)

Country.	Lignites and sub-bitum-inous coals.	Bituminous coals.	Anthracite and semi-anthracite.	Totals.
Germany .	13,381	409,975	..	423,356
Gt. Britain and Ireland	2	178,176	11,357	189,535
Russia . .	1,658	20,849	37,599	60,106
Austria-Hungary	18,174	41,095	..	59,269
France . .	1,632	12,680	3,271	17,583
Belgium .	..	11,000	..	11,000
Spain . .	767	6,366	1,635	8,768
Spitzbergen .	..	8,750	..	8,750
Holland. .	..	4,082	320	4,402
Balkan States	921	75	..	996
Italy . .	99	..	144	243
Sweden, Denmark and Portugal .	50	114	20	184
				784,192

contains the cannel coals and highly bituminous coal with volatile matter of from 30 to 40 per cent. Class D contains brown coals and lignites, generally younger in age than the carboniferous coals.

Estimated Reserves—Europe

In the following survey of the coal resources of the world, I am taking the figures directly from the Report and they are reproduced as there stated, that is to say, in metric tons of 2,205 lbs. each, instead of statute tons of 2,240 lbs., which is the unit I have generally used throughout this book.   If the figures were to be taken as statute tons the error would be less than the probable error in many of the estimates to be quoted.

The table on p. 774 gives the total, actual and probable reserves of the various European countries in order of quantity.

The two German coalfields are those of Westphalia and of Upper Silesia, and their reserves, stated as follows, each approximate to the total resources of Great Britain :—

In millions of metric tons.

	Actual Reserves.	Probable Reserves.	Total.
Westphalia (and Rhine Prov.) .	56,344	157,222	213,566
Upper Silesia . .	10,325	155,662	165,987

Whilst all the above, and some 30,000 million tons in the Saar, Lower Silesia, and other small coalfields, are of Carboniferous age, Germany has small deposits of coal of Permian and Cretaceous age, and large deposits of Tertiary age, mainly in Prussia, the North German States and Saxony, and totalling 13,883 million tons.   The Westphalian coalfield is becoming highly developed, with many very large mines, most of which are united for selling purposes in the well-known Westphalian Coal Syndicate. It is a *kartell* or combine for the sale of coal, which regulates output, the mines remaining in the ownership and control of individuals and companies. An extensive iron and steel industry and many kinds of manufactures have grown up at Essen, Dortmund and elsewhere in this great coalfield, which has an output considerably exceeding that of our greatest coalfield (Yorks — Derby — Nottingham). The total output of Germany in 1913 was 191,511,000 metric tons, or only about two-thirds the output of Great Britain ; but the percentage rate of increase of output is more rapid in Germany than with us, so they are catching us up, and will probably equal our output in thirty years' time.

The British reserves are described elsewhere in this book ; so I pass to a few words about those of Russia and Austria-Hungary, both of which are most extensive in comparison with the present small output. By far the largest Russian coalfield is the Donetz basin in the south of Russia, just north of the Sea of Azov. This field alone is estimated to contain 55,600 million tons, the coals being of many different kinds, including anthracite. The other coalfields are small and widely scattered. The coalfields of Hungary are small in area, but those of Austria are very extensive. They occur mainly in the Northern, Central and Southern Alps. Besides the Carboniferous coal, there is Triassic and Jurassic coal ; and a considerable part of Austria's coal reserves are brown coal or lignite of Tertiary age, which are extensively mined, the brown coal being useful for gas.

The coalfields of Belgium and the North of France

are geologically continuations of the Kent coalfield, and the coals have much the same character, except that they are often very steeply inclined through intense disturbance of the strata. There is an extraordinary number of small coal basins in France, from which the total output in 1912 was 40,300,000 metric tons. It will be observed that the outputs of France and Belgium bear a larger proportion to the available reserves of those countries than is the case with Great Britain or Germany, Russia or Austria, as is shown in the following table :—

	Proportion of Annual Output to Reserves.[1]
France	·229 per cent.
Belgium	·209 ,,
Great Britain and Ireland[2] . .	·147 ,,
Russia	·045 ,,
Germany	·041 ,,
Austria-Hungary . . .	·029 ,,

The low rate of output in proportion to reserves which characterises the last three centuries is not to be taken as evidence that their reserves will last a correspondingly long time. It rather shows what room there is for growth of the coal mining industry, and indicates that the output may be expected to expand rapidly during the next few years. Comparison of the figures show, indeed, that countries

[1] The outputs are those of 1912, except in the case of the United Kingdom, for which the average 1911–13 is used, as the 1912 output was affected by the Miners' National Strike.

[2] The reserve is taken at Mr. Strahan's estimate in *The World's Coal Resources*

with the lowest percentage of output to reserves are those in which the output is increasing most rapidly, and vice versa.

North America

In North America the reserves of coal are so enormous as to quite put into the shade the whole resources of Europe. The richest coalfields of the United States are in the Appalachian region—the anthracite, semi-bituminous and bituminous fields of Pennsylvania, stretching down into Maryland, West Virginia, Virginia, North Carolina and Tennessee. In these States alone it is estimated that there is the amazing quantity of 310,777 million tons at a depth less than 3,000 feet. This includes 19,056 million tons of anthracite and semi-anthracite in Eastern Pennsylvania. The semi-bituminous or high-class steam coal occurs in about four fields along the eastern margin of the Appalachian region in Pennsylvania, Maryland, Virginia and West Virginia. The reserve is 42,365 million tons. The great bulk of the merchantable coal of the United States, as elsewhere, is of the bituminous class. Besides the very extensive bituminous areas of the Appalachian fields in the States above named, there are three fields with 61,300 million tons reserve in Alabama, and a great field in Ohio with reserve of 85,270 million tons. The fields so far mentioned all belong to the eastern province ; and they consist entirely of coals of Carboniferous age. The interior province includes in the East the coalfields of Indiana,

Illinois and Kentucky, with a combined reserve of 280,000 million tons, and in the west the coalfields of Iowa, Missouri, Kansas and Oklahoma. All of the coal of this province (479,377 million tons) is bituminous and of Carboniferous age.

Throughout the western part of the United States the Cretaceous and Eocene were pre-eminently the periods when coal-forming conditions prevailed; and there are vast deposits of these periods stretching from Texas, New Mexico and Arizona, northwards on both sides of the Rocky Mountains, to Montana, Idaho and Washington State. These younger coals are nearly all lignites and brown coals; or, if black coal, are soft, and contain much moisture and volatile matter, being known as sub-bituminous. In some areas, however, earth movements have so affected the Cretaceous coals as to produce hard bituminous coals, and even anthracite in certain small areas affected by igneous intrusions. In the following summary of the reserves of the United States the coals are classified by kinds; and much the greater part of the younger coals comes in the form of lignites and sub-bituminous coals.

The total reserve is colossal; and it is interesting to note that considerably more than half of it is of Cretaceous or Eocene age. All of the foregoing coal is in seams of more than 14 inches thickness, and at less than 3,000 feet, these being the limits previously used by the Geological Survey of the United States. Put upon the same basis of 12 inches and 4,000 feet as used for other countries,

COAL RESERVES OF THE UNITED STATES

(IN MILLIONS OF METRIC TONS)

Province.	Lignites and sub-bitum-inous coals.	Bituminous coals.	Anthracite and semi-anthracite.	Totals.
Eastern . .	..	457,543	62,238	519,781
Interior (Missis-sippi Valley)	..	479,377	1,476	480,853
Northern Great Plains . .	1,134,026	41,337	..	1,175,363
Gulf (Texas, Ark.)	20,953	..	..	20,953
Rocky Mountains	643,696	325,371	464	969,531
Pacific Coast .	48,511	10,381	21	58,913
Totals .	1,847,186	1,314,009	64,199	3,225,394

there would be a further considerable increase of the estimates. The coal in the United States lying at depths between 3,000 feet and 6,000 feet is estimated, for areas where any estimate has been made, at 604,900 million tons.

Canada also has vast coal resources, mainly of Carboniferous age in the East, but of Cretaceous and Eocene age in the West, thus corresponding with the United States. Far the greatest reserves are in Alberta and British Columbia, as shown by the following summary :—

COAL RESERVES OF CANADA
(IN MILLIONS OF METRIC TONS)

	Seams over 1 ft. above 4,000 ft.	Seams over 2 ft., 6,000 to 4,000 ft.
East—		
Nova Scotia	7,080	2,639
New Brunswick	151	..
Middle—		
Ontario	25	..
Manitoba	160	
West—		
Saskatchewan	59,812	..
Alberta	1,059,927	12,700
British Columbia	73,875	2,160
North—		
Yukon	4,940	..
N.W. Territories	4,800	..
Arctic Islands	6,000	..
	1,216,770	17,499

Grand Total   1,234,269 million metric tons.

It is possible that very extensive further deposits will be found in the north of British Columbia, in Athabasca and the other northern regions.

A large amount of coal is found in Alaska, chiefly of Cretaceous and Tertiary age. Much of it is soft coal, but there are some extensive beds of good quality, including anthracite. The estimate printed in the Report is 19,593 million tons, occurring in seams of 3 feet thick and upwards, and at a depth of not more than 3,000 feet. As coal has been discovered in some thirty widely separated localities, and the estimate does not include any hypothetical extensions, it is quite probable that the coal reserves will ultimately be found to be at least ten times, and possibly twenty times, as great as that figure, when taken on the usual basis of 1 foot and 4,000 feet.

In Mexico there are some extensive coalfields of Triassic and Cretaceous age, the coals differing much in quality and value. No estimate of the reserves was made. The output in 1910 was 2,418,000 statute tons; but the revolutions have interfered with the industry subsequently.

South and Central America

The southern continent of America is not blessed with large coal reserves, so far as is yet known; but the geological explorations are hardly extensive or accurate enough for certainty upon this point, nor sufficient to give accurate estimates of the known coalfields. The estimates presented in the report, including both actual and probable reserves, are as follows :—

Country.	Millions of metric tons.
Colombia	27,000
Chili	3,048
Peru	2,039
Argentina	5
Venezuela	5
Honduras	5
Total . . .	32,102

The estimate for Argentina is probably a gross under-estimate ; and there are supposed to be small reserves which were not estimated, in Equador, Bolivia and Uruguay. Brazil is believed to have somewhat larger supplies. Some lignite and sub-bituminous coals have been found in Guatemala, Salvador, Costa Rica, Panama, Trinidad and Santo Domingo, mostly of poorish quality.

Asia

This vast continent contains, as might be expected, several large coalfields with big reserves. Probably as many more whole coalfields remain to be discovered, and the figures given for the known fields are stated in the Report to be probably under-estimated, as many of the Siberian and Chinese coalfields have not been sufficiently explored to admit of an estimate of their reserves being made. The coalfields of importance in India and China belong to the Permian or late Permo-Carboniferous period ; but in Japan the important coal-bearing deposits are of Miocene age, and in Siberia coal is found in formations varying in age from Lower Carboniferous to Recent. China is wonderfully rich in coal, the greatest fields being those of Shansi in the north, and Hunan-Szechuan in the south. The coal is extensively mined, both under European and Chinese control ; and the output may be about 17,000,000 tons per annum. There are no complete statistics.

The following table indicates the reserves and

the class of coal in the better known coalfields
of Asia.

COAL RESERVES OF ASIA

(IN MILLIONS OF METRIC TONS)

Country.	Lignites and sub-bitum-inous coals.	Bituminous coals.	Anthracite and semi-anthracite.	Totals.
China . .	600	607,523	387,464	995,587
Siberia . .	107,844	66,034	1	173,879
India . .	2,602	76,399	..	79,001
Indo-China .	..	..	20,002	20,002
Japan . .	778	7,130	62	7,970
Persia . .	..	1,858	..	1,858
Manchuria .	..	1,140	68	1,208
Korea . .	27	14	40	81
Totals .	111,851	760,098	407,637	1,279,586

Large unestimated reserves are reported in Siberia.
Small amounts are reported in the Federated Malay
States, Siam, and Asia Minor.

Africa

So far as is at present known, there are no important coalfields in the North of Africa ; and the largest
known deposits are in various parts of South Africa,
where they occur in the Karoo formation, which is
mostly of Permo-Carboniferous age.   The following
is a summary table :—

COAL RESERVES OF AFRICA
(IN MILLIONS OF METRIC TONS)

Country.	Lignites and sub-bituminous coals.	Bituminous coals.	Anthracites and semi-anthracites.	Totals.
Belgian Congo	900	90	..	990
Southn.Nigeria	80	..	..	80
Rhodesia .	74	493	2	569
Transvaal .	..	36,000	..	36,000
Rest of South Africa	..	8,540	11,660	20,200
Totals .	1,054	45,123	11,662	57,839

There are also large unestimated reserves in Southern Nigeria ; a moderate amount in Nyasaland, and small reserves in Madagascar, the East African Protectorate, the Soudan, and Abyssinia.

Australia and Oceania

The coals of Australia are mainly of Permo-Carboniferous age. There are also a few beds of coal of inferior quality in the true Carboniferous ; and seams of soft coal in the Triassic, Jurassic and Tertiary strata. The principal coalfields are in New South Wales, where there is a considerable mining industry, and in Queensland and Victoria. The following table gives a summary of the coalfields and reserves so far as known not only in Australia, but also in New Zealand and the other islands of the Pacific.

COAL RESERVES OF AUSTRALIA AND OCEANIA

(IN MILLIONS OF METRIC TONS)

Country.	Lignites and sub-bituminous coals.	Bituminous coals.	Anthracites and semi-anthracites.	Totals.
New Sth. Wales	..	118,439	..	118,439
Victoria .	31,114	52	..	31,166
Queensland .	866	13,693	659	15,218
Tasmania .	..	66	..	66
W. Australia .	653	..	..	653
New Zealand .	2,475	911	..	3,386
British N. Borneo	..	75	..	75
Netherlands India	1,071	240	..	1,311
Philippines .	61	5	..	66
Totals .	36,240	133,481	659	170,380

There is a possible further reserve in the Philippine Islands, and small reserves of lignite are found in South Australia, Queensland, Tasmania, and Netherlands India.

The Shackleton Polar expedition discovered a very large coalfield in the Antarctic Continent, which is reported on by Professor T. W. E. David, of Sydney (N.S.W.). He indicates a very large area, most of it covered by ice. Assuming mining to be possible as far as 15 or 20 miles from the coast, I should think, from the data given, that the available reserve would probably be about 15,000 million tons.

The World's Total Reserves

Having completed our survey of the continents we may add up the grand total, as follows :—

THE WORLD'S COAL RESERVES

(IN MILLIONS OF METRIC TONS)

Continent.	Lignites and sub-bitum-inous coals.	Bituminous coals.	Anthracite and semi-anthracite.	Totals.
Europe . .	36,682	693,162	54,346	784,190
Nth. America .	2,811,902	2,239,682	21,842	5,073,426
Sth. America .	4	31,398	700	32,102
Asia . .	111,851	760,098	407,637	1,279,586
Africa . .	1,054	45,123	11,662	57,839
Australia and Oceania	36,270	133,481	659	170,410
Totals .	2,997,763	3,902,944	496,846	7,397,553

This vast total of 7,000,000 millions of tons is doubtless far below the real total if all the known coalfields had been fully investigated, and if the wholly unknown coalfields were included. The reserves of Mexico, for example, are not included ; and those of the Argentine and Alaska are probably grossly under-estimated. Large deposits of coal known to occur in Afghanistan and in the Antarctic Continent are not included, whilst it is almost certain that further large reserves will be discovered in Siberia and other parts of Asia as well as in Africa and, possibly, in South America. For all these contingencies, bearing in mind that at least one-half of the land surface of the earth cannot have been geologically explored, except in the most superficial manner, it will not, I think, be exaggerating the probabilities of the matter if we add 25 per cent. of the estimated reserves, thus bringing the grand total up to 9,240,000 millions of tons. This total is so colossal that it is difficult for the imagination to grasp it.

Duration of the World's Coal Resources

It may be thought that the world's store of coal is so vast as to be practically inexhaustible for all time ; at any rate, during any future period which can be compared with the history of civilised man. With a present production of less than 1,300 million tons per annum it looks as if the world's coal supplies would last at least 7,000 years.

Such a view fails to take into account the extraordinary rapidity with which the world's production of coal is growing ; and I believe that I can give good reasons for supposing that the condition of serious exhaustion due to consumption of three-fourths of the world's total store may well occur at no later date than 400 or 500 years hence.

In England the coal-mining industry is entering upon a mature period of its existence ; and the production, which is already large in proportion to the population, is naturally not showing so large a percentage rate of increase as it did a century ago.

In most other countries, excepting only France, Belgium and Germany, the coal industry is no more advanced relatively to the resources than was the mining industry in Britain a century ago. There is no reason for surprise, therefore, when we examine the world's production of coal, and find it to be increasing with extraordinary rapidity, the total output doubling itself about every eighteen years. This rate of increase shows at present no signs of falling off.

The following are the figures for the world's coal output given in the *Report on the World's Coal Resources*, with the addition of such figures for 1913 as are yet available.

It is not possible entirely to judge of the development of the coal mining industry in other countries in the future by comparison with the past in Great Britain, for two very good reasons. In the first place the technique of mining has made enormous strides ; and large-scale mining, with electrical appliances for cutting, hauling and winding the coal, have completely altered the prevailing practices, and have rendered the removal of the coal a far more rapid operation than it used to be. In the kind of pit sunk fifty years ago it would be impossible to remove all the coal from a mine of large area and numerous seams in less than two hundred years ; but with the large shafts, often three in number, now sunk upon a taking the whole of the coal from ten or twelve seams could be worked out in less than a hundred years. The thicker seams would

TABLE OF ANNUAL COAL PRODUCTION OF PRINCIPAL COUNTRIES IN THE WORLD.

(In Million Tons.)

Name of Country.	1865.	1870.	1875.	1880.	1885.	1890.	1895.	1900.	1905.	1910.	1913.
Australia	..	..	..	..	..	..	4·01	6·48	6·83	10·00	12·50
New Zealand	..	..	..	..	..	..	0·76	1·11	1·41	2·23	2·40
China	..	..	..	..	..	..	..	..	..	14·59	..
India	..	..	..	..	..	..	2·65	6·22	7·92	12·09	15·49
Japan	..	..	..	..	..	..	4·84	7·43	11·89	14·79	20·92
S. Africa	..	..	..	..	..	..	1·40	0·76	3·22	5·50	9·82
Canada	..	..	..	..	..	..	3·19	5·09	7·96	13·01	13·50
United States	24·79	29·95	48·20	66·83	102·18	141·62	177·59	243·41	351·12	445·81	504·52
Mexico	..	..	..	..	..	..	..	..	..	2·45	..
United Kingdom	99·76	112·24	135·49	149·38	161·96	184·59	194·35	228·77	239·89	264·50	287·41
Spain	0·45	0·66	0·61	0·85	0·94	1·18	1·77	2·58	3·20	3·55	3·60
France	11·84	13·30	16·95	19·36	19·51	26·08	28·24	33·40	36·05	38·57	40·19
Belgium	11·84	13·69	15·01	16·88	17·44	20·37	20·41	23·46	21·84	23·13	22·50
Germany	28·33	34·88	48·53	59·12	73·67	89·29	103·96	149·79	173·66	221·98	273·65
Austria-Hungary	2·03	8·36	13·06	14·80	20·43	26·10	27·25	39·03	40·72	38·00	51·58
Italy	..	..	..	..	..	..	0·25	0·48	0·31	0·40	..
Sweden	..	..	..	..	..	..	0·20	0·25	0·33	0·21	..
Russia	0·33	0·69	1·17	3·27	4·24	7·00	9·10	14·76	17·12	24·57	29·87
Other Countries	2·71	4·04	6·26	9·28	12·45	16·89	1·75	2·90	4·55	8·00	..
Total	182·08	217·81	285·30	339·37	412·82	513·12	581·12	765·92	928·02	1143·38	1321·00

usually be worked first, and afterwards the thinner seams, which would be removed all the faster in proportion to their thinness.

In the second place, there are now large aggregations of capital available to be invested in sinking mines upon a very large scale, wherever market conditions render the investment profitable. Consequently, in the remotest coalfields, if the quality of coal is good enough, the cost of mining low enough, and the local market big enough, European and American capital will soon discover it and sink big mines. The only limits to the rapidity with which such mines can grow are the demand for coal and the necessary skilled and semi-skilled labour. Now the demand for coal depends partly upon the numbers of the local population, partly on the climate, partly on the degree of their civilisation and the development of general manufactures, mechanical transport and so forth, and partly upon the amount of international shipping. In the Asiatic countries, like China and India, which have already enormous populations, one may expect the conditions to favour a rapidly increasing demand for coal. In the more sparsely populated countries such as Australia, the West of Canada, and the United States, the growth of demand for coal must depend upon the growth of the population. Here, however, there is a reflex action, because the existence of the coal would favour the establishment of industries and the growth of manufacturing and mining populations—particularly so if the adoption of free-trade stimulates international commerce.

The suggestion may indeed be hazarded that with the highly developed conditions of modern industry which can be and are being applied to all parts of the world, no one coalfield can be expected to last more than about 300 years from the time that it is first seriously attacked by mining of the capitalist order. The technique of manufactures and of mining which has been developed in Europe and the United States is applicable to any part of the world where there is a sufficient demand for the products, and the necessary labour can be obtained. As soon as there is a fully settled agricultural population, there arises a demand for manufactured goods and for coal. For twenty or thirty years these may be brought long distances ; but as soon as large enough local markets are established, capital begins to arrive so as to produce coal or goods on the spot. The stage of mining in a small way will usually be short. After twenty or thirty years will come the period of mining by large companies owning many square miles of minerals. To each one of them the volume of output, to the limit the market can stand, is a sure road to big dividends. The growth of total demand will then regulate the number of large mining concerns which come into existence.

It may be suggested that the first century after large scale mining is commenced will see its gradual extension, until nearly the whole area of the coal-

field is taken up, with the exception of the portions
the most remote from markets, or where the best
seams lie deepest. During the next century the
first of the large scale mines will be working themselves out, whilst all the mines over the larger part
of the area will pass through maturity, and will
work out all the best seams by the end of the century.
The third century will see the mines near markets
working the thinnest and deepest seams and those
remote from the markets exhausting their better
seams. At the end of the third century there
would be nothing left but very thin seams at great
depths in parts remote from the principal markets.
In this country, probably only the Durham and
Staffordshire coalfields can be said to have been
already worked to an extent which I am assuming
for the first century of their duration. The South
Wales and Lanarkshire fields might be regarded
as having run for sixty and seventy years respectively ; allowance for the earlier mining being made,
and for the fact that mining methods of fifty years
ago were much slower than those which are now
applied to a new coalfield.

I cannot help thinking that the suggested duration
of three centuries for any new coalfield is unduly
liberal in an already civilised country ; for it is
difficult to believe that the Kent coalfield and the
concealed coalfield of South Yorkshire, Nottinghamshire and Lincolnshire will either of them last
300 years. Where the civilisation of the region is
now mainly agricultural, however, as in Western

America, China, and so forth, the growth of population and demand necessary to secure the taking up of a whole coalfield must be slower, and 300 years would appear a probable estimate.

These considerations would appear to foreshadow an extraordinary growth of the population of the world. In fact we have to recognise that the Malthusian check to the increase of population operates in a totally different manner when the people of any region change by the aggregation of capital and spread of education from a purely agricultural community to a manufacturing nation with an organised modern commerce. A great stimulus to such a change will come wherever coalfields exist; and the population will thrive and develop upon the coalfields of the world, almost like flies multiply upon honey.

I do not think that I am exaggerating the power of populations to grow under conditions of western civilisation and industry, if I am right in assuming that, like the Japanese and other oriental nations, they will assimilate European methods of government, sanitation and education. Consequently, I look for a continuous and very rapid expansion of the world's total output of coal. A hundred years hence it will probably have reached the neighbourhood of 10,000 million tons per annum, or more, and will be still rapidly increasing, so that 300 years hence probably not far from half the world's total coal resources will have been exhausted. Not only our own and other western European coalfields

will be practically exhausted, but also the vast
fields of the western United States; whilst the coalfields of western America, China, India and Australia
will be far on towards exhaustion.

The world will then be looking for its principal
supplies to the coalfields of Siberia and Central
Asia, of Central Africa, South America, and the
Polar regions. I have visions of the miners of the
future spending the arctic winter entirely underground in spacious chambers elaborately warmed and
lighted with abundance of fresh air and of the comforts of life. And so from these most remote coalfields, where now there is no population, the world
may derive its supplies for yet another 200 years.
When this space of time has passed, in the year 2400,
or thereabouts, the civilisation of the world will have
faced, and will probably have solved the question
of finding a substitute for coal as its main source of
power. The substitute will not be water power,
for if all the waterfalls of the world, the Himalayas
and Andes included, were harnessed to the uses
of industry, they would not give one-hundredth
part of the world's requirements. Nor can we think
of oil relieving the situation; for the fields of natural
oil will probably be near exhaustion in 100 years,
and our principal source of oil will be from the
distillation of coal.

There are two great sources of natural energy which
man could even now harness if it were commercially
worth while; and which can yield a never-failing,
and so far as we can think, a permanent supply

of power sufficient for all the purposes of man even with twenty times the present population of the world : I refer to the tides and to the direct radiant heat of the sun. The harnessing of the tides is by no means an impossible engineering feat, by damming up deep estuaries, bays and channels, where swift and powerful currents run twice daily. The only trouble is the huge capital outlay required for the enormous dams and machinery necessary, and for the locks necessary to make provision for shipping. Coal is at present a much cheaper source of power.

The radiant heat of the sun is a tremendous source of power in the desert regions of the tropics. In Colorado and in Egypt steam-engines driven by sun heat have already been employed for pumping operations. An array of mirrors is made to concentrate sunbeams upon a series of boilers which are preferably tubular. Though at present still in the experimental stage, it is clear that boilers to generate several hundred horse-power can be thus constructed. We may think of big factories built in the tropical desert regions with their roofs, and perhaps much of the surrounding ground, bristling with rows of mirrors ; or perhaps the power will be generated in these regions and transmitted hundreds of miles in the form of electricity to the densely populated coastal regions. There is also another method by which the sun's heat may be utilised which is applicable to the wetter and cloudy regions of the tropics—I mean by cultivating those tropical plants which grow with extraordinary rapidity and luxuriance, producing great bulk of stem and leaves. Every year several crops might be obtained from many thousands of square miles, and this vegetable material could be carbonised on the spot, producing valuable gas, oils, and other products of distillation, and a carbon residue which would be converted into briquettes. Both gas and briquettes could equally well be transmitted long distances to the centres of population.

It will not be in temperate regions of the earth that the great aggregates of population will be situated some four or five hundred years hence, but rather in the tropics. The population will tend to multiply more rapidly there in the coming era of peaceful government and with the extension of modern industrial methods. Whilst the coal resources of this country last, that is, for the next two centuries, I do not think we shall suffer much, if any, abatement of our industrial energies even from the position which several generations of continued progress will bring them to. But there is likely to be, I believe, a progressive concentration of the cruder and coarser manufacturing processes, and also much of the production of bulky goods, in tropical regions. As the natives of tropical countries progress under European guidance, and ultimately under their own government, in education, skill and enterprise, they will undertake in their own countries tasks which the more refined Europeans will only do for high wages.

When the coal of northern countries is nearing exhaustion, and recourse is had to sun-heat, probably in both the ways I have named, it is the tropics which will have the advantage for manufactures requiring much power. In England we shall become more than ever a commercial people owning and controlling the factories and plantations of the tropics, buying and selling their goods, and carrying them in our ships. A very large number of people in this country may be living, too, directly or indirectly, simply upon the interest on the vast amounts of capital we shall invest in tropical regions. Although the population of Great Britain will cease to increase, and may even slightly decrease, when our coal is nearly exhausted, we need not picture a future of want and misery. The tides will supply much of the electrical power we require, whilst we shall probably for a century or more import coal, and later on charcoal briquettes, such as I have described—all of which presupposes that no completely new source of energy from radium or other elements is discovered and found to be serviceable for commercial needs. So far we have only been able to develop other kinds of chemical energy by first of all burning coal or using water power; but though, personally, I think this condition will continue, I may be wrong, and scientific discovery in the future may yield some now unthought-of means of turning to the use of man the boundless energy stored in the molecules of matter on every side of us.

APPENDIX I

THE following are amongst the largest and best known colliery companies, each employing several thousand men :—

Durham

John Bowes & Partners, Ltd.
Bell Brothers, Ltd.
Bolckow, Vaughan, & Co., Ltd.
Consett Iron Co., Ltd.
Harton Coal Co., Ltd.
Horden Collieries, Ltd.
James Joicey & Co., Ltd.
Lambton & Hetton Collieries, Ltd.
Londonderry Collieries, Ltd.
North Bitchburn Coal Co.
Charles Perkins & Partners.
Pease & Partners, Ltd.
Priestman Collieries, Ltd.
South Hetton Coal Co., Ltd.
South Moor Colliery Co., Ltd.
Stella Coal Co., Ltd.
Weardale Steel, Coal & Coke Co., Ltd.
Wearmouth Coal Co., Ltd.

Northumberland

Ashington Coal Co., Ltd.
Backworth Collieries, Ltd.
Bedlington Coal Co., Ltd.
Cowpen Coal Co., Ltd.
Cramlington Coal Co., Ltd.
Seaton Delaval Coal Co., Ltd.
Throckley Coal Co., Ltd.

APPENDIX II

COLLIERIES OF THE YORKSHIRE COALFIELD

THE following are amongst the largest and best known colliery companies in Yorkshire :—

Ackton Hall Colliery Co.
Henry Briggs & Son, Ltd. (also Steel-makers).
Brodsworth Main Colliery Co., Ltd.
John Brown & Co., Ltd. (also Steel-makers).
Carlton Main Colliery Co., Ltd.
J. & J. Charlesworth, Ltd.
Dalton Main Collieries, Ltd.
Denaby & Cadeby Main Collieries, Ltd.
Dinnington Main Coal Co., Ltd.
Glass Houghton & Castleford Collieries, Ltd.
Hickleton Main Colliery Co., Ltd.
Houghton Main Colliery Co., Ltd.
New Monckton Collieries, Ltd.
Newton, Chambers & Co., Ltd.
Rother Vale Collieries, Ltd.
South Kirby, Featherstone, & Hemsworth Collieries, Ltd.
Tinsley Park Colliery Co., Ltd.
Wheldale Coal Co., Ltd.

APPENDIX III

Welsh Colliery Companies

The following is a list of some of the largest and best known colliery companies in the South Wales coalfield :—

Steam Coal Collieries

(Many produce also some Gas and House Coals)

	Number of Men employed.
Albion Steam Coal Co., Ltd. . .	2,196
Bedwas Navigation Colliery Co., Ltd. .	1,000
Burnyeat, Brown & Co., Ltd. . .	4,035
Blaenavon Co., Ltd.	2,485
Baldwin's, Ltd.	1,110
Bute, Marquis of	1,813
Cardiff Collieries, Ltd. (Llanbradach) .	2,832
Consolidated Cambrian, Ltd. (D. A. Thomas)	12,035
Cory Brothers & Co., Ltd. . . .	5,776
Crawshay Bros., Ltd., Cyfarthfa . .	2,796
Cwmavon Coal Co., Ltd. . . .	2,260
D. Davis & Sons, Ltd. . . .	8,793
Ebbw Vale Steel & Iron Co., Ltd. .	7,630
Fernhill Collieries, Ltd. . . .	1,867
Ffaldau Collieries, Ltd. . . .	1,300
Glenavon Garw Collieries, Ltd. . .	1,306

	Number of Men employed.
Guest, Keen & Nettlefolds, Ltd. . .	10,487
Great Western Colliery Co., Ltd. . .	4,461
Hill's Plymouth Co., Ltd. . . .	2,878
Insoles, Ltd.	3,069
International Coal Co., Ltd. . .	1,045
John Lancaster & Co., Ltd. . .	5,084
Lancaster's Steam Coal Collieries, Ltd.	2,986
Lewis' Merthyr Consolidated Collieries, Ltd.	7,352
Locket's Merthyr Collieries (1894), Ltd.	3,000
Main Colliery Co., Ltd. . . .	1,847
Newport Abercarn Black Vein Steam Coal Co., Ltd.	1,873
Nixon's Navigation Collieries (1889), Ltd.	5,160
Oakdale Navigation Collieries, Ltd. .	1,860
Ocean Coal Co., Ltd.	9,496
Partridge, Jones & Co. . . .	4,316
Penrikyber Navigation Colliery Co., Ltd.	1,974
Powell Duffryn Steam Coal Co., Ltd. .	13,611
Powell's Tillery Steam Coal Co., Ltd. .	3,516
Rhymney Iron Co., Ltd. . . .	3,886
Tir Pentwys Black Vein Steam Coal Co., Ltd.	1,422
Tredegar Iron & Coal Co., Ltd. . .	6,480
United National Coal Co., Ltd. . .	2,965
Windsor Steam Coal Co., Ltd. . .	2,246
Ynyshir Steam Coal Co., Ltd. . .	1,502

Anthracite Collieries

Glamorgan—

Gwaun-cae-Gurwen Colliery Co., Ltd.   1,270

Breconshire—

	Number of Men employed.
Abercrave Collieries Co., Ltd. . .	380
Gurnos Anthracite Colliery Co., Ltd. .	300
International Anthracite Colliery Co. .	320
South Wales Anthracite Colliery Co., Ltd.	958

Carmarthenshire—

Ammanford Colliery Co., Ltd. . .	1,454
Blaina Colliery Co., Ltd. . . .	600
Caerbryn & Empire Collieries, Ltd. .	460
Emlyn Anthracite Co., Ltd. . .	700
Garnant Anthracite Collieries, Ltd. .	450
Gellyceidrim Collieries Co., Ltd. .	583
Great Mountain Colliery Co., Ltd. .	862
New Cross Hands Collieries, Ltd. .	·700
Pentremawr Colliery Co., Ltd. . .	496
Ponthenry Colliery Co., Ltd. . .	358
Rhos Colliery Co., Ltd. . . .	550
Tirydail Colliery Co., Ltd. . .	460
Williams, Thomas & Sons (Llangennech)	500

Pembrokeshire—

Bonville's Court Coal Co., Ltd. .	324

APPENDIX IV

SALE CONTRACTS

The following is a form of Contract for the Sale of Coal, which was fixed by a Committee representing all the chief buyers and sellers in South Wales and Monmouthshire, and which is generally used :—

WELSH COAL CONTRACT, 19. F.O.B.

MEMORANDUM OF AGREEMENT entered into at......
......this............day of............19....
between.................of ....
hereinafter called the " Purchasers " and.........
.........hereinafter called the " Vendors."

1.—*Quantity*

The Purchasers agree to buy and the Vendors agree to sell..............tons of............
Large............Coal on the following conditions :—

2.—*Delivery*

The Purchasers will provide tonnage to take delivery of the Coal from the............to the ....in as nearly as possible equal proportions per calendar month. Such delivery shall be into Ship at one of the following Docks............

................as ordered, on application to the Vendors, before Ship's arrival.

3.—*Loading*

The time for Loading to be mutually agreed between the Purchasers and the Vendors when each Vessel is placed on stem, and being subjected in the case of Steam Vessels to the conditions and exceptions of Clauses 2 and 3 of the Chamber of Shipping Welsh Coal Charter, 1896, and in the case of Sailing Vessels to the Vendors' usual printed Form of Guarantee. In the event of the Purchasers and Vendors not being able to mutually arrange a stem, the Purchasers shall have the right to place a Vessel on stem for the portion of the month's deliveries then due on giving seven days' notice to the Vendors, in which case the Vendors shall be allowed customary hours for loading.

4.—*Trimming*

The Trimming shall be done by Trimmers who shall be some duly qualified persons selected by the Vendors or their Agents, and shall be appointed by the Purchasers at the Tariff Rates of the Port. The Vendors shall not be responsible for fire, explosion, or accident in connection with the Trimming, or for any act, default, or negligence of the Trimmers or Foremen.

5.—*Wharfage*

The Purchasers undertake to pay the Vendors or the Dock Company the usual Wharfage of 2d. per ton.

6.—*Quantity and Weights*

The Purchasers shall inspect the Coal in the Waggons on the tip-road at the place of shipment, and any objection to quality or condition shall be raised before shipment or be deemed to be waived. The returns of weight as ascertained at the Loading Tips by the Dock or Railway Company shall be final and conclusive for all purposes whatsoever.

7.—*Price*

The purchasers shall pay the Vendors the price of—
.... per ton for Colliery Screened Coal.
.... per ton if Single Screened at time of shipment.
.... per ton if Double Screened at time of shipment.

8.—*Payment*

The Purchaser shall make payment as follows
....

9.—*Strike and Accidents*

In the event of a stoppage or partial stoppage of the Vendors' Pits, or any of them, or on the Railway or Railways over which the Vendors' traffic is usually carried between any of the Vendors' Pits and the place of shipment, or at the Dock named as the place of shipment, or of a suspension of work by Trimmers, or Dock, Railway, or other hands connected with the working, delivery, or shipment of the said Coal, or from any cause whatever, whether or not of the same nature, the Vendors shall not be called upon to deliver any Coal during a total stoppage, and in the event of a partial stoppage the Vendors shall be entitled to a reduction in the deliveries proportionate to the reduction in their output. No liability shall attach to the Vendors for any such default of shipment, notwithstanding the fact that during such period Coal may have been shipped by them. The time within which the above named quantities are to be taken and delivered shall be extended for a period at least equal to that during which deliveries are suspended, or as otherwise arranged by mutual agreement.

10.—*Insolvency*

In case of any default on the part of the Purchasers in making any payment on the dates specified, the Vendors may suspend deliveries until such payment is made, and it shall be at their option whether they will afterwards make up any deficiency in deliveries so caused ; or if the Purchasers shall have been declared bankrupt, called any meeting of their creditors, or made any acknowledgment that they are unable to pay their debts in full, it shall be at the Vendors' option to consider the contract null and void.

11.—*Damage for failure to take or deliver*

Unless otherwise mutually arranged, Purchasers agree in the event of their failing to charter and

stem tonnage to take each month their regular
monthly quantity in accordance with the terms of
this contract to pay to the Vendors as and for
liquidated damages the difference between the
current price on the last day of such month and the
contract price on the quantity they have so failed
to take, and the Vendors agree that in the event of
their declining to accept tonnage to take each month
the regular monthly quantity in accordance with
the terms of this contract, to pay to the Purchasers
as and for liquidated damages the difference between
the current price on the last day of such month and
the contract price on the quantity for which they
have so declined to accept tonnage.

12.—*Prohibition of re-sale in the United Kingdom*

The Purchasers undertake, the whole of the Coal
herein named being purchased by them for *bona fide*
exportation, that no part of it will be sold by them,
either directly or indirectly, to any Export Merchants
or other person or persons in the United Kingdom, and
that none of this Coal shall be sent to the Vendors'
excepted ports, which are......................

The Purchasers agree to pay to the Vendors as
liquidated damages two shillings per ton for every
ton sold in violation of this clause, or the Vendors
may, at their option, decline to make further
deliveries under this contract.

13.—*War*

In the event of the United Kingdom being at
war with any European Power or any prohibition

being made by the British Government of the
export of Welsh Coal, the quantity due for
delivery under this Contract during the period over
which such war or prohibition extends shall be
cancelled.

APPENDIX V

EXTRACTS FROM COAL MINES ACT, 1911

PROVISIONS AS TO SAFETY

Ventilation

" **29.** (1) An adequate amount of ventilation shall be constantly produced in every mine to dilute and render harmless inflammable and noxious gases to such an extent that all shafts, roads, levels, stables, and workings of the mine shall be in a fit state for working and passing therein, and in particular that the intake airways up to within one hundred yards of the first working-place at the working-face which the air enters shall be normally kept free from inflammable gas. . . .

" (3) For the purposes of this section, a place shall not be deemed to be in a fit state for working or passing therein if the air contains either less than nineteen per cent. of oxygen or more than one-and-a-quarter per cent. of carbon dioxide, and an intake airway shall not be deemed to be normally kept free from inflammable gas if the average percentage of inflammable gas found in six samples of air taken by an inspector in the air current in that airway at intervals of not less than a fortnight exceeds one quarter : . . .

Safety Lamps

32.—(1) No lamp or light other than a locked safety lamp shall be allowed or used—

    (*a*) in any seam, where the air current in the return airway from any ventilating district in the seam is found normally to contain more than one half per cent. of inflammable gas ;

or

    (*b*) in any seam (except in the main intake airways within two hundred yards from the shaft) in which an explosion of inflammable gas causing any personal injury whatever has occurred within the previous twelve months, unless an exemption is given by the Secretary of State on the ground that, on account of the special character of the mine, the use of safety lamps is not required ;

    (*c*) in any place in a mine in which there is likely to be any such quantity of inflammable gas as to render the use of naked lights dangerous ;

    (*d*) in any working near to or approaching a place in which there is likely to be an accumulation of inflammable gas ;

    (*e*) in any place where the use of safety lamps is required by the regulations of the mine : . . .

34.—(1) In any mine or part of a mine in which safety lamps are required by this Act or the regulations of the mine to be used—

(i) A safety lamp shall not be used, unless it has, since last in use, been thoroughly examined at the surface by a competent person appointed in writing by the manager for the purpose and found by him in safe working order and securely locked, and a record shall be kept of the men to whom the several lamps are given out :

(ii) A competent person appointed in writing by the manager for the purpose shall also examine every lamp on its being returned, and if on such an examination any lamp is found to be damaged, he shall record the nature of the damage in a book to be kept at the mine for the purpose, and the damage shall be deemed to have been due to the neglect or default of the person to whom the lamp was given out, unless he proves that the damage was due to no fault of his own and that he immediately gave notice of the damage to the fireman, examiner, or deputy, or some other official of the mine appointed in writing by the manager for the purpose :

(iii) A safety lamp shall not be unlocked except at an appointed lamp station (which shall not be in a return airway) by a competent person appointed in writing by the manager for the purpose, nor, except in the case of electric hand lamps, shall it be relighted except by such a person at an appointed lamp station after examination by him, and no person other than such person as aforesaid shall have in his possession any contrivance for relighting or opening the lock of any safety lamp :

(iv) No part of a safety lamp shall be removed by any person whilst the lamp is in ordinary use.

Prohibition against Possession of Lucifer Matches

35.—(1) In any mine or part of a mine in which safety lamps are required by this Act or the regulations of the mine to be used, no person shall have in his possession any lucifer match nor any apparatus of any kind for producing a light or spark except so far as may be authorised for the purpose of shot firing or relighting lamps by an order made by the Secretary of State, or any cigar, cigarette, pipe, or contrivance for smoking.

(2) The manager of a mine in which, or in any part of which, safety lamps are required by this Act or by the regulations of the mine to be used, shall, for the purpose of ascertaining before the persons employed below ground in the mine or in the part of the mine, as the case may be, commence work whether they have in their possession any lucifer match or such apparatus as aforesaid or cigar, cigarette, pipe, or contrivance for smoking, cause either all those persons, or such of them as may be selected on a system approved by the inspector of the division, to be searched in the prescribed manner after or immediately before entering the mine or that part of the mine. . . .

Electricity

60.—(1) Electricity shall not be used in any part of a mine where, on account of the risk of explosion of gas or coal dust, the use of electricity would be dangerous to life.

(2) If at any time in any place in the mine the percentage of inflammable gas in the general body of the air in that place is found to exceed one and a quarter, the electric current shall at once be cut off from all cables and other electrical apparatus in that place, and shall not be switched on again as long as the percentage of inflammable gas exceeds that amount : . . .

(4) The use of electricity in any mine shall be subject to general regulations under this Act.

Explosives

61.—(1) The Secretary of State may, by order of which notice shall be given in such manner as he may direct, regulate the supply, use, and storage of any explosives at mines or any class of mines, and may, by any such order, prohibit the use of any explosive which appears to him of a kind to be or to be likely to become dangerous in mines or any class of mines, either absolutely or subject to such conditions as may be prescribed by the order.

(2) No explosives shall be taken into or used in any mine except explosives provided by the owner, and the price, if any, charged by the owner to the

workman for any explosives so provided shall not exceed the actual net cost to the owner.

Prevention of Coal Dust

62. In every mine, unless the floor, roof, and sides of the roads are naturally wet throughout,—

(1) arrangements shall be made to prevent, as far as practicable, coal dust from the screens entering the downcast shaft ; and, in the case of a mine newly opened after the passing of this Act, no plant for the screening or sorting of coal shall be situated within a distance of eighty yards from any downcast shaft unless a written exemption is given by the inspector of the division ;

(2) the tubs shall be so constructed and maintained as to prevent, as far as practicable, coal dust escaping through the sides, ends, or floor of the tubs, but any tub which was in use in any mine at the date of the passing of this Act may, notwithstanding that it is not so constructed, continue to be used in that mine for a period of five years from the said date ;

(3) the floor, roof and sides of the roads shall be systematically cleared so as to prevent, as far as practicable, coal dust accumulating ;

(4) Such systematic steps, either by way of watering or otherwise, as may be laid down by the regulations of the mine shall be taken to

prevent explosions of coal dust occurring or being carried along the roads ;

(5) The roads shall be examined daily and a report (to be recorded in a book kept at the mine for the purpose) made on their condition as to coal dust and on the steps taken to mitigate arising therefrom.

Inspections as to Safety

63. For the purpose of the inspections before the commencement of work in a shift hereinafter mentioned, one or more stations shall be appointed at the entrance to the mine or to different parts of the mine, as the case may require, and no workman shall pass beyond any such station until the part of the mine beyond that station has been examined and reported to be safe in manner hereinafter mentioned.

64. (1) The firemen, examiners or deputies of a mine shall, within such time not exceeding two hours immediately before the commencement of work in a shift as may be fixed by the regulations of the mine, inspect every part of the mine situated beyond the station or each of the stations, and in which workmen are to work or pass during that shift, and all working places in which work is temporarily stopped within any ventilating district in which the men have to work, and shall ascertain the condition thereof so far as the presence of gas, ventilation, roof and sides, and general safety are concerned.

3 F

(2) Except in the case of a mine in which inflammable gas is unknown, the inspection shall be made with a locked safety lamp, and no other light shall be used during the inspection.

(3) A full and accurate report specifying whether or not, and where, if any, noxious or inflammable gas was found, and whether or not any, and, if any, what defects in roofs or sides and other sources of danger were observed, shall be recorded without delay in a book to be kept at the mine for the purpose and accessible to the workmen, and such report shall be signed by, and, so far as the same does not consist of printed matter, shall be in the handwriting of, the person who made the inspection.

(4) For the purpose of the foregoing provisions of this section, two or more shifts succeeding one another so that work is carried on without any interval are to be deemed to be one shift.

65. A similar inspection shall be made twice at least in the course of each shift of all parts of the mine situated beyond the station or each of the stations aforesaid and in which workmen are to work or pass during that shift, but it shall not be necessary to record a report of the first of such inspections in a book : Provided that, in the case of a mine worked by a succession of shifts, no place shall remain uninspected for an interval of more than five hours.

APPENDIX VI

RULES OF THE MINERS' FEDERATION OF GREAT
BRITAIN.

Name and Place of Business

1.—This Federation shall be called " The Miners'
Federation of Great Britain," its office or place of
business shall be at 925, Ashton Old Road, Manchester, or at such other place as may at times be
most convenient. It shall consist of Federations
and Districts who are eligible to join, by paying an
Entrance Fee of One Pound per Thousand Members,
or fractional part thereof, but no section of a County,
where a Federation or County Association exists,
shall be eligible to join.

Objects of the Federation

(1) To provide funds to carry on the business of
this Federation, the same to be disbursed
as provided in the following rules.

(2) To take into consideration the question of
Trade and Wages, and to protect Miners
generally.

(3) To seek to secure Mining Legislation affecting
all Miners connected with this Federation.

(4) To call Conferences to deal with questions affecting Miners, both of a Trade, Wage, and Legislative character.

(5) To seek and obtain an eight hours' day from bank to bank in all mines for all persons working underground.

(6) To deal with cases of accidents, and attend inquests upon persons killed, in and about mines, whenever the Executive Committee considers it necessary to do so.

(7) To assist all Federations and Districts, in law cases, where they may have to appeal, or are appealed against, on decisions of the lower Courts.

(8) To provide funds to pay for the election expense of Labour Candidates and support Members who may be returned to the House of Commons.

Officials and Executive Committee

3.—There shall be in connection with this Federation a President, Vice-President, Secretary, and Treasurer, and an Executive Committee of not less than twelve members exclusive of officers.

Election of Officers

4.—The Officials of this Federation shall be elected or re-elected annually at the Yearly Conference. Each Official and Committee shall be duly nominated by a financial Federation or District.

5.—No person shall be eligible for election or re-election whose Federation or District is more than one month in arrears in contributions or levies.

Auditors

6.—There shall be two Auditors appointed annually, at the Annual Conference, to Audit the Accounts of the Federation.

Trustees

7.—There shall be three Trustees appointed who must be Members of the Federation, and shall continue in office during the pleasure of the Federation. In the event of any of such Trustees dying or being removed from office, at the first meeting of the Federation after the vacancy has been reported to Districts, another or others shall be appointed to supply such vacancy. They shall do and execute all the functions required of them by the Federation. They shall attend at the Audit, and examine the Treasurer's Bank Books, and Bonds, and Documents of Investments.

Annual Conference

8.—There shall be an Annual Conference of this Federation held in the month of October each year.

9.—Special Conferences shall be called by the President and Secretary when necessary.

10.—That all the Officials' Salaries shall be fixed at the Annual Conference.

Voting at Conference

11.—That the voting at all Conferences shall be by show of hands, but, in the event of a District or Federation claiming a vote by numbers, the voting shall be One Vote for every One Thousand financial members or fractional part thereof.

12.—No Official shall be allowed to vote at any Conference except the Chairman, and then only when there is a tie he shall give the casting vote.

Committee Meetings

13.—The Executive Committee shall meet when required. The President and Secretary shall call all Committee Meetings.

14.—The Executive Committee shall in the absence of a Conference, take into consideration all questions affecting the Mining interests, appoint Delegates to attend Coroners' Courts, Appeal Cases, inquire into Disputes on General Wage Questions, and the watching of legislation affecting the Members of the Federation. In cases of emergency the President and Secretary shall have power to appoint representatives to attend an inquiry.

Contributions

15.—That the ordinary contributions to this Society shall be at the rate of One Penny per Quarter per member, and that the same be paid the first week in each quarter to the Treasurer. The quarters to commence January, April, July and October in each year. The Executive Committee shall have power to call levies when necessary.

16.—That any Federation or District allowing its contributions or levies to fall into arrears shall not be entitled to any financial support until three months after all arrears have been paid up ; neither shall arrears be paid up at any time for the purpose of obtaining support in any case from the funds of this Federation.

When Support shall be Given

17.—That whenever any stoppage of collieries occurs, arising out of any action taken by a Conference, a Special Conference shall be called to determine whether support shall be given to any Federation or District.

18.—The Conference shall, after duly considering each dispute, have power to raise by levy upon the members of the Federation such sum as will meet the requirements.

19.—No Federation or District shall receive support unless more than 15 per cent. of the said Federation or District is out of work, consequent upon any action taken by a General Conference.

20.—Members shall be supported, in accordance with Rules 17 and 18, who may have been out of work twenty-one clear days from the commencement of the dispute ; and that pay at the rate of 7s. 6d. per week per member shall cómmence on the first day after the expiration of fourteen days from the commencement of dispute.

Defensive Action

21.—That whenever any Federation or District is attacked on the Wages Question, or the conditions of labour, or with the approval of the Conference especially called for that purpose, has tendered notice to improve the conditions of labour or to obtain an advance in wages, a Conference shall be called to consider the advisability of joint action being taken.

Standing Orders

The Conference (after first day) meet each day at ten a.m., adjourn at one, meet again at two p.m., and adjourn again at four.

The movers of resolutions be allowed for speaking ten minutes, and subsequent speakers five minutes. Replies to be confined to movers of resolutions only.

E. EDWARDS, President.
THOS. ASHTON, Secretary.

APPENDIX VII

VARIATIONS IN PERCENTAGE ADDITIONS TO STANDARD WAGES IN DIFFERENT DISTRICTS[1]

	Perc'ntage of wages above Standard at end of 1896.	Percentage Increase (+) or Decrease (−) on or off Hewers' Standard wages in years.						Perc'ntage of wages above Standard at end of 1912.
		1897-1900.	1901-1905.	1906-1907.	1908-1909.	1910-1911.	1912.	
Northumberland	3¾	+57½	−46¼	+32½	−17½	−2½	+11¼	38¾
Durham	15	+50	−37½	+26¼	−8¾	−6¼	+7½	46¼
Cumberland	30	+30	−22½	+17½	−7½	..	+7½	55
Federated Districts	30	+20	−10	+15	−5	..	+5	55
Sth. Staffs. & East Worcestershire	30	+20	−10	+15	−5	..	+5	55
Forest of Dean	15[2]	+35	−25	+20	−10	+5	..	35
Bristol	17½ / 22½[3]	+22½	−10	+15	−5	..	+5	45 / 50
Somersetshire (Radstock)	15	+27½	−15	+17½	−5	..	+5	45
South Wales & Mon.	10	+63¾	−43¾	+30	−12½	+2½	+7½	57½
Fifeshire & Clackmannan	4[4]	+97½	−60	+50	−37½	..	+18¾	68¾
West Scotland	12½	+81½	−62½	+50	−37½	..	+18¾	68¾

[1] From Board of Trade Report, "Changes in Wages and Hours of Labour" (Cd. 7080).

[2] At certain collieries the percentage above standard was 5 per cent. more.

[3] The lower percentage was on the Gloucestershire side of the district, the higher on the Somersetshire side.

[4] At Standard

APPENDIX VIII

SOUTH WALES MINERS' FEDERATION
WAGES AGREEMENT OF 1910 (TO REMAIN IN FORCE
UNTIL MARCH 31, 1915).

Memorandum of Agreement

made this Eighth day of April, one thousand nine
hundred and ten, BETWEEN the undersigned
Owners' Representatives duly authorised to act on
behalf of the Owners of Collieries in Monmouthshire
and South Wales, whose names or titles are set
forth in the Schedule hereto (hereinafter called " the
Owners ") of the one part, and the undersigned
Workmen's Representatives, duly authorised to act
on behalf of the Workmen (excepting Enginemen,
Stokers, and Outside Fitters) now employed at the
Collieries of the said Owners (hereinafter called
" the Workmen ") of the other part, whereby it is
mutually agreed as follows :—

1.—That a Board of Conciliation shall be established to determine the general rate of wages to be
paid to the Workmen, and to deal with disputes at
the various Collieries of the Owners subject to the
conditions hereinafter mentioned.

2.—The title of the Board shall be " The Board

of Conciliation for the Coal Trade of Monmouthshire and South Wales " hereinafter called " The Board."

3.—The Board shall consist of 24 duly authorised Owners' Representatives and 24 duly authorised Representatives of the Workmen employed at the Collieries of the Owners, and when dealing with questions relating to general advances or general reductions in the rates of wages also of a Chairman from outside who shall not be financially interested in any Coal Mine in the United Kingdom, and who shall have a casting vote only.

4.—The first Chairman of the Board from outside as aforesaid shall be Viscount St. Aldwyn, who is hereinafter called " the Chairman."

The Representatives on the Board of the Owners and Workmen respectively shall be appointed, and notice of such appointment given to the Secretaries, on or before the 30th day of April instant.

Wages

There shall be elected from the Members of the Board two Presidents, one elected by the Owners' Representatives and the other by the Representatives of the workmen.

Whenever a vacancy on the Board occurs from any cause (except in the office of Chairman) such vacancy shall be filled by the body which appointed the Member whose seat has become vacant, but during such vacancy the Board may transact the business of the Board. Intimation of such appointment shall be at once sent to the Secretaries. When and so often as the Office of Chairman becomes vacant the Board shall endeavour to elect a Chairman, and should they fail to agree will ask the Lord Chief Justice of England for the time being or, in case of his refusal, the Speaker of the House of Commons, to nominate one.

5.—The parties to this Agreement pledge their respective constituents to make every effort possible to avoid difficulties or disputes at the Collieries, and in case of any unavoidable difference the Owners or their Officials together with their Workmen or their Agent or Agents shall endeavour to settle all matters at the Collieries, and only in case of failing to effect a settlement shall a written appeal setting forth clearly the facts of the dispute and the contention of the parties making the claim be made to the Board by either or both of the parties concerned in the dispute to consider the same, and no notice to terminate contracts shall be given by either Owners or their Workmen before the particular question in dispute shall have been considered by the Board, and it shall have failed to arrive at an Agreement. The Board shall have power to refer such questions to a Committee consisting of one or more Owners' Representatives and an equal number of Workmen's Representatives, all of whom shall be members of the Board, to consider, and if so directed, with power to settle, and in all cases to report to the Board either a settlement or a failure to agree within three calendar months from the date of the reference to such Committee, and should the Board then fail to arrive at an agreement within one month or any extended period that may be agreed upon by the Board either party may give notice to terminate contracts. Any notices wrongfully given to terminate contracts on any question shall be withdrawn before the Board or any Committee thereof shall consider such question.

Both parties hereby respectively undertake to make every effort possible to secure the loyal observance by the Owners and workmen respectively of any Award made by representatives of the Board on any questions which may have been referred to them by the Board.

6.—Rules of procedure for the conduct of the business of the Board are set forth at the end hereof, and the same shall be deemed to be incorporated with and to form part of this Agreement.

7.—The Mineral to be gotten is clean large coal only as hereinafter described.

The cutting prices to be paid to the Collier shall be the several standard prices prevailing and paid at the Collieries of the Owners respectively.

Such standard cutting price shall be paid upon the weight of the large coal to be ascertained in manner hereinafter appearing, and includes all services in respect of the small coal necessarily produced in filling the large coal, in conveying it from the working places to the screen at the surface, and in the process of screening, that price being equal to the value of all the services involved in getting such large coal and small coal, and being more than the value of the services rendered in respect of the large coal only.

The respective weights of such large coal and small coal for the purpose of paying the Collier shall be ascertained as follows :—

After each tram of coal is brought to the weighing machine it shall be weighed, and the tare of the tram shall be deducted from the gross weight. The coal shall then be tipped over the screen in use at the Colliery to separate the small coal passing through the screen from the large coal passing over it.

The small coal which shall pass through the screen shall be weighed, and that weight shall be deducted from the gross weight of the coal in the tram in order to ascertain the weights of such large screened coal and small coal respectively, and the cutting price paid to the collier upon the weight of the large screened coal as aforesaid shall, during the continuance of this Agreement, be deemed to be the value of the services rendered in respect of both the large screened coal and small coal, the weights of which respectively shall be ascertained as aforesaid.

8.—It is distinctly understood that Clause 7 in this Agreement is not intended to change the system of weighing and screening the coal as it at present exists, but the Owners shall be at liberty to adopt such improved methods of screening and cleaning as they may consider necessary, provided that any methods so adopted shall not in any way prejudicially affect the wages of the workmen.

9.—Clause 7 shall not apply to or alter or in any way interfere with any agreements now existing or hereafter to be made for payment for through and through coal or where small coal is now separately paid for.

10.—The Board shall at the meetings held under Rule 6 of the said Rules of Procedure determine the general rate of wages to be paid for the three months commencing on the first day of the month following the dates of such meetings, but should neither party desire to vary the rate of wages, the then prevailing rate of wages shall continue until the same shall be varied in accordance with the said Rules of Procedure.

(a) All standard rates and prices shall be the Standards known as the Standards of December 1879 and 1877 respectively.

(b) The wages payable to the workmen shall until the same is advanced or reduced be 50 per cent. above the several rates actually paid at the respective Collieries under the Standard of December, 1879.

(c) During the continuance of this agreement the rate of wages shall, subject to subsection (d) hereof, not be less than 35 per cent. above nor more than 60 per cent. above the December, 1879, Standard of wages paid at the respective Collieries. The minimum of 35 per cent. above the December, 1879, Standard of wages shall, subject to subsection (*d*) hereof, be paid when the average net selling price of large coal is at or below 12s. 5d. per ton f.o.b. When the nett selling price of large coal reaches 14s. and does not exceed 14s. 9d. per ton f.o.b., the rate of wages shall, subject to sub-section (*d*) hereof, be 50 per cent. above the rates paid under the Standard of December, 1879, and when the nett selling price exceeds 14s. 9d. per ton f.o.b. the workmen shall be entitled to claim advances in the general rate of wages in excess of the 50 per cent. and up to the said maximum of 60 per cent., but in cases of claims to advances above 50 per cent., 50 per cent. shall be taken to be the equivalent of 14s. 9d. per ton f.o.b., and in the case of claims to reductions 50 per cent. shall be taken to be the equivalent of 14s. per ton f.o.b. The average nett selling prices shall be taken as for large Colliery screened coal delivered f.o.b. at Cardiff, Barry, Newport, Swansea, Port Talbot, and Llanelly.

(*d*) At Collieries where the Standard or basis upon which wages are now regulated is the rate of wages paid in the year 1877 the percentage payable thereat shall be 15 per cent. less than at the Collieries where the 1879 Standard prevails, and in cases where workmen have hitherto been paid net rates of wages or fixed or other percentages whether upon the 1877, 1879 Standards, or any other existing Standards, they shall continue to be paid such net rates fixed or other percentages only.

11.—At the Collieries under this Agreement all wages due to the workmen shall be paid once in each fortnight, provided that at those Collieries where wages are now paid weekly such practice shall continue in force.

12.—The hours of labour of workmen employed below ground at the said Collieries respectively shall be such as are authorised by the Coal Mines Regulation Act, 1908, except that such workmen shall not be under any obligation to work the extended hours mentioned in Section 3 (1) of the said Act, and that notwithstanding the limitation of hours to be worked under the said Act no alteration shall be made in the Standard rates and prices hitherto paid to such workmen during the continuance of this Agreement.

13.—Where payment of six turns for five worked by night has hitherto been paid the same shall continue. As under the said Act no overtime can be worked other than in cases of emergency as defined in Section 1 (sub-section 2) payment for overtime will cease. In cases of dispute as to whether any overtime or extra turns paid previous to the coming into operation of the said Act were in part paid for work done in the workmen's ordinary hours of work the same shall be referred to a joint Committee of the Conciliation Board with power to settle, and if the decision is in favour of the workman it shall date back to the time the dispute is

3 G

placed on the Agenda. In cases of failure to settle either side may determine the Contracts of the workmen affected by a month's notice to be given on the first day of the following month.

14.—The Owners will not press for double shift in the face, but shall be given an effective afternoon shift of such number of workmen as are required by the Owners for clearance purposes, repairing, double shift in headings and places that require to be pressed on for opening the Collieries.

Where six shifts for five are now paid at night the Owners shall pay six shifts for five in the afternoon shift, and where six shifts are paid for six shifts worked at night, the Owners shall only be required to pay six shifts for six in the afternoon.

15.—An overlapping shift shall be worked where required by the Owners, such shift shall start not earlier than 6 a.m. and not later than 9 a.m. On Saturdays this shift shall start and finish at the same time as the first shift.

The two sides of the Board shall unite in procuring an amendment of the Eight Hours Act making this early starting on Saturday legal.

16.—That where Sunday night shifts are worked they shall be eight hours shifts, only one shift to be paid.

17.—The mealtime for underground day wage-men (day and night) shall be twenty minutes, which shall be so arranged as not to interfere with the haulage and general working of the Colliery ; and in the case of all workmen connected with the winding, whether employed upon the surface or below ground, the mealtime shall be twenty minutes, and shall be so arranged as to secure the continuous winding of coal without interruption during the shift.

18.—Workmen on the surface engaged in handling the coal shall work half an hour per day beyond the coal winding time, either starting fifteen minutes before coal winding and working fifteen minutes afterwards, or, at the option of the Owners, working the half hour after winding, it being agreed that the hours of working of such workmen shall be $8\frac{1}{2}$ hours per day. The only workmen intended to come under the operation of this clause are those who handle the trams between the cage and the tipplers, the screenmen, slag pickers, and wagonmen. Banksmen are excluded.

19.—Where serious but non-fatal accidents occur an agreed number of men (but not more than 20) to be selected by the Management may accompany the injured workman out of the pit.

20.—No stoppage for funerals shall take place except by arrangement with the Management.

21.—The workmen shall be entitled to twelve General Holidays in each year, which shall include all Bank Holidays and Federation Demonstration Holidays, the dates to be agreed upon by the Board ; and the workmen's representatives agree that in arranging such holidays they will issue instructions to the workmen requesting them to resume work immediately after such holidays have terminated, and will use every effort to see that such instructions are complied with.

At any Collieries where " Mabon's Day " now exists, the custom of keeping such a holiday shall be forthwith abolished.

22.—The prices to be charged to workmen entitled to house coal for their own domestic purposes shall be the same prices as are fixed to be paid by the award of Sir David Dale, Bart., of the 11th day of July, 1903.

23.—During the continuance of this Agreement all notices to terminate individual contracts on the part of the Owners as well as on the part of the workmen shall be given on the first day of any calendar month and shall terminate upon the last day of the same month, provided that if the first day of any calendar month fall on a Sunday, the notice shall be given on the previous Saturday.

24.—Subject as aforesaid the Owners and Workmen at the respective Collieries shall be bound to observe and fulfil and shall be subject to all customs, provisions and conditions existing in December, 1899, at the Collieries respectively, and no variation shall be made therein by the Owners or workmen except by mutual arrangement at the Collieries respectively, or by a decision of the Board after a reference thereto in accordance with the provisions of Clause 5 of any proposal for a variation.

25.—Subject to the provisions of Clause 10 (c) hereof nothing in this Agreement or in the Rules of Procedure is to preclude either party bringing any matter before the Board or independent chairman which they consider as factors bearing upon the General Wage Question, but any evidence brought forward as to the selling price of large coal shall be confined to the price of large coal delivered f.o.b. at the shipping ports named in Clause 10 (c) hereof in the three calendar months immediately preceding the first day of the month prior to the month in which the meeting is held to consider any proposal to vary the General Wage Rate.

26.—This Agreement shall continue in force from the 1st April, 1910, until the 31st March, 1915, and thenceforth until either party gives to the other three calendar months' notice terminating the same, such notice to be given to the Secretary of such other party in writing, or left at his usual or last known address. Upon the termination of this Agreement all contracts of service between the Owners respectively and their workmen respectively shall cease.

27.—A copy of this Agreement shall be placed in a Contract Book at each Colliery, which shall be signed by or on behalf of the Owners of such Colliery and also by each workman employed thereat, as one of the terms of the engagement between the Owners and the said workmen.

Rules of Procedure

1.—The Constituents of the Board, *i.e.*, Owners' Representatives and Workmen's Representatives, are for brevity herein referred to as "the Parties."

2.—The Meetings of the Board shall be held at Cardiff or such other place as the Board may from time to time determine.

3.—Each of the parties shall appoint a Secretary, and shall give notice of such appointment when made to the other party, and such Secretaries shall remain in office until they shall resign, or be withdrawn by the parties appointing them. The Secretaries, or their respective deputies for the time being, shall attend all Meetings of the Board, and be entitled to take part in the discussion, but they shall have no power to move or second any resolution or vote on any question before the Board.

4.—The Secretaries shall conjointly convene all Meetings of the Board and record the names of the persons present thereat, and at all Meetings held under Rule 6 of the Rules of Procedure full minutes of the proceedings thereof shall be taken under the conjoint supervision of the Secretaries by an official shorthand writer to be mutually agreed on by the parties, which minutes shall be transcribed into duplicate books, and each such book shall be signed by the Presidents or other persons presiding at the Meetings at which such minutes are confirmed. One of such minute books shall be kept by each of the Secretaries, such minutes to be for the private use of the Board and not for publication. The Secretaries shall also conduct correspondence for their respective parties and conjointly for the Board.

5.—The Board shall meet once at least in each month for the purpose of dealing with difficulties or disputes arising at the several Collieries, and referred to in Clauses 5 and 24 of the foregoing Agreement, and the same shall be dealt with by the Board without reference to the Chairman. The Secretaries shall give to each Member of the Board seven days' notice of the intention to hold any such Meeting and of the business to be transacted thereat, and, except by mutual agreement, no subject shall be considered which has not been placed on the Agenda to the Notice convening the Meeting.

6.—Should there be a desire by either party to vary the rate of wages the Board shall meet to consider the same on the 10th day of the months of February, May, August, and November in every year (except when the 10th day of any of the said months falls on a Sunday, when the Meeting shall be held on the following day), to determine the general rate of wages to be paid for the three months commencing on the first day of the month next following the date of such Meetings. Either party intending to propose at such Meetings any alteration in the general rate of wages shall, ten days before the said 10th day of the months of February, May, August, and November for holding such Meetings, give to the

Secretary of the other party notice in writing of the proposition intended to be made, and of the grounds thereof, and the Secretaries shall enter such intended proposition and the grounds thereof on the Agenda to the notice convening the Meeting. The Secretaries shall send to each member of the Board seven days' notice of each such Meeting and of the business to be transacted thereat.

At all such last mentioned Meetings the questions to be dealt with thereat shall, in the first instance, be considered by the Board, it being the desire and intention of the parties to settle any differences which may arise by friendly conference if possible. If the parties on the Board cannot agree then the Meeting shall be adjourned for a period not exceeding twelve days, to which adjourned Meeting the Chairman shall be summoned, and shall attend and preside thereat, when the questions in difference shall be again discussed by the parties, and in the event of their failing to arrive at an agreement with regard thereto, the Chairman either at such Meeting or within five days thereafter shall give his casting vote on such questions, and the parties shall be bound thereby.

7.—Both Presidents shall preside at all Meetings (other than at Meetings at which it shall be the duty of the Chairman to preside in accordance with Clause 6 of these Rules), but if either or both of them shall be absent then a member or members of the Board shall be elected by the respective parties to preside

at such Meetings according as such Presidents who
shall be absent shall represent the Owners or workmen. The Presidents or other persons presiding
shall vote as representatives, but shall have no other
votes.

8.—All questions submitted to the Board shall be
stated in writing and may be supported by such
verbal, documentary, or other evidence and explanation as either party may submit, subject to the
approval of the Board.

9.—Each party shall pay and defray the expenses
of its own Representatives, Secretary, and Accountant ; but the costs and expenses of the Chairman,
Official Shorthand Writer, Joint Auditors (if any)
and of the stationery, books, printing, and hire of
rooms for Meeting shall be borne by the respective
parties in equal shares.

APPENDIX IX

Extracts from Coal Mines (Minimum Wage) Act, 1912

Minimum Wage for Workmen Employed Underground in Coal Mines

1.—(1) It shall be an implied term of every contract for the employment of a workman underground in a coal mine that the employer shall pay to that workman wages at not less than the minimum rate settled under this Act and applicable to that workman, unless it is certified in manner provided by the district rules that the workman is a person excluded under the district rules from the operation of this provision, or that the workman has forfeited the right to wages at the minimum rate by reason of his failure to comply with the conditions with respect to the regularity or efficiency of the work to be performed by workmen laid down by those rules ; and any agreement for the payment of wages in so far as it is in contravention of this provision shall be void.

For the purposes of this Act, the expression " district rules " means rules made under the powers given by this Act by the joint district board.

(2) The district rules shall lay down conditions as respects the district to which they apply, with respect to the exclusion from the right to wages at the minimum rate of aged workmen and infirm workmen (including workmen partially disabled by illness or accident), and shall lay down conditions with respect to the regularity and efficiency of the work to be performed by the workmen, and with respect to the time for which a workman is to be paid in the event of any interruption of work due to an emergency, and shall provide that a workman shall forfeit the right to wages at the minimum rate if he does not comply with the conditions as to regularity and efficiency of work, except in cases where the failure to comply with the conditions is due to some cause over which he has no control.

The district rules shall also make provision with respect to the persons by whom and the mode in which any question, whether any workman in the district is a workman to whom the minimum rate of wages is applicable, or whether a workman has complied with the conditions laid down by the rules, or whether a workman who has not complied with the conditions laid down by the rules has forfeited his right to wages at the minimum rate, is to be decided, and for a certificate being given of any such decision for the purposes of this section.

(3) The provisions of this section as to payment of wages at a minimum rate shall operate as from the date of the passing of this Act, although a

minimum rate of wages may not have been settled, and any sum which would have been payable under this section to a workman on account of wages if a minimum rate had been settled may be recovered by the workman from his employer at any time after the rate is settled.

Settlement of Minimum Rates of Wages and District Rules

2.—(1) Minimum rates of wages and district rules for the purpose of this Act shall be settled separately for each of the districts named in the Schedule to this Act by a body of persons recognised by the Board of Trade as the joint district board for that district.

Nothing in this Act shall prejudice the operation of any agreement entered into or custom existing before the passing of this Act for the payment of wages at a rate higher than the minimum rate settled under this Act, and in settling any minimum rate of wages the joint district board *shall have regard to the average daily rate of wages paid to the workmen of the class for which the minimum rate is to be settled.*

(2) The Board of Trade may recognise as a joint district board for any district any body of persons, whether existing at the time of the passing of this Act or constituted for the purposes of this Act, which in the opinion of the Board of Trade fairly and adequately represents the workmen in coal mines in the district and the employers of those

workmen, and the chairman of which is an independent person appointed by agreement between the persons representing the workmen and employers respectively on the body, or in default of agreement by the Board of Trade.

The Board of Trade may, as a condition of recognising as a joint district board for the purposes of this Act any body the rules of which do not provide for securing equality of voting power between the members representing workmen and the members representing employers and for giving the chairman a casting vote in case of difference between the two classes of members, require that body to adopt any such rules as the Board of Trade may approve for the purpose, and any rule so adopted shall be deemed to be a rule governing the procedure of the body for the purposes of this Act.

(3) The joint district board of a district shall settle general minimum rates of wages and general district rules of their district (in this Act referred to as general district minimum rates and general district rules), and the general district minimum rates and general district rules shall be the rates and rules applicable throughout the whole of the district to all coal mines in the district and to all workmen or classes of workmen employed underground in those mines, other than mines to which and workmen to whom a special minimum rate or special district rules settled under the provisions of this Act is or are applicable, or mines to which and workmen

to whom the joint district board declare that the general district rates and general district rules shall not be applicable pending the decision of the question whether a special district rate or special district rules ought to be settled in their case.

(4) The joint district board of any district may, if it is shown to them that any general district minimum rate or general district rules are not applicable in the case of any group or class of coal mines within the district, owing to the special circumstances of the group or class of mines, settle a special minimum rate (either higher or lower than the general district rate) or special rules (either more or less stringent than the general district rules) for that group or class of mines, and any such special rate or special rules shall be the rate or rules applicable to that group or class of mines instead of the general district minimum rate or general district rules.

(5) For the purpose of settling minimum rates of wage, the joint district board may subdivide their district into two parts or, if the members of the joint district board representing the employers agree, into more than two parts, and in that case each part of the district as so subdivided shall, for the purpose of the minimum rate, be treated as the district.

(6) For the purpose of settling district rules, any joint district boards may agree that their district shall be treated as one district, and in that case those districts shall be treated for that purpose

as one combined district, with a combined district committee appointed as may be agreed between the joint district boards concerned, and the chairman of such one of the districts forming the combination as may be agreed upon between the joint district boards concerned, or, in default of agreement, determined by the Board of Trade, shall be the chairman of the combined district committee.

Revision of Minimum Rates of Wages and District Rules

3.—(1) Any minimum rate of wages or district rules settled under this Act shall remain in force until varied in accordance with the provisions of this Act.

(2) The joint district board of a district shall have power to vary any minimum rate of wages or district rules for the time being in force in their district—

(a) at any time by agreement between the members of the joint district board representing the workmen and the members representing the employers ; and

(b) after one year has elapsed since the rate or rules were last settled or varied, on an application made (with three months' notice given after the expiration of the year) by any workmen or employers, which appears to the joint district board to represent any considerable body of opinion amóngst

either the workmen or the employers concerned ;

and the provisions of this Act as to the settlement
of minimum rates of wages or district rules shall, so
far as applicable, apply to the variation of any such
rate or rules.

APPENDIX X

DISTRICT RULES FOR THE DISTRICT OF SOUTH WALES
(INCLUDING MONMOUTHSHIRE)

1.—The following Rules shall apply to the work
ing of all Coal Mines (Minimum Wage) Act, 1912
hereinafter called " the Act " within South Wales
and Monmouthshire.

2.—In these Rules the word " Workman " means
any person to whom the Coal Mines (Minimum
Wage) Act, 1912, applies, the word " pay " means
the period in respect of which the workman's wages
are for the time being payable, and the word "day"
means a colliery working day.

3.—A workman who has reached 63 years of age
shall be regarded as an aged workman, within the
meaning of the Act, and shall be excluded from the
right to wages at the minimum rate. A workman who
from physical causes is unable to do the work ordinarily done by a man in his position in the mine or
who is partially disabled by illness or accident shall
be regarded as an infirm workman within the meaning
of the Act and shall be excluded from the right to
wages at the minimum rate. Where there is no
disagreement as to whether a workman has reached
the age of 63 years or is infirm or partially disabled

3 H

by illness or accident, a Certificate signed by the workman affected and the Manager of the Mine shall be conclusive evidence in reference thereto. Provided that in a case of a workman partially disabled by illness or accident such Certificate shall only apply during the period of such partial disablement.

4.—A workman shall forfeit his right to wages at the minimum rate on any day on which he delays in going to his working place or work at the proper time, or leaves his working place or work before the proper time, or fails to perform throughout the whole of the shift his work with diligence and efficiency and in accordance with the reasonable instructions of the Official having charge of the district in which such workman shall be engaged.

5.—A workman shall regularly present himself for work when the Colliery is open for work, and shall forfeit his right to wages at the minimum rate during any pay in which he has not worked at least five-sixths of his possible working days, unless prevented from working by accident or illness. Provided that a workman in a colliery open for work for less than 6 days in any pay shall not forfeit his right to wages at the minimum rate by absence from the Colliery not caused by accident or illness for one day during that pay, if he has worked on every day during the previous pay on which the Colliery was open for work and he was not prevented by accident or illness from working. In case of accident or illness the workman shall if required submit himself to the examination of a duly qualified medical man to be appointed by the employer ; and in case he shall refuse to do so he shall forfeit his right to wages at the minimum rate during that pay.

Every Collier or Collier's helper shall at all times work, get and send out the largest possible quantity of clean coal contracted to be gotten from his working place and shall perform at least such an amount of work as, at the rates set forth in the price list or other agreed rates applicable, would entitle him to earnings equivalent to the minimum rate. If at any time any workman shall in consequence of circumstances over which he alleges he has no control be unable to perform such an amount of work as would entitle him under the price list or other agreed rates to a sum equal to the daily minimum rate, then and in such case he shall forthwith give notice thereof to the official in charge of the district in which he shall be engaged, and if such official shall not agree that the workman cannot earn at the work upon which he shall be engaged a sum under the price list or other agreed rates equal to the daily minimum rate, then the matter shall be decided in the manner provided by Rule 8. The Management shall be at liberty to remove the workman to some other part of the Colliery.

If any workman shall act in contravention of this rule he shall forfeit the right to wages at the minimum rate for the pay in which such contravention shall take place.

6.—If a case of emergency in or about or connected with the Colliery shall render a workman's services for the time being unnecessary, and such workman shall be informed of such emergency when or before he reaches the pit bottom or a station within 300 yards therefrom, then such workman shall forthwith return to the surface (facilities being given) and shall not be entitled to any payment in respect of that shift. If the workman travels to his working place and is there informed or discovers that something has happened to prevent him working in his place and is offered but refuses other work which he may properly be called upon to perform, he shall not be entitled to claim any wages in respect of that shift. In the event of any interruption of work during the shift of any workman due to an emergency over which the management has no control whereby he shall be prevented from working continuously until the end of his shift, then he shall be entitled only to such a proportion of the minimum rate for the shift as the time during which he shall have worked shall bear to the total number of hours of such shift. Facilities shall be given to enable him to ascend the mine as soon as practicable.

7.—(1) In ascertaining whether the minimum wage has been earned by any workman on piecework the total earnings during two consecutive weeks shall be divided by the number of shifts and parts of shifts he has worked during such two weeks.

Upon the average earnings of any workman for two weeks being ascertained in accordance with this rule, the wages of such workman shall be adjusted and the amount found to be due to or from him ascertained and paid or debited to him as the case may be, and in the latter event the amount debited shall be deemed to be a payment on account of wages to become subsequently due to him.[1]

(2) In cases where workmen are working as partners on shares and pooling their earnings no member of such partnership shall be entitled to be made up to the minimum rate if the average earnings per day of the set over the whole week shall amount to the minimum rate.

(3) In ascertaining the earnings of workmen employed upon piece-work for the purposes of the minimum wage there shall not be deducted from the gross earnings for the helper more than the actual wages paid to the helper by the workman. All rates of wages so paid to the helper by the workman shall be registered with the Management. No workman on piece-work shall, without the consent of the Management, fix the wage paid to his helper at more than a Standard rate of 6d. per day, plus percentage, above the minimum Standard rate fixed for the class of helper in Schedule 1.

8—Should any question arise as to whether any particular workman employed underground is a workman to whom the minimum rate is to apply, or whether a workman has failed to comply with any

---

[1] It has been decided by the Court of Appeal that this portion of Rule 7 is *ultra vires*.

of the conditions contained in these rules, or whether by non-compliance with any of these rules such workman has forfeited his right to the Minimum rate, such question shall be decided in the following manner :—

(*a*) By agreement between the workmen concerned and the Official in charge of the Mine.

Failing agreement, by two officials of the Colliery representing the employer on the one side and two members of the Committee of the local lodge of the Workmen's Federation (or not more than two representatives appointed by them) on the other side.

Again failing agreement, by the Manager of the Mine and the District Miners' Agent.

(*b*) Still failing agreement, by an umpire to be selected by them (or if they disagree in the selection, by lot) without delay from one of the Panels constituted as hereinafter provided.

Three panels of persons having a knowledge of Mining to be prepared by the two Chairmen of the Employers and Workmen's Representatives on the Joint District Board. One of such panels shall be constituted for questions arising in the Newport District, one for questions arising in the Swansea District (including Pembrokeshire), one for questions arising in the Cardiff District. In case of difference as to the constitution of any panel, such panel shall be settled by the independent chairman of the Joint District Board. The Newport District shall consist of Collieries situated to the east of the Rumney

River. The Swansea District shall consist of Collieries situated westwards of the Llynfi River and of a line drawn from the top of that river into the Neath River at Ystradfellte. The Cardiff District shall consist of Collieries situated between the Newport and the Swansea District

If required by either employer or workman, a panel may be revised at the end of every twelve months from the constitution thereof. For the determination of any question arising under this rule the employers and workmen respectively shall be entitled to call such evidence as they may think proper before the person or persons who may have to determine such question, and such person or persons may make such inspections of workings as he or they may deem necessary for the proper determination of the matter in question.

Any questions that may arise for determination under paragraph (a) of this rule shall be determined within a period of three clear days from the date upon which the question to be determined first arose ; and any question to be determined by the Umpire shall be determined within seven clear days from the said date, or such further time as the Umpire shall appoint in writing. The colliery representative and the District Miners' Agent shall be entitled to attend and represent the employers and workmen respectively before the Umpire.

9.—A certificate in writing of any decision by any person or persons under the last preceding rule shall be given by such person or persons to both or either of the parties when requested, and such certificate shall be conclusive evidence of the decision. Any certificate so given as to the infirmity of a workman may be cancelled or varied on the application of either party after the expiration of six weeks from the date of the certificate. Any application to cancel or vary such certificate shall be determined as a question under the last preceding rule. The expenses and charges of the Umpire shall be paid by the Joint District Board and apportioned in the same manner as the expenses of the Joint District Board.

10.—Except as expressly varied by these rules, all customs, usages and conditions of employment existing at the respective Coal Mines to which these rules are applicable shall remain in full force unless altered by a mutual agreement.

11.—Overmen, Traffic Foremen, Firemen, Assistant Firemen, Bratticemen, Shotfirers, Master Hauliers, Farriers, and persons whose duty is that of inspection or supervision, are not workmen to whom the Coal Mines (Minimum Wage) Act applies.

12.—In the event of any question arising as to the construction or meaning of these rules, it shall be decided by the independent chairman of the Joint District Board.

(*Signed*)   ST. ALDWYN.

July 5th, 1912.

Revised Dec. 26th, 1913.

BIBLIOGRAPHY

DESCRIPTIVE

Berry, T. W. "Story of a Coal Mine." Pitman, 1914.

Boyd, R. N. "Coal Pits and Pitmen." Whittaker, 1895.

Cooke, Arthur O. "A Visit to a Coal Mine." Oxford Industrial Readers.

Davies, Henry. "A Mining Reader." Cardiff, 1904.

Holmes, F. M. "Miners and their Work Underground." London, 1896.

GENERAL

Arber, E. A. N. "Geology of the Kent Coalfield." Institute of Mining Engineers, 1914.

Boulton, Prof. W. S. "The Geology of the Coal Measures. Section I. of Practical Coal Mining." Gresham Press.

Gibson, Dr. Walcot. "Geology of Coal and Coal Mining." Arnold, 1908.

Harrison, W. J. "The Search for Coal in the South-East of England." Dulau & Co.

Hodgetts, E. A. Brayley. "Liquid Fuel for Mechanical and Industrial Purposes."

Hull, E. "The Coalfields of Great Britain." Stanford, 1905.

Jones, Atherley, and Bellot, Hugh L. "Miners' Guide to the Law Relating to Coal Mines." Methuen, 1914.

Lapworth, Charles. "An Intermediate Text Book of Geology. Blackwood.

Lewes, V. B. "Oil Fuel." Collins, 1913.

Martin, E. A. "The Story of a Piece of Coal. London, 1896.

Meade, Richard. "Coal and Iron Industries of the United Kingdom." Crosby, Lockwood & Co.

Meldola, A. " Coal and what we get from it." Romance of
     Science Series, 1891.

North, Sidney H.   " Oil Fuel."   Griffin & Co.

North, S. H. " Petroleum : Its Power and Uses." London,
     1905.

Redwood, Sir Boverton. " A Treatise on Petroleum."   2 vols.
     Griffin & Co.

Redwood, Iltyd L. " A Practical Treatise on Mineral Oils and
     their By-Products."   E. & F. Spon.

Sommermeier, E. E. " Coal."   New York, 1912.

Stokes, Ralph S. G. " Mines and Minerals of the British
     Empire."   E. Arnold, 1908.

Thorpe, Prof. (Editor). " Coal : Its History and Uses."
     Macmillan, 1876.

Tonge, Jas. " Coal." Constable, 1907.

Watts, W. W. " Geology for Beginners."   Macmillan, 1912.

White, D. " Origin of Coal."   U.S. Bureau of Mines, 1914.

Williams, Archibald. " Romance of Modern Mining."   1907.

Wilson, Francis H. " Coal."   Pitman, 1913.

HISTORICAL

Archer, M. " Sketch of the History of the Coal Trade of North-
     umberland and Durham."   London, 1897.

Boyd, R. N. " History of the Coal Trade."   London, 1892.

Galloway, R. L. " Annals of Coal Mining and the Coal Trade."
     2 vols. Colliery Guardian Co., 1898 and 1904.

Galloway, R. L. " Papers Relating to the History of the Coal
     Trade."   Colliery Guardian Co., 1906.

Galloway, R. L. " History of Coal Mining in Great Britain."
     London, 1892.

Report of Coal Commission, 1871. Vol. I. contains historical
     sketch of coal-mining down to 1870.

Report of Royal Commission on Coal Mines, 1907-10.   Contains
     a short sketch of Mining Legislation.

Vincent, J. E. " John Nixon, Pioneer of the Steam Coal Trade
     in South Wales."   London, 1900.

Wilkins. " The South Wales Coal Trade."   Cardiff, 1888.

SOCIAL

De Rousiers, Paul. "The Labour Question in Britain." Macmillan & Co., 1896.

Duland, H. "Among the Fife Miners." London, 1904.

McVail, John C. "Report on the Housing of Miners in Sterlingshire and Dumbartonshire." Glasgow, 1911.

Robb, Alex. "Report on the Housing Conditions of Miners in the Counties of Midlothian and Linlithgow." Edinburgh, 1912.

Small, A. "The Cry of the Miner."

TRADE UNIONS AND WAGES

Ashley, W. J. The "Adjustment of Wages." Longmans & Co.

Board of Trade Labour Department. Report on Trade Unions.

Evans, D. "Labour Strife in the South Wales Coalfield, 1910-11."

Hallam, William. "Miners' Leaders, 30 Portraits with Biographical Sketches, 1904.

Longmans. "The Coal Mine Workers: A Study in Labour Organisations." 1906.

Miners' Federation of Great Britain: Minutes and Verbatim Reports of Annual and Special Conferences. Printed for private distribution. T. Ashton, Manchester.

Miners' Federation of Great Britain: Various publications dealing with Minimum Wage Act, etc.

Munro, J. C. "Sliding Scales in the Coal Industry." Manchester, 1885 and 1889."

Percy, M. "Miners and the Eight Hours' Movement."

Smart, W. "Miners' Wages and the Sliding Scale, 1894."

Thomas, D. Lleufer. "Labour Unions in Wales." Article in "South Wales Labour Annual." Swansea, 1902.

Webb, Sidney and Beatrice. "A History of Trade Unionism."

Wilson, Ald. J. "History of Durham Miners' Association." Durham, 1907.

Wilson, J. "Memoirs of a Miners' Leader." Fisher Unwin, 1910.

TECHNICAL

Beard, J. T. "Mine Gases and Explosions." Chapman & Hall, 1908.

Boulton, Prof. W. S. (Editor). "Practical Coal Mining." 6 vols., Gresham Publishing Co., 1913.

Bulmore, F. H. "Colliery Working and Management." London, 1896.

Bulmore, F., and Redmayne, R. A. S. "Colliery Working and Management." Crosby Lockwood, 1912.

Burns, Daniel. "Electrical Practice in Collieries." Griffin & Co., 1909.

Burns, Daniel. "Safety in Coal Mines." Blackie, 1912.

Burns, D., and Kerr, G. L. "Modern Practice in Coal Mining." Whittaker & Co.

Byron, T., and Christopher, T. E. "Modern Coking Practice." Crosby Lockwood, 1910.

Cockin, T. H. "Elementary Class Book of Mining." Crosby Lockwood, 1909.

Forster, George. "Safety Lamps and the Detection of Fire Damp in Mines." Routledge & Sons, 1914.

Frazer, J. C. W. "Inflammability of Coal-Dust." U.S. Bureau of Mines, 1914.

Futers, T. Campbell. "The Mechanical Engineering of Collieries." "Colliery Guardian," 1909.

Galloway, W. "Course of Lectures in Mining." Cardiff, 1900.

Hagger, J. "Coal and the Prevention of Explosions and Fires in Mines." Reid, 1913.

Hughes, A. W. "A Text Book of Coal Mining." C. Griffin, 1904.

Kerr, G. L. "Practical Coal Mining." C. Griffin, 1905.

Lupton, Arnold. "A Short Treatise on the Getting of Minerals." Longmans, 1907.

Lupton, A., and Parr, G. D. A. "Electricity as Applied to Mining." Crosby Lockwood, 1906.

Mauchline, R. "The Mine Foreman's Handbook." E. & F. N. Spon.

McTrusty, J. W. "Mine Gases and Testing." Wall, 1913.

Pamely, Caleb. "The Colliery Manager's Handbook." Crosby Lockwood, 1904.

Peel, R. "Elementary Text Book of Coal Mining." Blackie, 1911.

Percy, C. M. "The Mechanical Equipment of Collieries." Manchester, 1905.

Redmayne, R. A. S. "Modern Practice in Mining." 5 vols. Longmans & Co.

Remer, J. "Shaft Sinking in Difficult Cases." C. Griffin, 1907.

Rice, G. S., and others. "Coal-Dust Explosion Tests in the Experimental Mine." U.S. Bureau of Mines, 1913.

Smyth, Sir W. W. "Coal and Coal Mining." Crosby Lockwood & Co.

Tonge, J. "Principles and Practice of Coal Mining." Macmillan, 1906.

Walker, S. F. "Coal-Cutting by Machinery." "Colliery Guardian," 1902.

Walner, R. "Ventilation in Mines." Scott Greenwood, 1903.

ECONOMIC

Carter, M.A., G. R. "The Tendency Towards Industrial Combinations." Constable, 1914.

Colliery Guardian Co. Various Pamphlets.

Hull, E. "Our Coal Reserves at the end of the Nineteenth Century." London, 1897.

Jevons, W. S. "The Coal Question." 3rd Edition, revised by A. W. Flux, M.A. Macmillan & Co., 1906.

Jevons, H. Stanley. "The Economics of Coal." Article in "Practical Coal Mining," Vol vi., pp. 403-436.

Jevons, H. Stanley. "Foreign Trade in Coal." Publication of Department of Economics and Social Science of the Univ. College of South Wales and Monmouthshire, 1909.

Lewy, E. "A Scheme for the Regulation of the Output of Coal by International Agreement."

Report on "The Coal Resources of the World" for the Twelfth International Geological Congress, 1913. 3 vols. and atlas.

Thomas, M.A., D. A. "The Growth and Direction of our Foreign Trade in Coal During the Last Half Century." Journal of the Royal Statistical Society, Vol. lxvi. (1903), pp. 439-533.

Walker, Francis. " Monopolistic Combinations in the German Coal Industry." American Economic Association, 1904.

COMMERCIAL

Evans, Charles E. " Hints to Coal Buyers. Business Statistics Co., Ltd., Cardiff.

Greenwell; T. A.; and Elsden, J. V. " Analyses of British Coals and Coke." Colliery Guardian, 1909.

Lawn, J. G. " Mine Accounts and Mining Book-keeping." Griffin & Co.

Kirkaldy, Prof. A. W. " British Shipping." Kegan Paul, 1914.

Mann, J. and Judd, Harold G. " Colliery Accounts." Gee & Co.

ANNUALS AND PERIODICALS

" The South Wales Coal Annual." 12 volumes, 1903 to 1915. Business Statistics Co., Cardiff.

" Business Prospects Year Book." Business Statistics Co., Ltd.

" Coal and Iron and By-Products Journal " (weekly).

" Iron and Coal Trades Review " (weekly).

" Northern Coal and Iron Companies " (annual). Business Statistics Co., Cardiff.

" South Wales Coal and Iron Companies " (annual).

" The Colliery Guardian " (weekly).

" The Colliery Manager's Pocket Book." Colliery Guardian Co.

" The North Country Coal and Shipping Annual." Business Statistics Co.

" The Mining Industry: World's Statistics." New York (annual).

GOVERNMENT PUBLICATIONS

" The Coal Tables." Parliamentary Paper."

" Coal Exports." Returns showing destinations from each port. Parliamentary Paper.

" Geological Survey." Memoirs and other Publications.

" Prices of Exported Coal." Showing prices and quantities from each port. Parliamentary Paper.

" Mines and Quarries : General Report and Statistics." Home Office. Part I., District Statistics ; Part II., Labour ; Part III., Output ; Part IV., Colonial and Foreign Statistics. (Paper by command.)

" Royal Commission on Coal Supplies, 1871." Reports, 3 vols.

" Royal Commission on Coal Supplies, 1903-05." 1st and 2nd Reports (3 parts each) and Final Report (13 parts).

" Royal Commission on Explosions from Coal Dust, 1891-94." Reports, 3 vols.

" Royal Commission on Mines, 1907-10." Reports, parts.

" Report of Committee on Causes and Prevention of Accidents Underground, 1909."

INDEX

A

Aberdare, 100-107, 323, 533, 536, 537
Abnormal places, 520-525
Abraham, Right Hon. W. ("Mabon"), 130, 468, 536
Accidents in mines, 122, 366-375
Adjustment of wages, 6
Admiralty, use of oil by, 695 703-705
    coal, 38, 294, 302-304
Alexandra Dock, 109
Allerton, Lord, 725
Amalgamations in coal trade, 314-330
Amaral, Sutherland & Co., 322
American competition, 765-767
Analysis of coal, 32-35, 234
Anglo-Persian Oil Co., 704
Anglo-Westphalian Syndicate, 168
Ankylostomiasis, 424
Anthracite, 33-35, 39, 97, 98, 115, 139, 148, 150, 232, 247-248, 507, 660-674
    collieries 668-669, 803-804
    trust, 324-326, 671
Anticline, 22, 24, 26
Arber, Dr. E. A. N., 159
Arley mine, 35
Ashton, Mr. T., 542
Askwith, Sir George, 553, 554
Asquith, Mr., 553, 555, 561, 600
Atherton, 619
Atkinson, Dr., 117
Austen Godwin, 161

3 I

B

Babcock & Wilcox, 46
Barry, 109-112
Basin, 23
Bêche, Sir Henry de la, 161, 378
Beehive coking ovens, 235
Beilby, Mr., 47
Bends, 29
Bituminous coal, 37-38
Black-band, 16, 146, 153
Blaina district of miners, 430
Blaina Valley, 101
Blyth, 61
Boreholes underground, 410
Borings in Kent, 19, 162
Boring processes, 177-181
Borings under London, 19
Boston, 68
Bovey Tracey, 80
Brace, M.P., Mr. Wm. 438, 561,
Bradbury, Judge J. K., 592
Bradford Dyers' Association, 314
Bramall, Mr. H., 728
Briggs Henry, Son & Co., Ltd., 318
    Mr. Currer, 737
Britannic Merthyr Coal Co., 321
Brown, Mr. W. Foster, 85, 765
Building clubs, 122, 129, 646-647
Burnyeat, Brown & Co., Ltd., 324
Burr, Mr. Arthur, 164-168, 175
Burt, Mr. Thos., 634
Bute, Marquis of, 99, 107, 108
Butty-gang system, 334, 410 454, 455-456

Buxton, Mr. Sydney, 554
Bye-products of coal, 9, 32, 56, 232-256

C

C. I. F., 295-297, 304
Caking and non-caking coals, 233
Calorific value, 35-36
Cambrian Combine, 320-322, 457, 802
    strike, 533-539, 564
Cannel, 149
Cardiff, 12, 36, 107, 109, 110, 111, 112, 114
Cawdor anthracite colliery, 35
Cement method of sinking, 192
Central Labour College, 635
Centralisation, 479
" Change," 627
Channel Tunnel, 162
Character of miners, 625, 629
Charters, coal-shipping, 308-309
Checkweighers, 71, 340, 384, 458, 464-465
Churchill, Mr. Winston, 705
Clanny, Dr., 376
Clark, Mr. G. T., 127-128
Clarke, Sir Edward, 579, 595, 596, 598
Cleat, 204
Coal, increasing cost of, 761-763
    increasing cost of mining, 718
    occurrence of, 2
    early mining operations, 3-4
    uses of, 4, 31-57
    composition, 32-35
    as a raw material, 9
    considerations affecting price of, 9, 36
    kinds of, 3, 37-40, 300
    nature of, 11, 12
    payment for, 298-300
    prospecting for, 18
    utilisation of small, 232
    measures, 14

Coalfields, possible new, 10, 78
    disposition of, 23
    Derbyshire, 14
    South Wales, 15, 17, 29-30, 93-137
    Kent, 18-19, 59, 155-175
    Yorkshire, 18-19, 29, 64
    Lancashire, 18, 72
    Midland, 29, 88
    Belgian, 29
    Durham and Northumberland, 59-64
    Cumberland, 75
    Bristol and Somerset, 77
    Devonshire, 80
    Forest of Dean, 82
    Ireland, 38-43
    Scotland, 143-154
    North Wales, 137-138
Coalfields, 2, 3, 58
    concealed, 3, 19, 58, 156
Coal-cutting machines, 62, 69, 210-214
Coal gas, 55-57
Coal Mines Acts, 7, 379-385, 401-444, 618-619, 811-818
Coal Resources of the World, 91, 171, 725, 735, 736, 772, 789
The Coal Question, 10, 718, 721, 722, 724, 741, 756, 757, 761, 765
Coaldust, 398-400, 422, 442
Coffering, 187-188
Coke, manufacture of, 9, 37, 233 *et seq.*
    manufacturing developments, 115
    shipments from South Wales, 113
Collieries, kinds of, 182
    of Northern coalfield, 799-800
    of South Wales, 802-804
    of Yorkshire, 801
Combinations. *See* Amalgamations.
Conciliation boards, 6, 87, 336-338, 498-519, 547

Conciliation districts, 506-512
Conferences of miners, 481-488.
  See also National Conferences.
Conservation of national capital,
  768-771
Consideration system, 290, 525
Consumption of coal, 40-43, 47-
  48, 742
Contracts for coal, 291-293, 805-
  810
Co-operative collieries, 459-460
  societies, 629, 631
Cornish boilers, 45
Cory Bros., 349
County average system, 356,
  358-362
Courrièrres disaster, 399
Cracking oil, 713
Crawford, Mr. W., 458
Creep, 26, 207
Crib (or curb), 183
Crossley, Mr., 53
Culm, 81
Cunningham, Sir Robert, 153
Curb. *See* Crib.
Cynon colliery, 322

D

Dalziel, Alexander, 103
Darby, Mr., 242
David, Prof. T. W. E., 786
Davies, M.P., Mr. D., 653
Davis, D. & Sons, Ltd., 323
Davis, Mr. F. L., 536
Davis-Calyx boring system, 180-
  181
Davy, Sir Humphrey, 376, 394
Deadwork, 29, 289, 342
Demand for coal, 257-260
  for labour, 275-283
Denudation, 24
Deputy gavellers, 83
Diamond system of boring, 180
Diesel engines.   *See* Internal
  Combustion Engines.
Diesel, Dr. Rudolph, 697, 698

Disputes, 512-519
Distillation, 31-32
Doncaster, 19, 65, 67, 69, 655,
  658
Double shifts, 63, 455, 759
Dover, 19, 162
Dowlais, 17, 99
Downcast, 200
Duffryn Rhondda colliery, 322
Duration of coal supplies, 118,
  753-756, 788-798
Durham Coal Sales Association,
  317
Dykes, 145, 146, 150

E

Earth-movements, 22, 24
Earth's crust, 22
Ebbsfleet syndicate, 168
Ebbw Vale Steel & Iron Co., 101,
  323
Ebbwy Valley, 101
Economics of coal trade, 257-290
Edinburgh Collieries Co., 320
Edwards, Mr. Enoch, 549, 561
Eight Hours' Movement, 63, 74,
  117, 125, 214, 338, 454, 455.
  462, 469, 483, 486, 532
Electrical safety lamps, 395-396
Electricity in mines, 61, 69, 402,
  416, 421
  underground, 421
Elliot washer, 228
Ellis, Sir T. Radcliffe, 542
Ely colliery dispute, 350, 515.
  533
  valley, 323
Emlyn anthracite colliery, 33
Enger, Dr., 710
Examiners, 408-409
Explosions, 387-403, 422
Exports of coal—
  Yorkshire ports, 67-68
  North-Eastern ports, 61
  Lancashire and Cheshire, 73-
  74

Exports of coal—
  anthracite, 665, 667
  from principal British coal-
    fields, 683
  from various countries, 677-
    679
  growth of foreign trade, 676,
    749-752
  sea-borne foreign coal trade,
    681
  South Wales ports, 111-114

F

F. O. B., 295-297
F. O. T., 295-296
Face, 5
Fan, 201, 412
Faraday, Sir Michael, 378
Faults, 27-30
Federation district, 475
  lodge, 472
Felspar washers, 229
Fernhill colliery, 322
Fife Coal Co., 319
Fireclay, 16
Firedamp, 388
Firemen (examiners or deputies),
  406-408, 422, 441
Foreign trade in coal, 675-693
Formation of coal, 2, 12, 20-23
Francis, Mr. J. O., 628
Freeminers, 83
Freight rates, 684-693, 714

G

Gales, 83-86
Galloway, Professor, 192
Galloway scaffold, 185
  water barrel, 186
Ganister, 16
Garden villages, 653-656
Garland, 186
Gateway, 199
Gavellers, 83

Gelli colliery dispute, 349
Geological formation of British
  coalfields, 26, 58, 59, 64, 72,
  75, 77, 81, 82, 88, 89, 90, 91,
  93-99, 137, 138, 143, 158, 663-
  664
  society, 161
Geology and coal-mining, 18
Geology of coal measures, 11-30
George, Mr. D. Lloyd, 554
German competition, 230
Gibson, Dr. Walcot, 144, 737
Glamorgan canal, 100
  Coal Co., 321
Goaf (or gob), 202, 204, 442, 611
Gob. See Goaf.
Gobert method of sinking, 191
Goole, 68
Grading of coal, 229-231
Great Western colliery, 33, 35,
  345
Grey, Sir Edward, 554
Grimsby, 68
Gueret, Ltd., 322
Guest, Keen & Nettlefolds, Ltd.,
  17
Guest, Mr. John, 17, 99
Gwendraeth, 97

H

Haase process of sinking, 190
Hade, 28
Halliday, Mr. T., 457
Hangers, 113
Hardie, Mr. J. Kier, 634
Hartley colliery, 381
Haulage of coal underground,
  213-218, 420
  roads, 420
Heading, 5
Hereford, Lord James of, 517
History of coal trade, 3-4, 99-
  109, 151-154
  of Kent coalfield, 161-168
  of miners' trade unions,
    445-471

Horden Collieries, Ltd., 318
Horizons, 19
Housing of miners, 63-64, 71-72,
122, 126-130, 175, 485, 637-659
Hull, 68
Prof., 140
Humboldt, 376
Hylton colliery, 193

I

Independent Labour Party, 135,
634
Inspection of mines, 411, 422,
426, 437, 439-441, 483, 485
Institutes, 352
Intake, 200, 419
Internal combustion engines, 51-
55, 421, 696-698, 715
Inventions Commission, 770
Iron ore imports, 16
Ironmasters of South Wales,
105-106
Ironstone, 16

J

Jevons, Prof. W. S., 10, 718, 720,
725
Jigger washer, 228
Joicey, Ltd., Sir James, 318
Joint district boards, 577
Jointing, 26
Jude, Mr. Martin, 451

K

Kartell, 326-330
Kelty Beach Colliery, 319
Kent Coal Concessions, Ltd.,
164, 166, 167, 173, 175, 656
Collieries Corporation, 163
Collieries Ltd., 163
Exploration Committee, 162
Light Railway Co., 167

Kibble, 184
Kind-Chaudron Method, 163,
191, 192
Kirkconnell, 656

L

Labour—
conditions underground, 63,
331-335, 608-617, 520-522
demand for, 275-283
foreign labour in mines, 623
skilfulness of South Wales
miners, 354
supply of, 283-287
Lancashire boilers, 45
Lens Mining Co., 192
Levels, 182
Lewes, Mr. Vivian, 717
Lewis, Sir W. T., 98
Lewis-Hunter crane, 11
Lignite, 80-82
Limitation of the Vend, 315-316,
327
Llanelly, 111
Llanerch colliery, 620
Lodge, 465, 472, 630
Lodgers, 643, 657
Londonderry, Lord, 449
Longwall, 202-209
Lusitania, 45
Lyell, Charles, 378

M

Mabon. See W. Abraham.
Macdonald, Alexander, 380, 451-
458, 462, 481, 482
Main and tail, 216-217
Management of mines, inade-
quacy of, 413, 428
Markham, Sir Arthur B., 655,
666
Markets for coal, 309-310
Mauretania, 45
McVail, Dr. J. C., 648

Measuring-up, 529
Mechanical stokers, 46, 47, 49
Medical clubs, 632
Mersey, Lord, 517
Merton washer, 227
Middlemen, 304-307
Midland Mining Commission, 378
Milbanke, Sir Ralph, 376
Millstone grit, 19, 20
Miners' achievements, 633
    Federation of Great Britain,
        87, 463-471, 480, 484-487,
        507, 520, 537, 819-824
Miners' Next Step, 554
Miners' societies, 629-633
Miners' Trade Unions, 6, 71, 107,
    130-135, 445-488
    Amalgamated Association of
      Miners, 457, 481
    Derbyshire Miners' Associa-
      tion, 71
    Durham Miners' Mutual
      Confident Association,
      358, 457-459, 469, 477-
      481
    early attempts to form
      unions, 444 *et seq.*
    Forest of Dean Miners'
      Association, 87
    history of miners' unions,
      445-471
    International Miners' Union,
      485
    Lancashire and Cheshire
      Miners' Federation, 479,
      480, 564
    Midland Counties' Federa-
      tion, 479, 480
    Midland Federation of
      Miners, 462, 468, 479, 480
    National Union of Miners,
      381, 448, 451, 454, 456,
      458, 463, 464, 474, 481,
      482
    North Staffs. Miners' As-
      sociation, 457
    Notts Miners' Association,
      71

Miner's Trade Unions—
    Scottish Miners' Federa-
      tion, 470, 479-481
    South Wales Miners' Feder-
      ation, 63, 71, 131, 356-
      358, 468, 478-481, 538
    Yorkshire Miners' Associa-
      tion, 71, 460, 462, 477,
      480
Minimum Wage Act—
    general district rules, 579-
      590, 849-856
    general district rates, 590-
      599, 602
    joint boards, 37, 577-579
    principles of, 569-572
    provisions, 562-577, 587,
      607, 842-848
    success of, 599-607
Minimum wage movement, 80,
    484-487, 522, 538-541, 560
Mining, methods of, 202-218
Mining communities—
    character of population,
      473, 620-623
    permanence of, 623-625
    standard of life, 7, 71, 351-
      353, 772
Mining Examinations Board, 405
    Law, 404-444, Appendix 5
Mixing of coals, 307-308
Mortality in mines, 366-375
Mather and Platt boring system,
    179

N

Nantyglo Ironworks Co., 101
National Coal Strike, 1912, 6, 133,
    504, 520, 539-586, 566-568
    conferences, 450, 452, 453,
      454, 455, 462, 463, 471,
      481-488, 545, 547, 548,
      550, 552
Nationalisation of mines, 486,
    771
    of railways, 771
Naval Colliery Co., 321, 533

Neath, 109, 110, 111
New Unionism, 466
Newcastle, 3, 61, 315, 451
    Vend, 315-316
Newcomen's steam engine, 153
Newport, 11, 109, 110
Newspaper quotations for coal, 293-295
Nixon, John, 108, 690
Normansell, John, 482
Nystagmus, 424

O

Oakdale, 656
Ocean Coal Co., 324, 531, 619, 653
Ogmore Valley, 107
Oil fuel, 54, 55, 694-717
Opening a mine, 264-270
Outcrop, 23

P

Packing deals, 183
Passbyes, 214
Patent fuel, 50, 113, 244-256, 673
Pay-tickets, 345
Pease & Partners, Ltd., 318
Penarth docks, 108, 111
Pendleton colliery, 73
Percussive boring, 177-179
Permitted explosives, 210, 421
Persons employed—
    Bristol and Somerset, 79
    Cumberland, 76
    Durham and Northumberland, 62
    Forest of Dean, 86
    Ireland, 143
    Lancashire and Cheshire, 74
    Midland, 92
    North Wales, 138
    South Wales, 98-99, 116, 117

Persons employed—
    Yorkshire and North Midland, 65, 68
Petroleum. See Oil Fuel.
Pickford, Justice, 589
Picking table, 225
Piecework, 288, 290, 335
Pillar and stall, 202, 206-208
Pillson & Co., 322
Pit-bottom arrangements, 194-196
Pit-head baths, 346, 424, 617-620
Pits, 182
Playfair, Lyon, 378
Plans of mines, 409-410
Poetsch method, 190
Politics of Welsh miner, 135-137
Population, growth of, 742
Port Talbot, 109, 110, 111, 251
Powell Duffryn Steam Coal Co., 323
Preparation of coal for market, 49-51, 219-231
Prestwich, Sir Joseph, 162
Price lists, 338-362, 491
Price-Williams, Mr., 743, 744, 745, 746, 748, 751
Prices of coal, 73, 82, 112, 150, 151, 294, 489, 511
Princess Charlotte, 700
Producer gas engines, 51, 52, 53, 162
Production of coal—
    Aberdare valley, 103
    anthracite, 665, 667
    at differing periods, 740
    Bristol and Somerset, 79
    chief coalfields of world (1912), 678
    comparative cost (1897 and 1900), 279
    Cumberland, 76
    factors in, 264-270
    Forest of Dean, 86
    growth of, 794
    in principal countries, 790
    Ireland, 141

Production of coal—
  Kent, 173
  Lancashire and Cheshire, 74
  Midland counties, 92
  North Wales, 138
  Northumberland and Dur-
    ham, 62-63
  Scotland, 149, 154
  South Wales, 116
  United Kingdom, 116
  Westphalia, 329
  Yorkshire, 65, 68, 69
Production of coke—
  foreign countries, 244
  United Kingdom, 243
Production of lignite—
  Devonshire, 81
  foreign countries, 82
Production of patent fuel—
  foreign countries, 251
  United Kingdom, 250
Production of petroleum—
  crude petroleum, 711
  growth of, 708
  Scottish oil shale works, 712
  world, 707
Profits on mining, 272-274
Prospecting for coal, 176-177
Punch prop, 183
Putters, 214

R

Railways—
  Brecon & Merthyr, 110
  Burry Port, 110
  Cardiff, 110
  Great Central, 67
  Great Northern, 67
  Great Eastern, 566, 700
  Great Western, 79, 86, 110
  Hull and Barnsley, 67
  in mining areas, 59
  Kent Light, 167, 173
  London & North Western,
    67, 73, 110, 700

Railways—
  Lancashire & Yorkshire, 67,
    73
  Midland, 67, 79, 86
  Neath & Brecon, 110
  North British, 61
  North Eastern, 61, 67
  Port Talbot, 110
  Rhondda & Swansea Bay,
    110
  South Eastern & Chatham,
    167, 173
  South Wales Mineral, 110
  Taff Vale, 100, 108, 110
Reciprocating engines, 45
Redmayne, Mr. R. A. S., 437
Remaux, M. E., 192
Reserves of coal—
  A. Strahan's estimate, 736
  Africa, 784-785
  America, 778-783
  Asia, 783-784
  Australasia, 785-786
  Author's estimate, 739
  Bristol and Somerset, 79
  Cumberland, 76-77
  Durham and Northumber-
    land, 62
  Estimate of Royal Commis-
    sions (1871 and 1904), 732,
    733
  Europe, 774-778
  Forest of Dean, 86
  Ireland, 141
  Kent, 171-172
  Methods of estimating, 726-
    732
  Midlands, 91
  North Wales, 138
  South Wales, 118-120
  World, 785-786
Reversed fault, 27
Rhenish Westphalian Coal Syn-
  dicate, 317, 326-330, 775
Rhondda Valley, 33, 35, 101,
  103, 104, 105, 107, 197, 198,
  536, 537, 545, 621

Rhymney, 101, 323
  Iron & Coal Co., 101, 324
Royal Commission of Mines,
  (1871), 383, 722, 726, 730, 737,
  739, 743, 753
    Commission on Coal Dust,
    (1891), 384
Royal Commission on Coal Supplies, 1904—
    advantages of coal-cutting
    machines, 211-213
    bye-product recovery ovens,
    241-242
    classes of coal in South
    Wales, 98
    coal reserves of South Wales,
    118-120
    economies in coal consump-
    tion, 44, 47
    estimate of coal reserves,
    733
    grading and cleaning coal,
    231
    Irish coal-mining, 140
    manufacture of briquettes,
    252-254
    preparation of coal for mar-
    ket, 255
    use of coal in powdered
    form, 50
    use of oil as fuel, 694
    use of producer gas engines,
    52-53
Royal Commission on Coal Supplies (1904), 161, 171, 385,
  618, 725, 728, 731, 733, 734,
  748
Royal Commission on Coal Supplies (1905), 40, 44, 47, 50, 52,
  62, 86
    coal reserves of North and
    Durham, 62
Royal Commission on Employment of Children in Mines,
  (1840), 377
Roberts, Judge Bryn, 531
Roberts, W. P., 451, 482
Robinson washer, 228

Rolls, 29
Romer, Mr. Robert, 517
Rotary boring, 179-181
  engines, 45
Royalties, 270
Rule, 8, 21, 539, 545
Ruskin College, 635

S

Safety lamps, 394-395, 414-416,
  438
  legislation, 365-403, 404
    et seq.
  strikes, 430, 432
St. Aldwyn, Lord, 577, 601
St. Briavels, 83
Sale of coal, 291-313
Sandstone, 15
Sauerbeck, Mr., 761, 762
Scottish Coalowners' Associa-
  tion, 471
Sea-coal, 3
Selandia, 702
Select Committee on Firedamp
  Explosions (1849), 517
    Committee on mine acci-
    dents (1835), 376
    Committee on Mines Venti-
    lation (1853), 380
    Committee on Working of
    Mines Acts (1865), 382
Self-trimming vessels, 11
Senghenydd, 123, 365, 366, 411,
  414, 425, 430, 432-436
Shackleton Expedition, 786
Shaft pillars, 196-198
Shaftesbury, Lord, 377
Shafts, 181-190, 417
Shale, 15
Shell Transport Co., 699
Shipping agents, 304-305
  charters, 308-309
  coal, cost of, 312
Shot-firing, 399-403, 613
Show-card days, 134
Signalling, 217-218, 421

Sill, 145, 146, 150
Sinking new mines, 176-201
Sirhowy, 101
Slant, 4, 182
Sliding scales, 87, 461-463, 489,
    491, 527
Sludger, 178
Small coal, 232-233, 254
Smillie, Mr. Robert, 484
Smith, Sir H. Llewelyn, 554
Snowdown Colliery Co., 167,
    170, 172
Socialism, 125, 135, 136, 466,
    627
South Shields, 61
Speculative middlemen, 305-307
Standard wage, 335-338, 507
Steam engines, 44-49
Stoking, 46, 49
Stonehead, 183, 184
Strahan, Mr. Aubrey, 14, 40, 91,
    159, 171, 725, 735, 736, 737
Stringing deals, 183
Subsidences, 197-198
Sump, 186
Sunderland, 61
Sunderland Association, 376
Sunken forests, 20
Supply of coal, 260-264
Supply of labour, 283-287
Swansea, 107, 111, 114, 245,
    251
Syncline, 22, 24, 26, 92
Syndicalism, 125, 135, 136, 554,
    627

                    T

Tables, list of—
    Analyses of coal, 33, 35
        of cost of production of
        coal, 279
    Annual coal production of
        principal countries, 790
    Anthracite coal production,
        665
        collieries, 668, 669

Tables, list of—
    Average home consumption
        of coal, 747
        market prices of Cardiff
        coal, 489
    British coal exports to
        foreign markets, 683
        coke production, 243
        patent fuel production,
        250
    Census housing statistics,
        640
    Coal, coke and patent fuel
        shipments from South
        Wales, 113
        consumption of France,
        42
        consumption of United
        Kingdom, 41, 42, 43
        cutting machines, 211
    Comparative increases of
        population and housing
        accommodation, 639
    Cumberland coal produc-
        tion, 76
    Death-rates from accidents
        in various coalfields,
        124
    Death-rates from accidents
        in foreign mines, 374
    Destinations of Cardiff and
        Swansea coal exports
        (1912), 114
    Estimate of coal reserves—
        A. Strahan (1912), 736
        Author (1914), 739
        Royal Commission
        (1871), 732
        Ditto (1904), 733
    Estimate of future popula-
        tion, 744
    Export of coal from Bristol
        Channel, 111
    Exports of anthracite coal,
        665, 667
    Foreign production of coke,
        244
        of patent fuel, 251

Tables, list of—

Freight rates from Cardiff to foreign ports, 685, 686

Future population, home consumption, export and total output, 752

Growth of export trade in coal, 675, et seq., 749-752

production of petroleum, 708

Housing statistics, Northern and South Wales coalfields, 643

Output of coal in Bristol and Somerset, 79

Outward freight rates from Cardiff (1863-1913), 692-693

Overcrowded coalfield areas, **641**

Percentage changes in wages, 337

Persons killed and injured in mines, 367-368

Produce of shale oil works, Scotland, 712

Production in Durham and Northumberland, 62

in Forest of Dean, 86

in Lancashire and Cheshire, 74

of the Midlands, 92

of Yorkshire, 68-69

Production and exports of principal coalfields of world, 678

coal during different periods, 740

coal in Ireland, 141

coal in North Wales, 138

coal in Westphalia, 329

crude petroleum, 711

lignite abroad, 82

Reserves of Africa, 875

of Asia, 784

of Canada, 781

Tables, list of—

Reserves of Australia and Oceania, 786

of Durham and Northumberland, 62

Europe, 774, 778

of the Midlands, 91

of South and Central America, 782

of United States, 780

of the world, 787

of Yorkshire, 70

Reserves of German coalfields, 775

Taff Vale railway mineral traffic, 100

Uses of European coal, 43

oil, 712

Variations in percentage additions to standard wages in different districts, 825

World's petroleum production, 707

seaborne foreign coal trade, 681

Temperature of mines, 615, 728

Thomas & Davey, 322

Thomas, Mr. D. A., 317, 320, 321, 322, 533, 538, 689

Three shift system, 759

Thrust, 27, 107

Tilmanstone colliery, 167, 169, 172, 173

Timbering, 420 421

Tipplers, 220 221

Town planning, 658

Trade Unions. See Miners' Trade Unions.

Tredegar Iron & Coal Co., 101, 324

Tremenheere, Mr., 378, 379

Trepan, 191

Trimming of coal, 311-312

Trough washers, 227

Truck system, 102, 446, 447, 454

Tubbing, 187-188

Turbines, 45

U

Under sea coal, 60, 76
United Collieries, Ltd., 319
United National Collieries, Ltd., 324
   States Steel Corporation, 314
Unmarked Bar Association, 314
Upcast, 200

V

Vale of Neath, 97, 664
Ventilation of mines, 200-201, 411-414
Volatile products of coking, 242

W

Wage agreements, 336, 826-848
   standards, 507
Wages, 288-289
   demand for new standard, 363-364
Wages of miners, 5, 458
   fluctuations in, 496-497
   Forest of Dean, 87
   Lancashire and Cheshire, 73
   methods of payment, 6, 330-364

Wages of miners—
   Northumberland and Durham, 63, 458
   South Wales, 104, 120-122
   Yorkshire, 70
Wages, percentage changes in, 337, 825
Walling of pit shafts, 184
Washeries, 225-234
Wash-out, 29
Water-tube boilers, 45
Watering mines, 400, 429
Weekly pays, 528
Welsh Iron & Coal Companies, 101 et seq., 802-804
Welsh miners' characteristics, 124-126
   politics, 135-137
Welsh Navigation Steam Coal Co., 323
   Town Planning and Housing Trust, Ltd., 653
West Hartlepool, 61
White-shirt parades, 135
Wilkins, Charles, 99
Williams, John, 723
Wilson, Mr. John, 634
Wire Nail Association, 314
Woodlands Village, 655
Workers' Educational Association, 628
Workmen examiners, 408, 430